Lecture Notes in Networks and Systems

Volume 116

The series "Lecture Notes in Networks and Systems" publishes the latest developments in Networks and Systems—quickly, informally and with high quality. Original research reported in proceedings and post-proceedings represents the core of LNNS.

Volumes published in LNNS embrace all aspects and subfields of, as well as new challenges in, Networks and Systems.

The series contains proceedings and edited volumes in systems and networks, spanning the areas of Cyber-Physical Systems, Autonomous Systems, Sensor Networks, Control Systems, Energy Systems, Automotive Systems, Biological Systems, Vehicular Networking and Connected Vehicles, Aerospace Systems, Automation, Manufacturing, Smart Grids, Nonlinear Systems, Power Systems, Robotics, Social Systems, Economic Systems and other. Of particular value to both the contributors and the readership are the short publication timeframe and the world-wide distribution and exposure which enable both a wide and rapid dissemination of research output.

The series covers the theory, applications, and perspectives on the state of the art and future developments relevant to systems and networks, decision making, control, complex processes and related areas, as embedded in the fields of interdisciplinary and applied sciences, engineering, computer science, physics, economics, social, and life sciences, as well as the paradigms and methodologies behind them.

**** Indexing: The books of this series are submitted to ISI Proceedings, SCOPUS, Google Scholar and Springerlink ****

More information about this series at http://www.springer.com/series/15179

Maitreyee Dutta · C. Rama Krishna ·
Rakesh Kumar · Mala Kalra
Editors

Proceedings of International Conference on IoT Inclusive Life (ICIIL 2019), NITTTR Chandigarh, India

Springer

Editors
Maitreyee Dutta
Department of IMCO
National Institute of Technical Teachers Training and Research
Chandigarh, India

C. Rama Krishna
Department of CSE
National Institute of Technical Teachers Training and Research
Chandigarh, India

Rakesh Kumar
Department of CSE
National Institute of Technical Teachers Training and Research
Chandigarh, India

Mala Kalra
Department of CSE
National Institute of Technical Teachers Training and Research
Chandigarh, India

ISSN 2367-3370 ISSN 2367-3389 (electronic)
Lecture Notes in Networks and Systems
ISBN 978-981-15-3019-7 ISBN 978-981-15-3020-3 (eBook)
https://doi.org/10.1007/978-981-15-3020-3

This Springer imprint is published by the registered company Springer Nature Singapore Pte Ltd.
The registered company address is: 152 Beach Road, #21-01/04 Gateway East, Singapore 189721, Singapore

Committee Members

Patron

Dr. S. S. Pattnaik, Director, National Institute of Technical Teachers Training and Research, Chandigarh, India

Conference Chair(s)

Dr. Maitreyee Dutta, Professor & Head, IMCO, National Institute of Technical Teachers Training and Research, Chandigarh, India
Dr. C. Rama Krishna, Professor & Head, Department of Computer Science & Engineering, National Institute of Technical Teachers Training and Research, Chandigarh, India

Conference Coordinator

Dr. Rakesh Kumar, Assistant Professor, Department of Computer Science & Engineering, National Institute of Technical Teachers Training and Research, Chandigarh, India
Dr. Mala Kalra, Assistant Professor, Department of Computer Science & Engineering, National Institute of Technical Teachers Training and Research, Chandigarh, India

Proceedings Editors

Dr. Maitreyee Dutta, Professor & Head, IMCO, National Institute of Technical Teachers Training and Research, Chandigarh, India
Dr. C. Rama Krishna, Professor & Head, Department of Computer Science & Engineering, National Institute of Technical Teachers Training and Research, Chandigarh, India
Dr. Rakesh Kumar, Assistant Professor, Department of Computer Science & Engineering, National Institute of Technical Teachers Training and Research, Chandigarh, India
Dr. Mala Kalra, Assistant Professor, Department of Computer Science & Engineering, National Institute of Technical Teachers Training and Research, Chandigarh, India

Local Organizing Committee

Dr. Maitreyee Dutta, Professor & Head, IMCO, National Institute of Technical Teachers Training and Research, Chandigarh, India
Dr. C. Rama Krishna, Professor & Head, Department of Computer Science & Engineering, National Institute of Technical Teachers Training and Research, Chandigarh, India
Dr. Rakesh Kumar, Assistant Professor, Department of Computer Science & Engineering, National Institute of Technical Teachers Training and Research, Chandigarh, India
Dr. Mala Kalra, Assistant Professor, Department of Computer Science & Engineering, National Institute of Technical Teachers Training and Research, Chandigarh, India
Ms. Shano Solanki, Assistant Professor, Department of Computer Science & Engineering, National Institute of Technical Teachers Training and Research, Chandigarh, India
Mr. Amit Doegar, Assistant Professor, Department of Computer Science & Engineering, National Institute of Technical Teachers Training and Research, Chandigarh, India
Mr. Amrendra Sharan, Jr. System Programmer, Department of Computer Science & Engineering, National Institute of Technical Teachers Training and Research, Chandigarh, India

Student Volunteers

Mr. Talvinder Singh, M.E. Student, National Institute of Technical Teachers Training and Research, Chandigarh, India
Mr. Aaqib, M.E. Student, National Institute of Technical Teachers Training and Research, Chandigarh, India
Ms. Aastha Sood, M.E. Student, National Institute of Technical Teachers Training and Research, Chandigarh, India
Ms. Empreet, M.E. Student, National Institute of Technical Teachers Training and Research, Chandigarh, India
Ms. Nikita Katnoria, M.E. Student, National Institute of Technical Teachers Training and Research, Chandigarh, India
Ms. Savneet Kaur, M.E. Student, National Institute of Technical Teachers Training and Research, Chandigarh, India
Mr. Kamal Deep, Research Scholar, National Institute of Technical Teachers Training and Research, Chandigarh, India
Ms. Manisha Malik, Research Scholar, National Institute of Technical Teachers Training and Research, Chandigarh, India

Preface

"You never change things by fighting the existing reality. To change something, build a new model that makes the existing model obsolete."

—Buckminster Fuller

It's a great privilege for us to present the proceedings of International Conference on IoT Inclusive Life (ICIIL) 2019 to the authors, delegates and general public. We hope that you will find it useful, exciting and inspiring.

Unlike previous conferences, this time the theme was the link between the IoT and various applications in day-to-day life. The power of Internet connectivity has now stepped beyond computers and smartphones. Every 'smart' device around us is now aiming to solve real-world problems with digital interventions. These are the real-life applications of IoT. Needless to mention, the buzz around IoT is immense. This disruptive technology is penetrating into various industries, developing new real-life applications of IoT and connecting every Internet-enabled device around us. But amongst the mad rush of 'newer' and 'better' IoT applications, some shine through more than the rest.

ICIIL 2019 aimed to provide a platform for discussing the issues, challenges, opportunities and findings in the area of IoT. The ever-changing scope and rapid development of IoT create new problems and questions, resulting in the real needs for sharing brilliant ideas and stimulating good awareness of this important research field.

This book focuses on the following sub-themes:

- Security and Privacy in IoT
- IoT Sensing, Monitoring, Networking and Routing
- Data Science and Computational Intelligence
- IoT Enabling Technologies

The book contains survey papers highlighting the challenges or manuscripts showing simulation or testbed-based experimental results. We hope this book will be quite useful for academicians, researchers and scientists in order to carry out further experimentation and technology enhancements.

Chandigarh, India

Maitreyee Dutta
C. Rama Krishna
Rakesh Kumar
Mala Kalra

Contents

Data Science and Computational Intelligence

IoT Enabling Technologies

About the Editors

Dr. Maitreyee Dutta completed her basic studies in ECE at Guahati University in 1993; her M.E. at PEC Chandigarh in 1999; and her Ph.D. at Punjab University in 2007. She joined the NITTTR Chandigarh in 2001. She is currently a Professor and Head of the institute's IMCO Unit. She has secured funding for two major projects from the Ministry of IT, New Delhi, and Ministry of Social Justice. She received the Rota Women's Distinction Award from the Rotary Club Chandigarh Midtown in 2013, and the Best Teacher Award from Bharatiya Shikshan Mandal, Uttar Pradesh, in the same year.

Dr. C. Rama Krishna completed his B.Tech. at the JNTU, Hyderabad; his M.Tech. at CUSAT, Cochin; and his Ph.D. at the IIT, Kharagpur, in the area of MANET. He is a Senior Member of the IEEE, USA. Since 1996, he has been working with the Department of CSE, NITTTR, Chandigarh, and currently holds the position of Professor and Head of Department, with more than 22 years of teaching experience. He has more than 100 research publications in international and national journals and conference proceedings to his credit, has implemented several projects funded by various government and private agencies in India, and holds one patent.

Dr. Rakesh Kumar received his B.Tech. in CSE from the IKGPTU, Jalandhar; his M.Tech. in IT from the GGSIPU, New Delhi; and his Ph.D. in Computer Engineering from the NIT Kurukshetra in 2004, 2008, and 2015, respectively. With more than 15 years of teaching experience, he is currently working as an Assistant Professor at the Department of CSE, NITTTR Chandigarh. He has several international/national conference/journal publications to his credit, and serves as a reviewer for many international journals/conferences.

Dr. Mala Kalra is an Assistant Professor at the Department of Computer Science and Engineering, National Institute of Technical Teachers Training and Research (NITTTR), Chandigarh, India. She received her B.E. in Computer Science and Engineering from the MDU, Rohtak, India, in 2001; her M.E. in CSE from PEC

University of Technology, Chandigarh, India, in 2006; and her Ph.D. in CSE from the PU, Chandigarh, in 2019. She has more than 15 years of teaching and research experience, and has published over 40 research papers in reputed international journals and conference proceedings.

Security and Privacy in IoT

Remotely Triggered Blackhole Routing in SDN for Handling DoS

Sugandhi Midha and Khushboo Tripathi

Abstract The DoS attack is one of the simplest yet most vulnerable attacks that is prominent in SDN. No doubt, SDN is leveraging several benefits like no vendor lock-ins, better fault tolerance, more opportune to innovations, etc., over traditional networks. Yet, SDN controller is a bigger target of security attacks. DoS makes the SDN controller unavailable for providing information/services. Remotely Triggered Black Hole technique is used to protect SDN from DoS attack. This technique has the ability to drop traffic coming from an undesired network element before it penetrates into the network. Our paper explains how this algorithm works and how it can be used to secure our SDN. We have tried to analyse, quantify and detect the impact of DoS in SDN. One key advantage of the proposed approach is its ability to accurately detect DoS with a nominal rate of failure. We simulated the system with our test bed of virtual machine with different attack scripted in Python.

Keywords Denial of Service (DoS) · Software-defined network controller · Remotely Triggered Black Hole (RTBH) routing · Access Control List (ACL)

1 Introduction

Software-Based Network [1] commonly called Software-Defined Network (SDN) is a new networking technology that has removed several shortcomings like high equipment cost, buggy product, complex infrastructure, vendor dependencies, etc., of a conventional network. SDN has made innovations easy and has brought flexibility in the networking environment. SDN has lowered down the cost and reduced the complexity of network devices.

SDN is broken down into three layers: Application Plane, Control Plane and Data Plane.

Application plane is how the user manages and uses applications like security, load balancing, SNMP, NETCONF, etc. Control Plane defines how SDN Controller

S. Midha (✉) · K. Tripathi
Amity University, Gurgaon, India
e-mail: mailmetech@gmail.com

M. Dutta et al. (eds.), *Proceedings of International Conference on IoT Inclusive Life (ICIIL 2019), NITTTR Chandigarh, India*, Lecture Notes in Networks and Systems 116,
https://doi.org/10.1007/978-981-15-3020-3_1

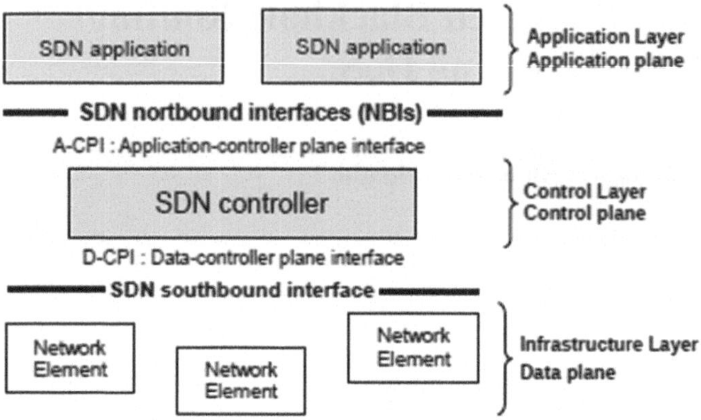

Fig. 1 SDN architecture [2]

handles the Network Elements (NEs). It dynamically provisions the NEs by setting up the flow rules on them. It improves automation by using common APIs and provides a mechanism for the rapid deployment of new services [1]. The Data Plane is merely a collection of physical devices that are responsible for the flow of packets among NEs (Fig. 1).

SDN is the sum of North Bound Interfaces (NBI) and South Bound Interfaces (SBI) which allows the creation of a powerful and dynamic virtual network that enables SDN Controller to configure NEs flow table and allows for more sophisticated traffic management [3].

2 DoS Attack

"**CIA** Triad" [1] is a key term in the security field which refers to three major goals of security:

Confidentiality: It deals with ensuring the privacy and secrecy of information, i.e. no unauthorised user gets access to information and resources.

Integrity: It ensures that no tampering has been done on the data either on the storage or communication medium.

Availability: It ensures that information is available to the intended/right user, i.e. the user is legitimate and has the right to access that information. Ensuring timely information to the legitimate user is the key goal of availability.

DoS is one of the key attacks on the SDN Controller that affects the availability of information to legitimate users. In the DoS attack, a malicious attacker consumes all the network capacity by flooding packets on SDN Controller that attacked SDN Controller is left with no capacity to reply to the legitimate user (Fig. 2).

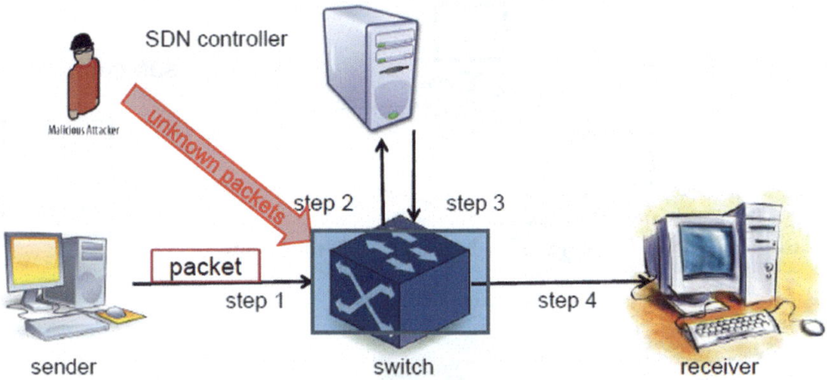

Fig. 2 DoS attack [5]

3 Remotely Triggered Blackhole Routing—Proposed Work

In this paper, the RTBH routing protocol has been proposed for SDN Controller with an aim that all packets from the malicious attacker will be dropped in no time by maintaining ACL (Access Control List). Both of these mechanisms, RTBH routing and ACL, can be applied as an edge on the SDN Controller for providing access to legitimate users. For defeating the DoS attack, SDN Controller can scale up to a level based on the size of ACL. It depends on how quickly a packet is accessed and filtered so that it can either be dropped or forwarded (Fig. 3).

RTBH routing algorithm triggers a discard route and maintains an ACL. ACL is updated from time to time to take care of the scalability issue too. The attacker ID is checked against an ACL. If the ID is not found in ACL, traffic is routed to discard

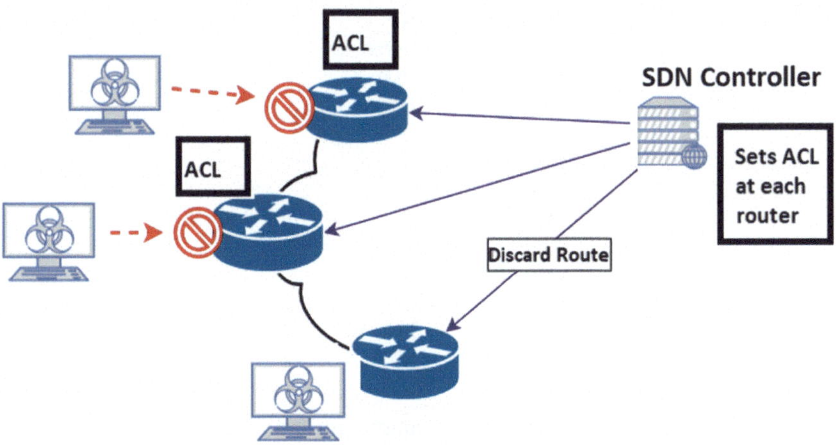

Fig. 3 Maintaining ACL and configuring discard route

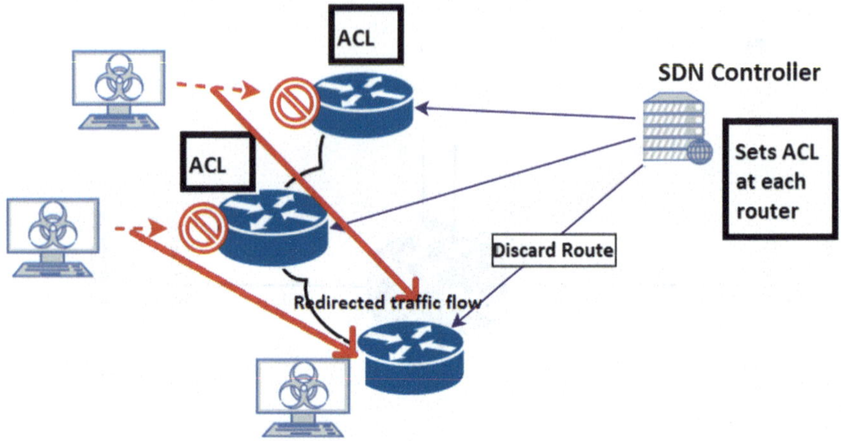

Fig. 4 Redirecting traffic to discard route

the route or else an entry is further checked into the flow table for forwarding the packet to a legitimate user (Fig. 4).

4 Analysis, Detecting and Quantifying DoS—Results and Discussion

A complete scenario is simulated on the virtual machine using a Python script. IP/TCP stack is used and the src—192.168.1.1 and destination 192.168.1.7 are the valid addresses.

```
version   = 4
ihl       = None
tos       = 0x0
len       = None
id        = 1
flags     =
frag      = 0
ttl       = 64
proto     = ip
chksum    = None
src       = 192.168.1.1
dst       = 192.168.1.7
\options   \
```

Figure 5 shows the Wireshark packet capture file at normal packet flow.

An attacker with the src—192.168.1.6 tries to flood the route with the packets so that legitimate user and controller do not connect.

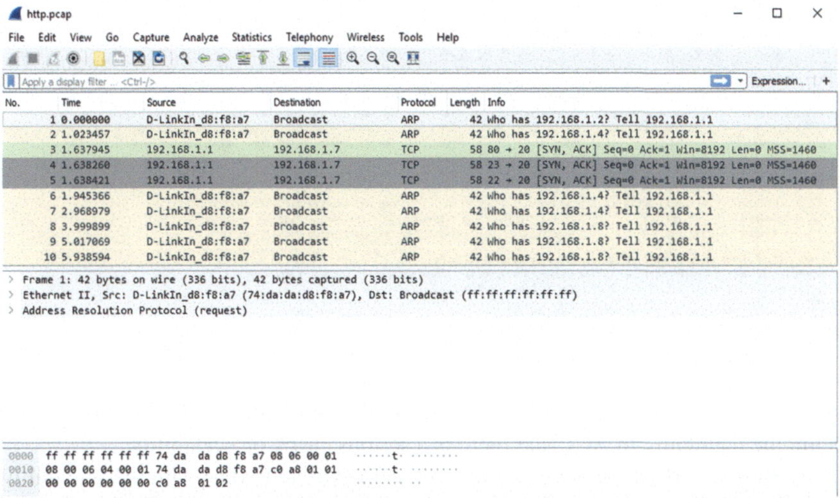

Fig. 5 Packet capture—normal flow

```
version   = 4
ihl       = None
tos       = 0x0
len       = None
id        = 1
flags     =
frag      = 0
ttl       = 64
proto     = ip
chksum    = None
src       = 192.168.1.6
dst       = 192.168.1.7
\options   \

Sent 21394 packets.
```

Figure 6 shows the Wireshark packet capture file at abrupt packet flow which identifies the source attacker and reroutes the packets to a configured discarded route.

5 Conclusion and Future Work

In this paper, we have discussed what a DoS attack is and what aims to consume network resources and hampers its performance. We highlighted the ways how it could be detected and analysed. We pointed out that in SDN, the control plane is at stake in the case of DoS. We mentioned and focused on Black Hole based routing

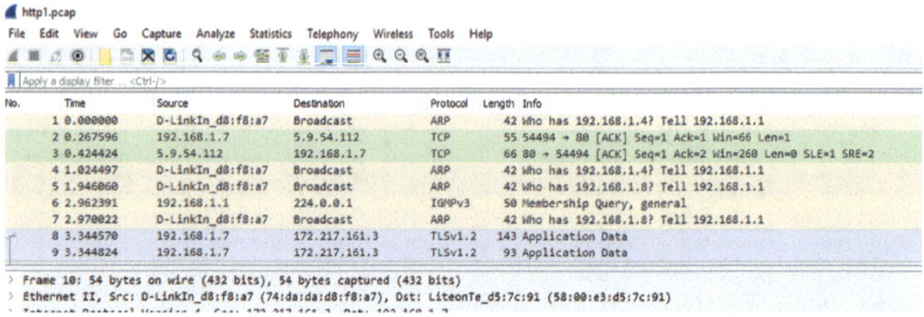

Fig. 6 Packet captures—DoS (Packets flooding)

which can serve as an edge to protect our SDN from DoS. No doubt, this routing technique maintains an ACL and there is a serious check on scalability.

For future work, there are many directions to explore to deal with the scalability issue. Our next set of experiments regarding this is in progress.

References

1. A. Owokade, How to configure remotely triggered black hole routing to protect from DDOS attacks (2017), https://sdn.ieee.org/newsletter/January-2016/sdn-in-the-cable-access-network
2. Open Networking Foundation, Software-defined networking: the new norm for networks. White Paper (2012), https://www.opennetworking.org/images/stories/downloads/sdn-resources/white-papers/wp-sdn-newnorm.pdf, pp. xvii, 14, 15, 53, 54
3. A. Mirchev, Survey of concepts for QoS improvements via SDN, in *Seminars FI/ IITM SS 15, Network Architectures and Services* (2015)
4. S. Hassas Yeganeh, Y. Ganjali, Kandoo: a framework for efficicient and scalable offloading of control applications, in *Proceedings of the first workshop on hot topics in software defined networks, HotSDN'12*, (New York, NY, USA), pp. 19–24; (ACM, 2012), pp. xvii, 16, 17
5. AdmelaJukan, Marcel Caria, Siquan Zhao, "Security in SDN", in proceedings of IEEE conference in IEEE, 2014
6. R. Sherwood, M. Chan, A. Covington, G. Gibb, M. Flajslik, N. Handigol, T.-Y. Huang, P. Kazemian, M. Kobayashi, J. Naous, S. Seetharaman, D. Underhill, T. Yabe, K.-K. Yap, Y. Yiakoumis, H. Zeng, G. Appenzeller, R. Johari, N. McKeown, G. Parulkar, Carving research slices out of your production networks with open ow. SIGCOMM Comput. Commun. Rev. **40**, 129, 130 (2010), xvii, 31, 32, 39
7. Open Networking Foundation, OpenFlow, White Paper, https://www.opennetworking.org/wp-content/uploads/.../openflow-switch-v1.5.1.pdf
8. J. Suh, H.-G. Choi, W. Yoon, T. You, T. Kwon, Y. Choi, Implementation of content-oriented networking architecture (CONA): a focus on DDoS countermeasure, in *1st European NetFPGA Developers Workshop* (2010), p. 34
9. R. Sherwood, G. Gibb, K.K. Yap, G. Appenzeller, M. Casado, N. McKeown, G. Parulkar, Flowvisor: a network virtualization layer. OpenFlow Switch Consortium, Technical report (2009)
10. Q. Yan, F. Richard Yu, Distributed denial of service attacks in software-defined networking with cloud computing. IEEE Commun. Mag. (2015)

11. K. Govindarajan, K.C. Meng, H. Ong, A literature review on software-defined networking (SDN) research topics, challenges and solutions, in *IEEE fifth international conference on advanced computing (ICoAC)* (2013)
12. A. Hakiri, A. Gokhalec, P. Berthou, D.C. Schmidtc, T. Gayraud, Software-defined networking: challenges and research opportunities for future internet. **75**(Part A), 453–471 (2014). Elsevier
13. D. Kerutz, P. Essteves, S. Azodolmolky, Software defined networking: a comprehensive survey. Proc. IEEE **103**(1) (2015)
14. A. Kucminski, A. Al-Jawad, P. Shah, R. Trestian, QoS-based routing over software defined networks, in *IEEE international symposium on broadband multimedia systems and broadcasting (BMSB)* (2017)
15. S. Pisharody, J. Natarajan, A. Chowdhary, A. Alshalan, A security policy analysis framework for distributed SDN-based cloud environments. IEEE Trans. Dependable Secur. Comput. **PP**(99) (2017)
16. I. Ahmad, S. Namal, M. Ylianttila, A. Gurtov, Analysis of deployment challenges of host identity potocol (IEEE, 2017)
17. Noxrep, POX: OpenFlow Controller (2015), https://www.opennetworking.org/wp-content/uploads/2014/10/Principles_and_Practices_for_Securing_Software_Defined_Networks_applied_to_OFv1.3.4_V1.0.pdf
18. Hengky Hank Susanto, Sing Lab, Introduction to SDN, www.cse.ust.hk/~kaichen/courses/spring2015/comp6611/.../SDN-presentation.pptx
19. B. Heller, N. Handigol, V. Jeyakumar, N. McKeown, D. Mazières, Where is the debugger for mySoftware-Defined Network? In *HotSDN* (2012)
20. Z. Hu, M. Wang, X. YAN, Y. YIN, A comprehensive security architecture for SDN, in *The IEEE proceedings of 18th international conference on Intelligence in Next Generation Networks* (IEEE, 2015)
21. S. Shin, G. Gu, Attacking software-defined networks: a first feasibilitystudy, in *Proceedings of the second ACM SIGCOMM workshopon Hot topics in software defined networking* (ACM, 2013), pp. 165–166
22. E. Al-Shaer, S. Al-Haj, Flowchecker: configuration analysis and verification of federated openflow infrastructures. in *Proceedings of the 3rd ACM workshop on assurable and usable security Configuration* (2010)
23. P. Porras, S. Shin, V. Yegneswaran, M. Fong, M. Tyson, G. Gu, A security enforcement kernel for OpenFlow networks, in *Proceedings of the first workshop on Hot topics in software defined networks* (ACM, 2012), pp. 121–126
24. S. Shin, G. Gu, CloudWatcher: network security monitoring using OpenFlow in dynamic cloud networks (or: How to provide security monitoring as a service in clouds?), in *20th IEEE international conference on network protocols (ICNP)*. (IEEE, 2012), pp. 1–6
25. D. Kordalewski, R. Robere, A dynamic algorithmfor loop detection in software defined networks (2012)
26. A. Khurshid, W. Zhou, M. Caesar, P. Godfrey, Veriflow: verifying network-wide invariants in realtime. ACM SIGCOMM Comput. Commun. Rev. **42**(4), 467–472 (2012)
27. H. Yang, S.S. Lam, Real-time verification of networkproperties using atomic predicates, in *ICNP, The IEEE international conference on network protocols* (2013)
28. Z. Hu, J. Luo, Cracking network monitoring inDCNs with SDN, in *2015 IEEE conference on computercommunications, INFOCOM 2015* (Hong Kong, China, 2015), pp. 199–207
29. F. Chen, B. Bruhadeshwar, A.X. Liu, Privacypreservingcross-domain network reachability quantification, in *2011 19th IEEE international conference on network protocols (ICNP)* (IEEE, 2011), pp. 155–164
30. M.J. Freedman, K. Nissim, B. Pinkas, Efficientprivate matching and set intersection, in *Advances in Cryptology-EUROCRYPT 2004* (Springer, Berlin, 2004), pp. 1–19
31. N. Chellani, P. Tejpal, P. Hari, V. Neeralike, Enhancing security in openflow, in *IEEE Proceedings* (2016)
32. O.I. Abdullaziz, Y.-J. Chen, L.-C. Wang, Lightweight authentication mechanism for software defined network using information hiding (IEEE, 2016)

33. S.M. Mousavi, M. St-Hilarie, Early detection of DDoS attacks against SDN controllers (EEE, 2015)
34. IBM Knowledge Center, An overview of the SSL or TLS handshake (2017), http://www.ibm.com/support/knowledgecenter/en/SSFKSJ_7.1.0/com.ibm.mq.doc/sy10660_.htm. Accessed 7 Feb 2017

An Improved Authentication and Key Agreement Protocol for Smart Healthcare System in the Context of Internet of Things Using Elliptic Curve Cryptography

Uddalak Chatterjee, Dipanwita Sadhukhan and Sangram Ray

Abstract Internet of Things (IoT) is heterogeneous in nature and has diverse application areas. Smart healthcare (s-health) is one of the most significant applications of IoT that handles sensitive health-related data, which can be manipulated to cause fatal safety consequences. Hence, security and privacy must be considered as the most important factors during the design of wearable s-health. Therefore, it is necessary to establish a secure communication channel between wearable sensors or devices and the remote server of the IoT network. In this work, we propose a protocol for authentication and key agreement using ECC (Elliptic Curve Cryptography) to protect the health-related data in s-health. In this work, trusted sensors take the responsibility of completing the complex cryptographic operations on behalf of resource-constraint sensors. The smartphones accelerometers data and the data received from sensors installed on the body are used to do the trust verification by finding data correlations. Moreover, through security and performance analysis we have shown that the proposed protocol is secure against all relevant cryptographic attacks and is lightweight due to the use of ECC.

Keywords Internet of things · Mutual authentication · Key management · Resource-constrained sensors · ECC (Elliptic curve cryptography) · Body area networks

U. Chatterjee · D. Sadhukhan (✉) · S. Ray
Department of Computer Science and Engineering, National Institute of Technology Sikkim, Ravangla 737139, Sikkim, India
e-mail: dipanwitasadhukhan2012@gmail.com

U. Chatterjee
e-mail: uddalak.udi@gmail.com

S. Ray
e-mail: sangram.ism@gmail.com

M. Dutta et al. (eds.), *Proceedings of International Conference on IoT Inclusive Life (ICIIL 2019), NITTTR Chandigarh, India*, Lecture Notes in Networks and Systems 116,
https://doi.org/10.1007/978-981-15-3020-3_2

1 Introduction

Importance and issues of s-health: IoT integrates physical objects irrespective of their resource and computing capabilities or type of networks [1] for seamless communication and transmission of data over the Internet. It involves devices with various sensing, measuring and data capture capability sensors, RFID, etc., to achieve intelligent identification and monitoring. One of such standalone IoT device is smart wearable devices that have become an essential part of our daily life. These have a wide area of implementation such as healthcare applications, smart home applications, sports and fitness tracking applications, etc. Wearable devices like Fitbit, smartwatches, etc., in health care have become very common in the present life. The main issues that arise in this context and considered in this paper are – *(i)* how the remote server and the resource-constrained sensors are authenticating each other to be paired; and *(ii)* how to balance users' privacy and usability. Moreover, the sensor nodes have very less computing capabilities and constrained resources in the area of the potential of processing, memory, communication bandwidth, and energy especially in s-health. Since the sensors are of different resource types in nature, the same encryption technique cannot be used for all sensors. Hence, security measures must be pre-embedded. To achieve high-security strength, it is necessary to establish secure key management among the communicating parties. In general, the symmetric encryption algorithms such as AES are well suited to achieve sound security of constraint resources with processing capability. However, due to the high communication overhead and huge storage requirements, the symmetric key encryption techniques are not suitable for this purpose [2]. Asymmetric key cryptography has also its own demerits like high computation and communication overheads which are the prime concern for sensors with less resource and computing capability. Other established security measures also could not be applied due to less resources, low computing capability, and heterogeneity of sensors implanted or worn in the body in IoT-based healthcare system. To prevent sensitive health data from malicious activity, end-to-end (E2E) data protection is crucial.

Our contribution: In this paper, the complex cryptographic operations which need high computing capability and resources are offloaded from less resourceful sensors to resource-rich correlated sensor nodes. Thus, we have proposed an enhanced and lightweight protocol using ECC for s-health in the context of IoT.

Organization of the paper: The structure of the rest of the paper is as follows. Section 2 highlights the summary of the related works; Sect. 3 describes the assumptions and preliminaries for selection; Sect. 4 describes the proposed key establishment protocol; Sect. 5 focuses on the security evaluations of the proposed protocol; Sect. 6 illustrates performance analysis, and finally Sect. 7 concludes the work with an overview of future work directions.

2 Related Work

Authentication and key agreement protocols are mainly motivated to design for establishing a secure key between the communicating entities for secure data transfer as well as to achieve the primitive security objectives such as protection of users privacy or confidentiality of the users credentials that are transmitted over the network; maintain the integrity of the transferred message as well as providing required services to be availed by the legitimate user.

Many different security measures for providing security in IoT devices and other remote servers, in the network have been proposed and implemented to achieve secure authentication between endpoints of the network. Among them, many have proposed public-key cryptography, but could not become successful due to significant computational and processing overheads. Therefore, researchers are aiming to reduce overheads which occur during key agreement and authentication handshake. In the case of resource-constraint sensors, the use of public-key cryptography in such a scenario will increase the overhead since the resources are limited. In [3], the authors have implemented an authentication scheme which is a two-way scheme for the IoT using PKI. X.509 certificates and RSA public keys with DTLS handshake are being used in this approach, which is not suitable as it generates a high amount of network traffic. Other existing protocols using compression of IPV6 header, extension headers, and UDP (User Datagram Protocol) header have been proposed which is standard in 6LoWPAN. To authenticate header and encapsulating security, 6LoWPAN compressions have been presented by the authors in [1] for IPsec payload header. In another interesting approach, the authors in [4] proposed a solution based on the proxy to move the heavy cryptographic operations from a less resourceful device to resource-rich sensor nodes. Although the authors [1] assumed the assisting nodes to be trustworthy, the authors have not described any trust determination technique. Moreover, in the scheme [1] the re-keying processes of compromised proxies are not described, which behaves irrationally or does not perform their assigned task. (n, k) threshold scheme and polynomial interpolation by Lagrange for reconstituting the sender's Diffie–Hellman's public key are considered in these protocols [1, 4] where reconstructing sender's DH public-key k polynomial shares are sufficient. However, the potential threat of revealing the secret key could not be avoided as they [1, 4] have overlooked the threat of revealing these secret keys through cryptanalysis although the shared symmetric keys are considered secure from brute force attacks. The requirement of IoT-based application is to have a secure end-to-end connection between devices. Both the remote server and resource-constrained sensor require having a secure E2E communication link between them that means both ends first need to authenticate each other and securely establish a secret session key to encrypt the transmitted data. In this regard in [2] Diffie–Hellman (DH) security protocol is used in their proposed key agreement and distribution protocol. The security of this key scheme lies in the difficulty of computing discrete algorithms that is very hard to break and also achieves the perfect forward secrecy property. DH key establishment protocol requires both ends to agree first on the appropriate prime (p) and generator (g) numbers. If remote server A wants to communicate with a highly resource-constrained sensor, an E2E key distribution protocol is needed to establish a secure channel between them for further communications. In accordance with this paper [2] the public keys for node B will be generated in server side and eventually the session key will be generated using the DH algorithm. In this research, an enhanced version of the protocol developed in [2] with the same network model explained in [2] is proposed with much lesser communication and computation overheads. The proposed scheme also reduced the existing limitations of the previous protocol [2].

3 Preliminaries

The essential preliminaries for the proposed protocol are discussed as follows:

A. *Finding correlation and selection of assisting sensor nodes*

The sensor nodes have a triaxial accelerometer which is the same as the smartphone to compare data directly for correlation. Accelerometers detect force acting in the opposite direction to the displacement vector. The magnitude of these three axes of accelerometers is used for data correlation. We can get the rate of change of speed over time for a particular sensor. We can find how two signals correlate in the frequency domain by finding coherence, which is the cross-spectral density of each individual signal [2]. After the correlation is found, the selection of assisting trusted sensors in the proximity of the resource-constraint sensor node is done by finding whether they are installed on the body or not. If any of the sensor nodes data is not correlated with the gateway nodes accelerometer's data, then it implies that the sensor is not trusted and therefore could not be assigned any task related to key agreement mechanism.

B. *Network Model*

The network model for the s-health system consists of different sensors with different resource limitations. They are accompanied by triaxial accelerometers which are wearable or implanted on the body of a person and a smartphone working as a gateway. The implanted or wearable sensors are highly resource constraint, therefore they are unable to perform cryptographic operations. Some sensors which are less resource-constraint are able to do heavy cryptographic computations for the key agreement protocol. Others are the devices or sensors with no constraint on resources like remote servers. Now if the remote server wants to communicate with a highly resource-constraint sensor then an E2E key distribution protocol is essential to establish a secure channel for further communications. In this regard, the resource-constraint sensor offloads its heavy computational task to the neighboring trusted sensors which are installed on the body and are correlated.

C. *Elliptic Curve Cryptography (ECC)*

The elliptic curve cryptosystem [5, 6] was initially proposed by Koblitz [7] and then Miller [8] in 1985 to design public-key cryptosystem and presently, it becomes an integral part of the modern cryptography. Let E/F_p denotes an elliptic curve E over a prime finite field F_p, can be defined by the following equation:

$$Y^2 = x^3 + ax + b, \tag{1}$$

where $a, b \in F_p$ and the discriminate D such that, $D = 4a^3 + 27b^2 \neq 0$.

The points on E/F_p together with an extra point O called the point at infinity used for additive identity form an additive group A as

$$A = \{(x, y) : x, y \in Fp, E(x, y) = 0\}u\{0\} \tag{2}$$

Let the order of A be n, which is very large and it can be defined as $n \times G \bmod q = O$, where G is the generator of A.

The A be a cyclic additive group under the point addition "+" defined as follows:

$$P + O = P$$

where $P \in A$. The scalar point multiplication over A can be defined as

$$tP = P + P + \ldots P\,(t\,times) \tag{3}$$

The hardness of ECC depends on solving the computational problems of ECC such as ECDLP in polynomial time.

4 Proposed Authentication and Key Agreement Protocol

In our approach, if a remote server R_s wants to communicate with resource-constraint node the gateway reports the *IDs* of neighboring assisting nodes installed on the body by the data correlation process [2]. The sensor is said to be compromised when the neighboring sensor accelerometer's data and gateway accelerometer's data fails to correlate. The asymmetric key agreement involves heavy cryptographic computations. Since the node R_{c_s} is resource constraint it distributes the heavy tasks to its neighboring trusted sensors installed on the body for assistance. In our proposed protocol for key agreement and distribution, ECC-based Diffie–Hellman (ECDH) protocol is used since ECC based scheme is much lighter and secure than the DH method [9–12, 15–19]. The proposed key agreement protocol is designed considering the following initial assumptions (Table 1).

Table 1 Interpretation of symbols and notations used

Symbols	Interpretation
R_s	Remote Server
R_{c_s}	Resource-Constraint Sensor
A_{t_s}	Assisting trusted sensor
ID_S	Remote Servers Identity
ID_R	Identity of resource-constraint sensor
K_{ECDH}	Session key
(s, P_s)	Private key, public-key pair of the remote server
N_i	Random Nonce
$(b_{i,}\ T_i)$	Private key, public-key pair of resource-constraint sensor
E/D	Symmetric Key Encryption, Decryption
$\|\|$	Concatenation symbol
K_{ir}	Pairwise keys between Resource-Constraint node and trusted sensors
F_p	A finite field
P	A point on Elliptic Curve
E/F_p	An elliptic curve defined on F_p with prime order n

A. *Assumptions*

All the assumptions that are required for executing the protocol are mentioned below.

i. Each sensor node has a pairwise secret key with its neighboring nodes/trusted devices as $K_1, K_2, K_3, \ldots K_n$ Each sensor nodes can discover a set of high-resource trusted nodes In their proximity by finding the data correlations using gateway nodes data and accelerometer
ii. The field size whose length is about 512 bits in general applications is either an odd prime or a power of 2.
iii. An elliptic curve E over F_p is $E : y^2 = x^3 + ax + b$ $E{:}y^2 = x^3 + ax + b$, where the two field elements a, b are members of F_p and $4a^3 + 27b^2 \ (\mathrm{mod}\ p) \cdot 0$; and all $x, y \in F_p$.
iv. P is a base point of order n over $E/(F_p)$, where n is a large prime.
v. "." is the multiplication operation in p, where $p\ p \in [2, n-2]$, p contains a set of integers such as master and private keys, identities and random numbers.
vi. $X(P)$ is the value of point P's x-coordinate, where P is an elliptic curve point order n over $E/(F_p)$.

B. *Key Agreement Protocol*

The proposed key agreement protocol is described below in a stepwise manner, where $X \to Y : M$ means sender X sends message M to receiver Y.

Step 1: $\mathrm{R_s} \to \mathrm{R_{c_s}} : ID_s\ N_i$

Remote server generates a data request along with its identity ID_s and a random number (Nonce N_i).

Step 2: $R_{c_s} \rightarrow A_{t_s} : E_{K_i}(b \cdot ID_R)$

After receiving the identity of remote servers the implanted resource-constrained sensor

R_{c_s} verifies ID_S and the message freshness by checking nonce N_i. Then, R_{c_s} divides its private key b into small fragments $b_1, b_2, b_3 \ldots b_i$ to assign the fragments to the neighboring trusted assisting sensors. The R_{c_s} encrypts the fragments along with ID_R using pairwise secret key K_i to send a to the message containing $E_{K_i}(b_i \cdot ID_R)$ assisting trusted sensors.

Step 3: $A_{t_s} \rightarrow R_s : T_1 \ldots T_i$, *Auth_Proof, Sign* T_i, Ts_1

Upon receiving $E_{K_i}(b \cdot ID_R)$, the trusted sensor decrypts the same and compute the public-key shares $(T_1 \ldots T)_i$ of the resource-constraint sensor.

$$T_1 = b_1 \cdot P$$
$$T_2 = b_2 \cdot P$$
$$\vdots$$
$$T_i = b_i \cdot P$$

Then it sends the public-key shares along with *Auth_proof, Sign* T_i and Timestamp Ts_1 to the remote server (Fig. 1).

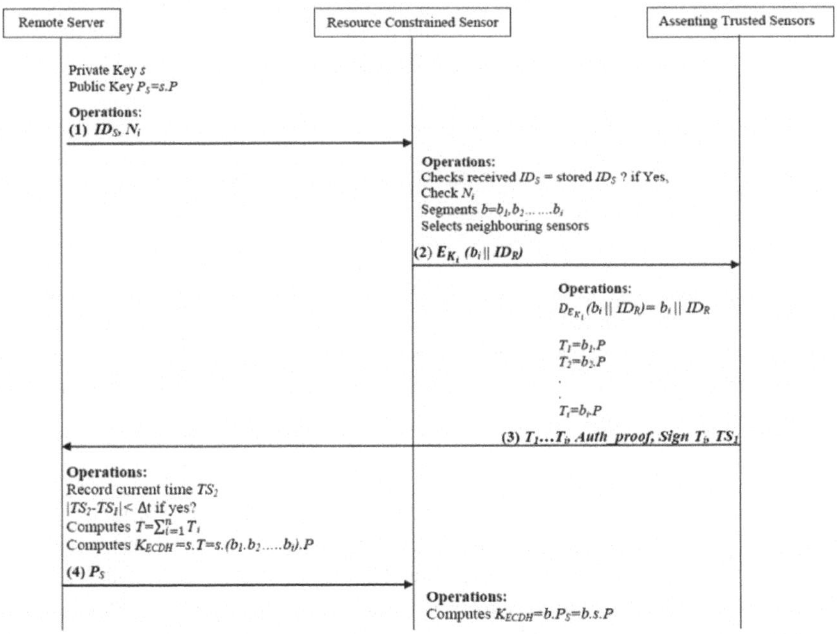

Fig. 1 Proposed key agreement protocol

Step 4: $\mathrm{R_s} \rightarrow \mathrm{R_{c_s}} : P_s$

Remote server first checks the timestamp with its recorded current time for freshness, then computes the public key *(T)* of resource-constraint sensor $T = \sum_{i=1}^{n} T_i$. Then remote server computes ECDH-based session key $K_{ECDH} = s \cdot T = s \cdot (b_1, b_2, \ldots b_i) \cdot P$. Finally, remote server sends its public key P_s to resource-constraint sensor.

Similarly, the resource-constraint sensor also computes ECDH-based same session key $K_{ECDH} = b \cdot P_s = b \cdot s \cdot P$. Then, it finally sends a message to the server for key agreement.

Now a secure communication channel is established between the remote server and resource-constraint sensor.

Proposition 1 *We compute the public key of resource-constraint sensor in remote server side as*

$$\begin{aligned} T = b \cdot P = &\quad (b_1 + b_2 + \cdots + b_i) \cdot P \\ = &\quad b_1 \cdot P + b_2 \cdot P + \cdots + b_i \cdot P \\ = &\quad T_1 + T_2 + T_3 + \cdots + T_i \end{aligned}$$

Proposition 2 : *Proof*: $s \cdot T = b \cdot P_s$:

$$L.H.S = s \cdot T = s \cdot \sum_{i=1}^{n} T_i$$

$$= s \cdot (b_1 \cdot P + b_2 \cdot P + \cdots + b_i \cdot P)$$

$$= s \cdot (b_1 + b_2 + b_3 + \cdots + b_i) \cdot s \cdot P$$

$$= b \cdot P_s = R.H.S$$

5 Security Analysis

The security of any key agreement protocol for the IoT environment is most crucial although it is very difficult to achieve [13, 14]. In this section, various existing security breaches applicable to the proposed scheme are considered and analyzed the robustness of the proposed scheme against those attacks in detail to guarantee that the proposed scheme is secure.

A. *Man-in-the-Middle Attack*:

The security of the proposed scheme largely depends on the trustworthiness of the assisting sensors. Initially, the pre-shared keys between the resource-constrained sensor and assisting trusted sensors are exchanged during the bootstrapping phase. Moreover, only the correlated trusted sensors can assist in calculating the key shares of the resource-constrained sensors. If any sensor is not correlated with the device means that the specific sensor is not trusted and also not involved in the key agreement. So, any adversary cannot pretend to be present as a neighboring trusted assisting sensor. Besides, if any man/adversary intends to be present in between the remote server and the resource-constrained node he/she cannot be able to generate the ECDH-based session key without the knowledge of the secret private key of the participating remote server. Due to the hardness of the ECC, it also cannot be calculated from the public key of the remote server. Key fragments of the private key of the resource-constrained sensors are exchanged over the channel encrypted using pre-shared key and fragments are individually computed by the assisting trusted sensors. Hence, it is can be said that no man/adversary present with the network can harm the protocol. So, the protocol is secure against man-in-the-middle attacks.

B. *Denial of Services (DoS)*:

In the proposed protocol, the communications are done through the trusted sensor nodes, and as without the approval of the user it is not possible to install sensors, and therefore this protocol stops the chances of DoS attack from any compromised or fake sensors. There is no chance that a malicious node will try and interrupt by sending redundant messages to the responder.

C. *Replay Attack and Integrity*:

The proposed key agreement protocol initially uses the nonce between the remote server and resource-constraint sensors to check the message freshness and then it uses timestamp between trusted sensors and the remote server for communication hence ensure message freshness and integrity and leaves out chances of any replay attack.

D. *Scalability*:

In the proposed network model, adding new sensors is allowed in the initialization phase to create a shared pairwise key with other sensors. After successful initialization new sensor is able to create a secure channel with a remote server which makes the protocol scalable.

6 Performance Analysis

The performance of the proposed protocol is analyzed and compared in this section with the previous key agreement protocol proposed by Iqbal et al. [2]. In the proposed protocol, the key size is chosen to be 512 bits which is smaller than the previous method (1024 bits) and provides better security, less computation and communication cost as well. Table 2 showing the time requirements for the cryptographic operations used in both the schemes. The standard size of the ID of the resource-constraint sensor is 128 bits. The comparison tables showing computational and communication overheads are given in Tables 3 and 4, respectively.

The test programs and computational operations in both the cases were executed on an intel i5 processor. From the above comparative performance analysis, we can

Table 2 Time requirement for cryptographic operations

Symbols	Explanation	Time *(ms)*
$T_{SE/D}$	Time for symmetric encryption/decryption	*3.8500*
T_{SM}	Time for point multiplication	*2.226*
T_{PA}	Time for point addition	0.0288
T_{ME}	Time for modular exponentiation(1024 bits)	*3.8500*
T_{MAC}	Time to compute MAC	*0.0046*
i	No of trusted sensors	–

Table 3 Comparison of computational cost

Protocol	Resource-constraint sensor	Asst. trusted sensor	Remote server
Iqbal et al. [2]	$i{*}T_{SE/D} + T_{MAC} + T_{ME} + i * T_{SE/D}$ $= (3.8546 + 2i{*}7.7)ms$	$3{*}T_{SE/D} + 2{*}\,T_{ME}$ $= 19.25\ ms$	$2{*}T_{ME} + T_{MAC} + T_{SE/D}$ $= 11.5546\ ms$
Proposed Scheme	$i\,{*}T_{SE/D} + T_{SM}$ $= (3.85{*}i + 2.226)ms$	$T_{SE/D} + T_{SM}$ $= 6.076\ ms$	T_{SM} $= 3.85\ ms$

Table 4 Comparison of communication cost

Protocol	Resource-constraint sensor
Iqbal et al. [2]	*1024 + 128*i bits*
Proposed Scheme	*512 + 128*i bits*

easily infer that our proposed protocol has less overheads than the previous protocol proposed by Iqbal et al. [2]. In the previous protocol, modular operations were being used which requires higher execution time and as a result, the total computational overhead becomes higher than our protocol, where point multiplication operation is being used. On the other hand, while comparing the communication overhead the previous protocol was using a key size of 1024 bits for RSA encryption. However, in our protocol, the key size of 512 bits is being used for ECC. We could have used 160-bit key size but to achieve even greater security the key size is chosen to be 512 bits. So the communication overhead is also minimized in our proposed protocol.

7 Conclusion

In this work, we have developed an improved scheme to establish a secure E2E communication channel for an s-health system using elliptic curve cryptography. The proposed scheme works by offloading the heavy cryptographic operations to correlated assisting trusted sensors. The security analysis and performance evaluation reveal that the proposed protocol is more secure as well as lightweight in terms of computation and communication overheads compared to the Iqbal et al. scheme [2]. Thus, the proposed scheme is efficient, reliable, sustainable, and practically implementable.

References

1. S. Raza, S. Duquennoy, T. Chung, D. Yazar, T. Voigt, U. Roedig, Securing communication in 6LoWPAN with compressed IPsec, in *2011 International Conference on Distributed Computing in Sensor Systems and Workshops (DCOSS)* (IEEE, 2011), pp. 1–8
2. M.A. Iqbal, M. Bayoumi, A novel authentication and key agreement protocol for internet of things based resource-constrained body area sensors, in *2016 IEEE 4th International Conference on Future Internet of Things and Cloud Workshops (FiCloudW)* (IEEE, 2016), pp. 315–320
3. T. Kothmayr, C. Schmitt, W. Hu, M. Brünig, G. Carle, DTLS based security and two-way authentication for the Internet of Things. Ad Hoc Netw. **11**(8), 2710–2723 (2013)
4. Y.B. Saied, A. Olivereau, D. Zeghlache, M. Laurent, Lightweight collaborative key establishment scheme for the Internet of Things. Comput. Netw. **64**, 273–295 (2014)
5. A. Menezes, SA (Scott Alexander) Vanstone, *Guide to Elliptic Curve Cryptography*. (Springer, New York, 2004)
6. N. Koblitz, Elliptic Curve Cryptosystem. J. Math. Comput. Jan. **48**(177), 203–2009 (1987)
7. V.S. Miller, Use of elliptic curves in cryptography, in *Conference on the Theory and Application of Cryptographic Techniques* (Springer, Berlin, 1985), pp. 417–426
8. S. Ray, G.P. Biswas, Establishment of ECC-based initial secrecy usable for IKE implementation, in *Proceedings of the World Congress on Engineering*, vol. 1 (2012)
9. J. Athena, V. Sumathy, K. Kumar, An identity attribute–based encryption using elliptic curve digital signature for patient health record maintenance. Int. J. Commun. Syst. **31**(2), e3439 (2018)
10. A.P. Haripriya, K. Kulothungan, ECC based self-certified key management scheme for mutual authentication in Internet of Things, in *2016 International Conference on Emerging Technological Trends (ICETT)* (IEEE, 2016), pp. 1–6
11. D. Sadhukhan, S. Ray, Cryptanalysis of an elliptic curve cryptography based lightweight authentication scheme for smart grid communication, in *2018 4th International Conference on Recent Advances in Information Technology (RAIT)* (IEEE, 2018), pp. 1–6
12. S. Ray, G.P. Biswas, M. Dasgupta, Secure multi-purpose mobile-banking using elliptic curve cryptography. Wireless Pers. Commun. **90**(3), 1331–1354 (2016)
13. M. Turkanović, B. Brumen, M. Hölbl, A novel user authentication and key agreement scheme for heterogeneous ad hoc wireless sensor networks, based on the Internet of Things notion. Ad Hoc Netw. **20**, 96–112 (2014)
14. M.S. Farash, M. Turkanović, S. Kumari, M. Hölbl, An efficient user authentication and key agreement scheme for heterogeneous wireless sensor network tailored for the Internet of Things environment. Ad Hoc Netw. **36**, 152–176 (2016)
15. R. Amin, N. Kumar, G.P. Biswas, R. Iqbal, V. Chang, A light weight authentication protocol for IoT-enabled devices in distributed Cloud Computing environment. Future Gen. Comput. Syst. **78**, 1005–1019 (2018)
16. Y. Yang, L. Wu, G. Yin, L. Li, H. Zhao, A survey on security and privacy issues in Internet-of-Things. IEEE Internet Things J. **4**(5), 1250–1258 (2017)
17. X. Li, J. Niu, M.Z.A. Bhuiyan, F. Wu, M. Karuppiah, S. Kumari, A robust ECC-based provable secure authentication protocol with privacy preserving for industrial Internet of Things. IEEE Trans. Industr. Inf. **14**(8), 3599–3609 (2017)

18. A.A. Alamr, F. Kausar, J. Kim, C. Seo, A secure ECC-based RFID mutual authentication protocol for internet of things. J. Supercomput. **74**(9), 4281–4294 (2018)
19. S. Kumari, M. Karuppiah, A.K. Das, X. Li, F. Wu, N. Kumar, A secure authentication scheme based on elliptic curve cryptography for IoT and cloud servers. J. Supercomput. **74**(12), 6428–6453 (2018)

Highly Scalable Block Cipher Encryption in MapReduce-Based Distribution System

Prashant Verma, Gurjit Singh Walia and Kapil Sharma

Abstract The paper presents the Scalable Modular Hadoop Encryption Framework (SMEH). High performance and low overhead encryption mechanisms are introduced to realise secure operations on sensitive data. The framework uses MapReduce to scale up the processing of linear encryption algorithms to the performance levels of distributed and parallel algorithms. Proposed distributed and modular approach to plugin not only optimally process a large amount of data but also solve the hurdle of memory management. In addition, the modular approach also allows for the algorithm and its implementation to be tuned according to the user's tolerances.

Keywords Big data, Hadoop · Security · Encryption · Hashed keys · Low overhead

1 Introduction

In today's digital world, data size is growing rapidly by means of social media, online shopping, smartphones, etc. The volume, velocity and veracity of big data play an important role, which can be explicitly revealed from the reality that in the year 2012–2013, there exists only a few dozen terabytes of data in a single dataset, which has increasingly been shifted to a large number of petabytes. Hadoop is an open-source framework that allows the distributed handling for large datasets on multiple nodes/machines, and therefore generates good results using MapReduce that is iterative in nature. Iterative MapReduce plays an important role in data analysis performance and this can be implemented on Hortonworks Data Platform Sandbox

P. Verma (✉) · G. S. Walia
SAG, DRDO, Delhi, India
e-mail: prashantperot@gmail.com

G. S. Walia
e-mail: gurjit.walia@gmail.com

K. Sharma
Department of Information Technology, Delhi Technological University, Delhi, India
e-mail: kapil@ieee.org

M. Dutta et al. (eds.), *Proceedings of International Conference on IoT Inclusive Life (ICIIL 2019), NITTTR Chandigarh, India*, Lecture Notes in Networks and Systems 116,
https://doi.org/10.1007/978-981-15-3020-3_3

known as Cloudera Hadoop [1]. We are present in the era of data and it is not easy to calculate the amount of data. The sources and ways of data generation are varied and keep changing but the technology built to utilise this data has largely entered around Hadoop and similar frameworks [2]. Hadoop is a framework that supports the running of big data applications and it is written in Java language [3]. As the centre of Hadoop are Hadoop Distributed File System and MapReduce and both surely lack comprehensive security architecture of framework which provides security to the data beyond access control. MapReduce and HDFS have been popular because of their affinity to highly scalable distributed programming or computing [2]. Due to their distributed nature, they are able to deal with very large amounts to data with relative ease as they follow a recursive divide and conquer approach to simplify the operations on large-scale data. Because of their distributed nature, they also enjoy immense processing capacity using thousands of nodes, simply deploying its Mapper and Reducer tasks. MapReduce algorithm provides highly parallel and distributed algorithms based on divide and conquer approach across a very large set of data. The notion is to divide a larger job into smaller jobs that are tackled by different machines. On the other hand, Hadoop breaks the input into inalterable pieces called input splits to a MapReduce job and creates one mapper job for each individual split that runs the user-defined mapper function for every record in the split. Therefore, the processing time for the whole input is very large as compared to the processing time for each individual split. HDFS runs two types of node one is a master called NameNode and the other is working/storage nodes called DataNodes of HDFS [4]. Data files are actually situated in Hadoop, known as DataNode providing the core storage. However, NameNode maintains an index of where the data is actually stored [5].

Hadoop framework was never built with secure application architecture. Generally, Hadoop was built to run on publicly accessible data over a public network with trusted job submissions from trusted users. Over time the use case of Hadoop has changed and so have the security requirements of Hadoop. Initially, there was no security model for Hadoop. The user was not authenticated and the user could impersonate and run potentially malicious code on the system. As Hadoop was designed in a way to run over a parallel distributed cluster, the code can be submitted by anybody without any prior authorization. Although file permissions were placed inside Hadoop, they can be easily circumvented by impersonation or unauthorised code execution [6]. Over time security systems have been placed inside Hadoop but most of them could not either tackle threats or provide fundamental security to the underlying confidential data [7]. Currently, a major security issue faced by Hadoop is the integrity of data stored in HDFS [8]. In sum, major Security Risks in Hadoop and associated technologies can be summarised as follows:

- Hadoop does not enforce mandatory user authentication thus unauthorised users can easily access the system using RPC or HTTP [9].
- Authentication and Access Control are non-existent at data nodes, and therefore data nodes are highly insecure create a serious lapse in data security.

- Super-User or System can execute any kind of malicious query without any checks and blocks creating a potential security hazard for the System.
- An attacker can pose as a Hadoop service and spoof other services to get access to sensitive data.

The rest of the paper is organised as follows. In Sect. 2, the proposed methodology for data at rest encryption is introduced. Experimental validation of the proposed framework is discussed in Sect. 3. Finally, conclusion and future direction are sketched in Sect. 4.

2 Proposed Methodology

The proposed work utilises technologies such as Hadoop and MapReduce to provide security for data at rest. The approach is based on fast encryption algorithms like PRESENT and Twofish to provide encryption of data within the map and reduce phases of the data to provide a high multi-level encryption to the data and also to create a strong decryption methodology to provide reclamation of data without very large overheads [4]. Figure 1 depicted the architecture of the proposed framework for the security of data at rest. In the proposed framework, the structured data splits and passes to the mapper where key, value paired data is extracted. The key is extracted from files as well as passes to the reducer where the hash of the keys has to be found. The combined key repository contains keys extracted from files. We obtained the semi-encrypted structured data from Reducer Phase and passed to the multi-MapReduce-based PRESENT algorithm instances to obtain multi-level encrypted data. The architecture of the proposed system contains two independent modules:

i. High-Performance Encryption and Decryption of Data.
ii. High-performance classifier algorithm that can help with piecewise decryption of data to reduce overheads in accessing the encrypted data.

The details of these modules are presented below:

2.1 Encryption Module

The encryption module is itself divided into three phases namely Scanning Phase, Extraction Phase and Encryption Phase and depicted in Fig. 2. These different phases can be used in various combinations and settings to provide differential levels of encryption as per the requirements of the user.

A. *Scanning Phase*

In the scanning phase, the data is stuffed into the machine for examining purpose. In order to avoid misplacement of data in the loading phase, each of mapper and reducer

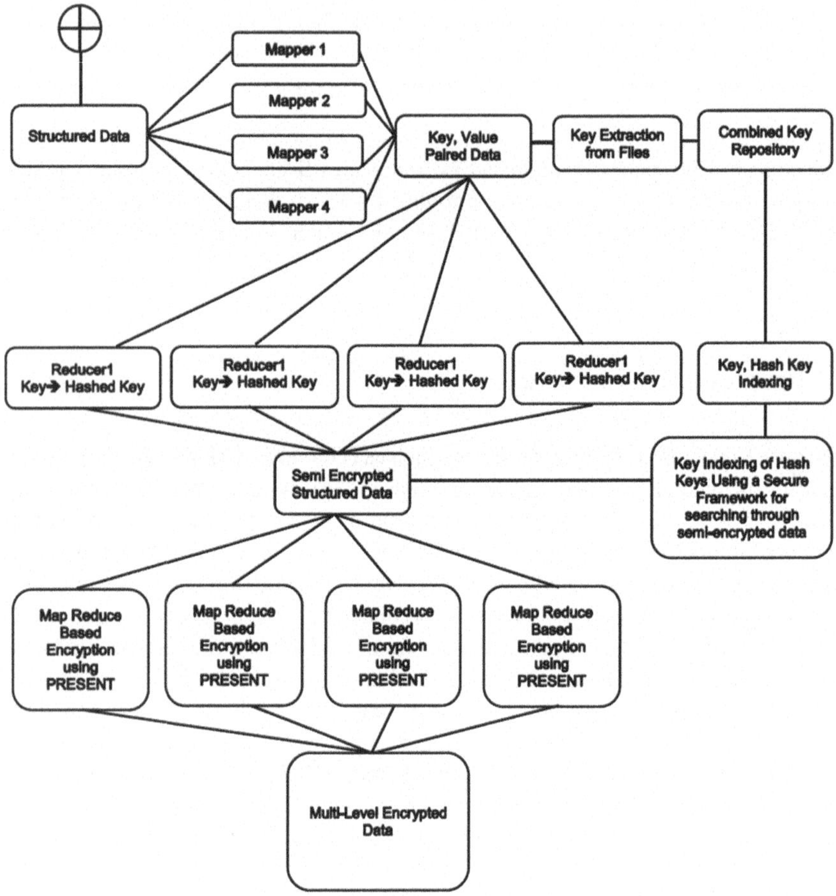

Fig. 1 Architecture of the proposed framework for security of data at rest

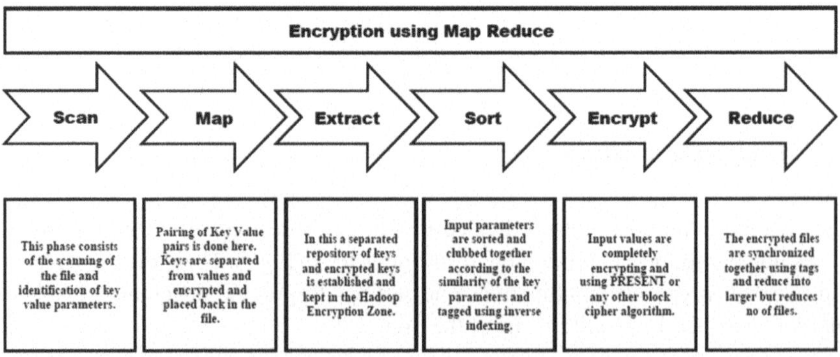

Fig. 2 Execution of the encryption mechanism

jobs reads the data which is pre-partitioned based on no of lines or memory chunk. In order to achieve fast processing, the mapper task transforms the data rows or tuples to a pair of key-value. The records are stored in the value part and rows or tuples are grouped based on key, for e.g. key, (value, value). The reducer task recognises the key parameters using a regex pattern for unstructured files or a predefined key offset for structured files [10]. The mapper and reducer having identical keys manage the same set of data between mappers and reducers. In order to achieve an optimal balance between resource usage and execution times, a threshold can be identified as an ideal size for the mappers and reducers. To reduce overheads key-value pairing would be done and irregularities would be removed from the data and the resultant data would be stored in a hash table or in a file system which is distributed in nature and to be reachable to the succeeding Map Reduce job in the Extraction Phase [11].

B. *Extraction Phase*

The extraction phase follows immediately after the scanning phase and runs on the machines in the same previous map and reduce configuration as the scanning phase. The target of this phase is to extract keys from the data and replace them with encrypted key parameters. In big data, the largest hurdle is to find the required data from the massive pool of data entries [12]. If the key identification parameter is removed the finding the required data becomes monumentally difficult and also introduces the element of partial privacy preservation. During extraction, the mapper reads the key-value-paired output of the Scanning Phase where the key parameters have already been identified and writes them to a different file and replaces the key with an encrypted key. Further, if the file is unstructured the mapper identifies the keys and replaces them with encrypted keys. An indexing database is maintained to match encrypted keys to the actual keys. This level of partial encryption provides passive security to the data by removing key parameters from the data. The process can be stopped at this step itself if a low overhead yet privacy-preserving version of the data is required. The original file is broken into several easily operable smaller files.

C. *Encryption Phase*

In the last phase, the file is broken down into smaller files. These files serve as the input for the encryption phase. Here a block cipher algorithm, namely PRESENT having high storage efficiency is used. Every small file generated in the last phase is used as an independent input for the encryption algorithms. This allows us to utilise the distributed nature of the system as a Mapper reads a file and prepares it into blocks to be processed and as many instances of the encryption algorithm are created as the number of reducers. The encryption algorithms work inside the reducer and process the input files in parallel [11]. The files are stored as separate outputs, which are later processed using the classification module.

2.2 *Classification Module*

The classification module will be used to classify similar records together for indexing. The system will only decrypt part of the data which is similar to the data that is needed. This can provide a low overhead way of providing encryption and decryption for Big Data.

This module classifies similar encrypted keys together and files pertaining to these keys are then tagged and appended together. In case decryption of a file corresponding to a key or multiple keys are required then only the part between two tags followed by the tag corresponding to the keys is only decrypted thus allowing for only partwise decryption and lowering operational overheads in the encrypted state [13]. In sum, the proposed architecture for the security of data at rest can be summarised as follows:

- First, we need to separate the identifying keys and attributes from the actual data so as to achieve the basic level of security by hiding confidential data in the vast pool of data.
- Replacement of the keys with hashed keys is needed to provide an identifying metric inside the data to facilitate further processing of the data.
- Key replacement should be done over MapReduce so as to access and process the records and keys in parallel and to reduce any possible overheads.
- Indexing method is devised so as to match the data to the hashed keys providing the first and second layers of security to the data.
- Finally, the data needs to processed over the MapReduce to completely encrypt it using a 128-bit or a 80-bit PRESENT algorithm [14].
- The encryption is performed using MapReduce Algorithm running inside the MapReduce Framework.

In parallel the classification is being trained by using the repository of keys and classification is being done to the keys mapped along with file tags. Further, the proposed framework is experimentally validated over the different design parameters to gauge the merit over the existing approach. The details of the experimented validation follow in the next section.

3 Experimental Validation

3.1 *Experimental Setup*

Using existing big data, we have evaluated the performance of the proposed algorithm. In comparison with existing methods, our method is much more robust and scalable. We have implemented the algorithm on Hadoop version 2.0.3 and jdk-7u3 in a cloud-enabled environment that is virtual and consisting of eight machines. In this, the first machine serves as both master and slave node and the rest seven machines serve only as slave nodes. All experiments are performed on nodes/machines with 2.64 GHz Intel Xeon CPU, 8 gigabytes main memory, and 1 GigaByte network bandwidth [6]. The test data for our testing phase is obtained from the data mining tools of IBM Quest. In an experimental evaluation, we performed scalability and sensitivity analysis over the dataset as discussed below. The first phase of evaluation is analysing the performance of the machine under different block sizes.

3.2 Scalability Analysis

The scalability of the proposed framework is determined through the integration of the PRESENT algorithm which is the state-of-the-art block cipher algorithm on the cloud computing environment. For this, the dataset needs to be very huge for latency and overhead testing. IBM's data contains around 8,00,000–12,800,000 records an ideal block size of 32 bits has been used [10]. The results are depicted in Fig. 3 and it reveals that the algorithm definitely provides good scalability and arranges satisfactorily as we increase the number of transactions in our test data. We can also see the drastic change in execution time over the increasing dataset. This leads to increased network traffic and higher memory consumption. The capability to tackle huge data sets depends upon the capacity of the server and also depends on the no. of commodity servers used.

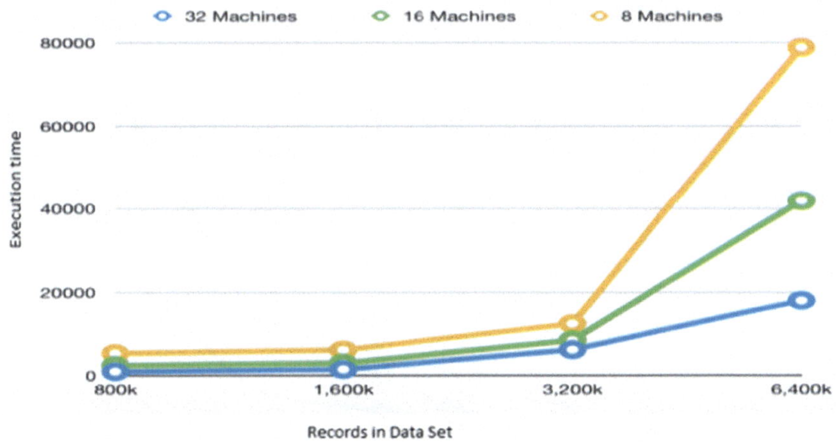

Fig. 3 Scalability of framework

3.3 Sensitivity Analysis

Results have shown that in IBM's dataset with an increased number of items results have drastically changed. Due to the minimum memory consumption and reduced network traffic, all the subsequent processes can be generally executed and evaluated in main memory only. As the value of N increases, the execution time also increases.

This is mainly attributed to the deployed PRESENT algorithm, which utilises cipher block chaining and thus file needs to broken into multiple smaller files and independent MapReduce phases should be initialised for each file. It takes a large time in the initialisation and consolidation process. Thus all these evaluations show that the algorithm has an upper hand [7]. In addition, during experiments, the performance analysis of various machines is conducted first. The number of machines used may vary from 2, 4, 8, 16 to even 32 machines to optimise the performance of the algorithm on varying data sets with transactions ranging from 600000 (six lakh) to 800000 (eight lakh). With block size assigned to 32, as shown in Fig. 3, the evaluation time drastically reduces as we increase the number of machines [9].

We can also conclude that execution time is affected by the attributes of the dataset. Number of MapReduce tasks: On investigating the results depicted in Fig. 4 we found that the larger the number of MapReduce rounds, the better the Present Encryption [14]. Thus all these will help in reducing the initialization cost and making enhanced and straightforward dataflow. The results also indicate that in general finding substantial block sizes will take more time, so the algorithm concentrates on a block size of 32 and learn it so that it can be further used to propagate on other blocks that are larger in size.

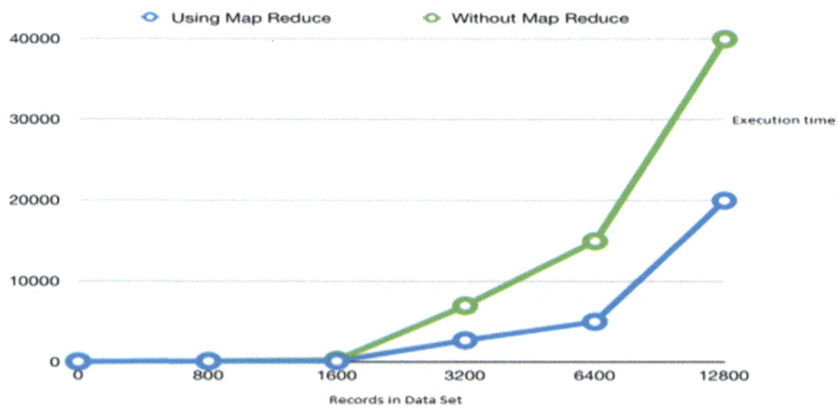

Fig. 4 Comparison with and without MapReduce

4 Conclusion and Future Direction

In this paper, we presented a modified version of a traditional block cipher algorithm on a parallel and distributed setup by using the Hadoop Framework and MapReduce algorithm. The overall work done can be stated as follows.

A stepwise system to divide, encrypt and combine the data set using MapReduce was created and tested. An elaborate setup was used for testing and benchmarking the algorithm to simulate a real-time scenario and partitioning and job achieving was modified to significantly reduce run times and high scalability and load balancing were achieved using MapReduce. In the future, we plan to investigate acceleration of such algorithms on GPUs and see how the algorithm behaves in a massively parallel environment with real-time a large amount of data and a better memory management scheme [15].

References

1. D. Das, O. O'Malley, S. Radia, K. Zhang, Adding Security to Apache Hadoop in Hortanworks
2. S. Park, Y. Lee, *Secure Hadoop with Encrypted HDFS* (Springer, Berlin, 2013)
3. J. Dean, S. Ghemawat, Map reduce: simplified data processing on large cluster, in *OSDI* (2004)
4. T. Kevin, T. Smith, Big Data Security: The Evolution of Hadoops Security Model Posted (2013)
5. M. Satyanarayanan, Integrating security in a large distributed sytem. ACM Trans. Comput. Syst. (1989)
6. H. Lin, S. Seh, W. Tzeng, B.P. Lin, Toward data confidentiality via integrating hybrid encryptionschemes and HDFS, in *26th IEEE International Conference on Advanced Information Networking and Applications* (2012)
7. P.P. Sharma, C.P. Navdeti, Securing big data hadoop: a review of security issues, threats and solution? (IJCSIT) Int. J. Comput. Sci. Inf. Technol. **5**(2) (2014)
8. K. Aoki, T. Ichikawa, M. Kanda, M. Matsui, S. Moriai, J. Nakajima, T. Tokita, Camaellia: a 128-bit block cipher suitable for multiple platforms design and analysis, in *Proceedings of SAC 2000* (Springer, Berlin, 2000), pp. 39–56
9. J. Cohen, S. Acharya, Towards a trusted hadoop storage platform, in *IEEE 10th International Conference on Ubiquitous Intelligence and Computing* (2013)
10. S. Sharma, Rise of big data and related issues, in *Proceedings of the 2015 Annual IEEE India Conference (INDICON)* (Delhi, India, 2015), pp. 1–6
11. R. Bobba, H. Khurana, M. Prabhakaran, Attribute-sets: a practically motivated enhancement to attribute-based encryption, in *Proceedings ESORICS* (Saint Malo, France, 2009)
12. J. Li, N. Li, W.H. Winsborough, Automated trust negotiation using cryptographic credentials, in *Proceedings ACM Conference (CCS)* (Alexandria, VA, 2005)
13. O. O'Malley, K. Zhang, S. Radia, R. Marti, Hadoop Security Design, TR (2009)
14. P. Priya, Sharma, Chandrakant P. Navdeti Securing Big Data Hadoop: A Review of Security Issues, Threats and Solution?(IJCSIT) International Journal of Computer Science and Information Technologies, Vol. 5 (2), 2014
15. B. Thuraisingham, Big data security and privacy, in *Procedings of the 5th ACM Conference on Data and Application Security and Privacy* (San Antonio, TX, USA, 2015), pp. 279–280

RansomAnalysis: The Evolution and Investigation of Android Ransomware

Shweta Sharma, Rakesh Kumar and C. Rama Krishna

Abstract Ransomware is not a Personal Computer (PC) problem anymore, but nowadays smartphones are also vulnerable to it. Various types of ransomware such as Android/Simplocker and Android/ Lockerpin attack Android OS to steal users' personal information. In this paper, we present the evolution of Android ransomware and coin a term—RansomAnalysis—to perform the investigation of samples to analyze the AndroidManifest.xml file for the extraction of permissions. We perform a comparison between permissions gathered by ransomware and benign apps. Besides this, we analyze the topmost permissions used by Android ransomware.

Keywords Android security · Malware · Ransomware · Analysis · Manifest · Permissions

1 Introduction

Smartphones play a vital role in modern day-to-day life by performing various tasks such as document writing, sending e-mails, online shopping, online food delivery, and so on. Nowadays, smartphones with internal memory ranging from 8GB to 128GB store users' personal information including photos, videos, documents, and so on. Out of all mobile operating systems, there are a vast majority of Android users with approximately 70% market share to date [1]. But this continual popularity of Android operating system (OS) makes it a primary target to attack. Thus, the developers introduced permissions in its security model to combat the threats on Android OS. The Android applications (apps) include the required permissions

S. Sharma (✉) · R. Kumar · C. R. Krishna
Department of CSE, NITTTR, Chandigarh, India
e-mail: shweta.cse@nitttrchd.ac.in

R. Kumar
e-mail: rakeshdhiman@nitttrchd.ac.in

C. R. Krishna
e-mail: rkc@nitttrchd.ac.in

M. Dutta et al. (eds.), *Proceedings of International Conference on IoT Inclusive Life (ICIIL 2019), NITTTR Chandigarh, India*, Lecture Notes in Networks and Systems 116,
https://doi.org/10.1007/978-981-15-3020-3_4

in AndroidManifest.xml file to take grant from users before getting installed. But the cybercriminals write malicious apps to overclaim the permissions by requesting unnecessary permissions from users. For example, Andro/LockerPin ransomware [2] asks BIND_DEVICE_ADMIN permission from users' to get administrative privileges to reset the lock screen PIN of the smartphone.

A report by Symantec antivirus [3] reveals that the number of cases of ransomware is more than other types of malware attacks on smartphones in the year 2018. Ransomware, in the form of a malicious app, needs permissions to be granted to spread the malign infections in the Android OS. This motivates us to investigate Android ransomware by performing analysis of the acquired permissions.

The rest of the paper is arranged as follows: Sect. 2 describes the related work on permissions overclaimed by various Android malware. Section 3 outlines the evolution of Android ransomware. Section 4 describes the collected datasets and data preprocessing techniques for analysis. Section 5 investigates the Android ransomware samples and compares the permissions gathered by it with benign apps. Finally, Sect. 6 concludes the paper.

2 Related Work

The permission feature in Android OS has been used to investigate malware to detect its abnormal patterns. The researchers observed that the permissions requested by malicious apps differ from benign apps. Thus, we perform a literature review of all types of Android malware overclaiming permissions from users to harm the Android OS.

Felt et al. [4] analyzed SMS and READ_PHONE_STATE permissions were commonly used by malicious apps but infrequently called by benign apps. The Android OS checks whether the developers mentioned the requested permission in AndroidManifest.xml file or not. But the Android OS does not perform any security check on the permissions asked by apps. Thus, the researchers [5] proposed a security framework (Kirin) to define the requirements and rules. The security rules of Kirin state that an application must not have INTERNET, SMS, PHONE_STATE, AUDIO, SHORTCUT, and ACCESS_FINE_LOCATION related permission labels. When a user installs a new app on Android OS, Kirin considers it as a malicious app if it does not fulfill the criteria. Wu et al. [6] proposed a framework (DroidMat) to extract permissions from the AndroidManifest.xml file. They examined that Android malware requires a large number of permissions as compared to benign apps to perform harmful actions.

Arp et al. [7] proposed a framework (DREBIN) to analyze the permissions used by various apps to investigate Android malware. They observed that a large number of malware samples use SEND_SMS permission to send SMS to premium-rate numbers where the network operator will cut the cost from users' account. Sanz et al. [8] and Qiao et al. [9] compared the frequency of permissions obtained by Android malware. They analyzed that INTERNET permission is the most frequently used permission by malware and goodware. They also observed that malicious apps are more interested in acquiring SMS-related permissions. But the authors did not analyze the permissions which are rarely used by Android malware. Li et al. [10] analyzed permissions which are frequently and rarely requested by Android malware. They ranked permissions according to its usage and used a priori association rules to extract significant permissions for classification of malign and benign apps. Diamantaris et al. [11] distinguished permissions required by the core functions of the Android apps and integrated by third-party libraries. They showed 30 topmost permissions used by core and third-party libraries.

Discussion: The background study reveals that the researchers performed analysis on permissions obtained by generic Android malware. However, there is a need to study and analyze the permissions obtained by ransomware on Android OS. But before analyzing the permissions, the next section presents the evolution of ransomware attack on Android OS.

3 Ransomware Attack on Android OS

Ransomware is a kind of malware, but it demands ransom from the victim to unlock the phone and release the seized material. Android ransomware can either hide malicious functions behind legitimate apps or it can masquerade the apps by taking the same name and icon. After installation, it sends users' information such as device model, IMEI number, device language, photos, and documents to Command and Control (C&C) server. Ransomware executes commands sent by the malware writer(s) which generates a botnet of infected Android devices under the attacker's control. In addition to device locking and encryption, it performs other harmful actions such as to send SMS to victim's contacts, steal phone numbers or messages, display ransom messages on the screen, enable or disable mobile data/Wi-Fi, stop antivirus processes by acquiring several permissions (discussed in Sect. 5) from users. After installation, ransomware kills processes of anti-malware and disable it so that its presence cannot be identified by the user.

3.1 Evolution of Android Ransomware

Android ransomware became visible in the form of fake antivirus (AV) software. These fake AVs scan and infect users' document and convince users to pay money to remove threats. Further, lock screen functionality has been added to it, where users have to pay ransom to unlock the screen. Figure 1 demonstrates the timeline of Android ransomware families since its beginning (2013) till date (2017). The behavior of each ransomware family is discussed in the following:

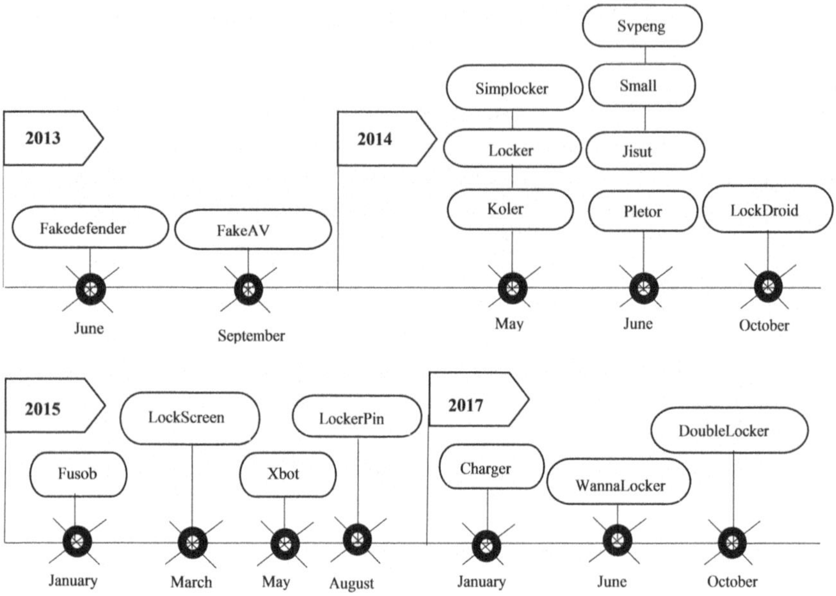

Fig. 1 Android ransomware family chronology

1. Android Defender Ransomware
   Discovered: June 2013
   Category: Fake Antivirus
   Behavior: It scans users' documents and displays files which are infected with malware by performing a fraud scan in users' phone. If the victim clicks on "CLEAR ALL THREATS" button, then it shows a pop-up with a warning message suggesting victim to remove this fake antivirus as it is infected with malware. This pop-up gives two options to the victim: Remove Threats Now or Stay Unprotected. The victim can choose the second option to stay safe and close the fake antivirus app. But the background services of fake antivirus display infinite malware warning pop-ups until the user chooses the first option. It also modifies the OS settings so that the victim is unable to do a factory data reset.
   Goal: It locks the smartphone device.
   Example: Fake antivirus "Avast".
2. Simplocker ransomware
   Discovered: May 2014
   Category: Fake application
   Behavior: It masquerades legitimate apps which are frequently used by teenagers as they cannot differentiate between legitimate and illegitimate app, for example, Grand Theft Auto: San Andreas, Car Racing application, and so on. It scans the memory card of smart-phone to find extensions such as JPEG, PDF, DOCX, AVI, and MP4 to encrypt these files using Advanced Encryption Standard (AES)

cipher. It clicks the victim's photo by the front camera and displays it along with the threatening message. It uses Extensible Messaging and Presence Protocol (XMPP) to send information about the victim's device to C&C server.
Goal: It locks the device and encrypts users documents.
Example: Android/Simplocker.

3. Xbot
Discovered: May 2015
Category: Fake banking app
Behavior: It steals banking credentials and credit card information by impersonating Google Play payment screen and login pages of various e-banking apps. It asks credit card number, expiration date, CVV number, card owner's name, billing address, and mobile number from the victim as shown in Fig. 2. This collected information will be transferred to the C&C server.
Goal: It locks the device and encrypts users documents.
Example: hxxp://market155[.]ru/Install.apk, hxxp://illuminatework[.]ru/Install.apk.
4. WannaLocker
Discovered: June 2017
Category: Plugin
Behavior: It spreads via Chinese game forums where it masquerades a plugin for the game, namely, King of Glory. After installation, it changes the wallpaper of mobile phone and encrypts files stored on the device's external storage using AES algorithm.
Goal: It locks the device and encrypts users documents.
Example: Android.WannaLocker.

4 Data Preprocessing

After discussing the ransomware attack on Android OS, this section discusses the available datasets of Android ransomware and benign apps followed by the data preprocessing tools for feature extraction.

4.1 Dataset

The Android ransomware samples have been collected from RansomProber dataset [12]. The benign samples have been collected from AndroZoo [13] dataset. The total number of collected datasets include approximately 2,050 ransomware APKs and 2,050 benign APKs. But the collected dataset is not in a readable format for which there is a need to perform preprocessing of the dataset by reverse engineering tools.

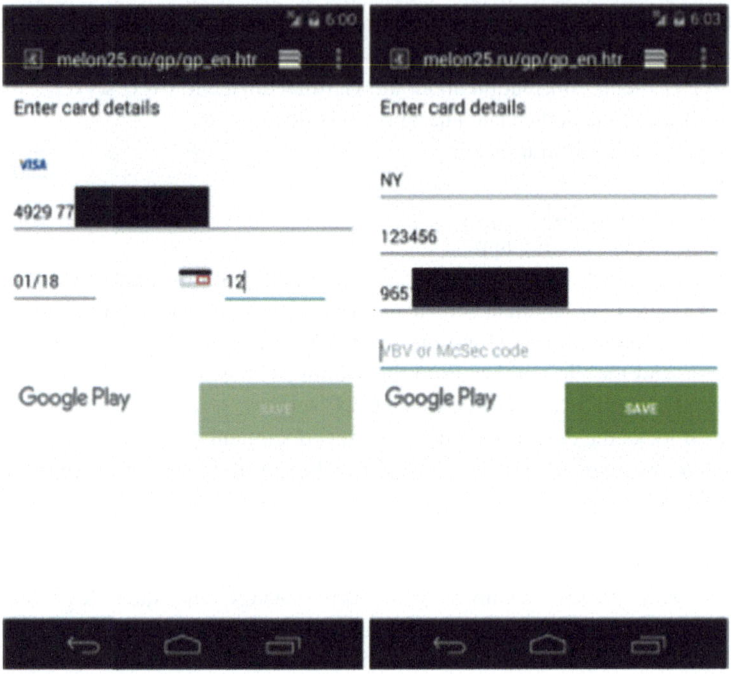

Fig. 2 Fake Google Play payment pages

4.2 Reverse Engineering

The APKTool [14] has been used to perform reverse engineering on collected raw datasets. The APKTool disassembles the APK file into Dalvik Executables (DEX) and eXtensible Markup Language (XML) formats. The extracted formats contain AndroidManifest.xml file, which includes the permissions acquired by an Android app. Thus, the AndroidManifest.xml file can be analyzed to extract permissions obtained by ransomware and benign apps.

5 Analysis of Android Permissions

During installation, the first and foremost step of Android ransomware is to obtain permissions from users to start its noxious operation. This makes permission an important feature to analyze for investigation of Android ransomware. Permission feature can identify ransomware only if the number of permissions obtained by it differs from benign apps. Thus, we coined a term “RansomAnalysis” for analysis of Android ransomware in which we analyzed the AndroidManifest.xml file present in ransomware and benign apps to extract permission feature using Python Programming.

The results are shown in Fig. 3 with a comparison graph between the frequency of permissions used by the ransomware and benign apps. The graph shows that the benign apps mostly use ACCESS_NETWORK_STATE, LOCATION, INTERNET, and VIBRATE permissions. Besides this, the graph shows that ransomware apps are more interested in other permissions such as RECEIVE_BOOT_COMPLETED, WAKE_LOCK, GET_TASKS, and so on. Table 1 shows the 10 topmost permissions used by ransomware with their malicious purpose to harm the Android OS. The frequency field shows the number of ransomware apps obtaining permissions from users. For example, 2049 out of 2050 ransomware apps obtained RECEIVE_BOOT_COMPLETED permission.

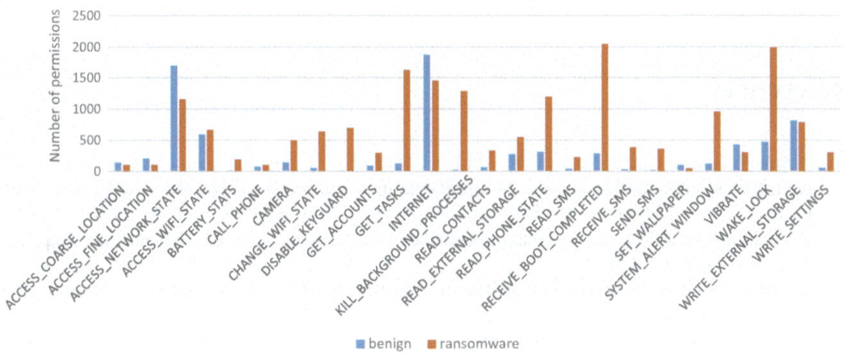

Fig. 3 Comparison of permissions mostly used by ransomware and benign apps

Table 1 Top 10 permissions asked by Android ransomware

Permission	Purpose	Frequency
RECEIVE_BOOT_COMPLETED	To check when the device boots up	2049
WAKE_LOCK	To keep the device screen turned on	1998
GET_TASKS	To get information about running tasks	1631
INTERNET	To open network sockets	1465
KILL_BACKGROUND_PROCESSES	To stop the antivirus process	1295
READ_PHONE_STATE	To get read access to phone state	1200
ACCESS_NETWORK_STATE	To access information about networks	1165
SYSTEM_ALERT_WINDOW	To create windows above all other apps	965
WRITE_EXTERNAL_STORAGE	To write to external storage	796
DISABLE_KEY-GUARD	To disable the keyguard	701

6 Conclusion

In this paper, we have presented the evolution of Android ransomware. We have used a term—RansomAnalysis—to perform the investigation of Android ransomware by analyzing the AndroidManifest.xml file to extract permission feature. We have compared the permissions and their frequency obtained by ransomware and benign apps. We analyzed that almost all ransomware samples from the collected dataset used RECEIVE_BOOT_COMPLETED permission to check the booting status of the device. On the other side, the benign apps are more interested in INTERNET permission. Thus, revoking these topmost permissions as discussed in Table 1 to get access by Android ransomware can stop Android ransomware at entry level. The future work involves analysis of other features such as system calls and network communication to investigate Android ransomware.

References

1. Netmarketshare, Operating System Market Share (2019), https://www.netmarketshare.com/operating-system-market-share.aspx?qprid=9&qpcustomb=1&qpcd=1. Accessed 5 Aug 2019
2. R. Lipovsky, L. Stefanko, G. Branisa, The Rise of Android Ransomware (2016), https://www.welivesecurity.com/wp-content/uploads/2016/02/Rise_of_Android_Ransomware.pdf. Accessed 31 May 2019
3. Symantec, Internet Security Threat Report (2019), https://www.symantec.com/content/dam/symantec/docs/reports/istr-24-2019-en.pdf. Accessed 20 July2019
4. A.P. Felt, M. Finifter, E. Chin, S. Hanna, D. Wagner, A survey of mobile malware in the wild, in *Proceedings of the 1st ACM Workshop on Security and Privacy in Smartphones and Mobile Devices* (Illinois, USA, Chicago, 2011), pp. 3–14
5. W. Enck, M. Ongtang, P. McDaniel, On lightweight mobile phone application certification, in *Proceedings of the 16th ACM Conference on Computer and Communications Security* (Illinois, USA, Chicago, 2009), pp. 235–245
6. D.-J. Wu, C.-H. Mao, T.-E. Wei, H.-M. Lee, K.-P. Wu, Droidmat: android malware detection through manifest and api calls tracing, in *Seventh Asia Joint IEEE Conference on Information Security* (Japan, Tokyo, 2012), pp. 62–69
7. D. Arp, M. Spreitzenbarth, M. Hubner, H. Gascon, K. Rieck, C. Siemens, Drebin: effective and explainable detection of android malware in your pocket. in *Network and Distributed System Security Symposium* (SanDiego, CA, USA, 2014)
8. B. Sanz, I. Santos, C. Laorden, X. Ugarte-Pedrero, J. Nieves, P.G. Bringas, G. Álvarez Marañón, Mama: manifest analysis for malware detection in android. Cybern. Syst. **44**(6–7), 469–488 (2013)
9. M. Qiao, A.H. Sung, Q. Liu, Merging permission and api features for android malware detection, in *International Congress on Advanced Applied Informatics* (IEEE, 2016), pp. 566–571
10. J. Li, L. Sun, Q. Yan, Z. Li, W. Srisa-an, H. Ye, Significant permission identification for machine-learning-based android malware detection. IEEE Trans. Ind. Inform. **14**(7), 3216–3225 (2018)
11. M. Diamantaris, E.P. Papadopoulos, E.P. Markatos, S. Ioannidis, J. Polakis, Reaper: real-time app analysis for augmenting the android permission system, in *Proceedings of the Ninth ACM Conference on Data and Application Security and Privacy* (ACM, 2019), pp. 37–48
12. J. Chen, C. Wang, Z. Zhao, K. Chen, R. Du, G.-J. Ahn, Uncovering the face of android ransomware: characterization and real-time detection. IEEE Trans. Inf. Forensics Secur. **13**(5), 1286–1300 (2018)

13. K. Allix, T.F. Bissyandé, J. Klein, Y. Le Traon, Androzoo: collecting millions of android apps for the research community, in *Proceedings of the 13th International Conference on Mining Software Repositories* (ACM, 2016), pp. 468–471
14. Kali, apktool (2019), https://tools.kali.org/reverse-engineering/apktool. Accessed 12 July 2019

On the Applicability of Certificateless Public Key Cryptography (CL-PKC) for Securing the Internet of Things (IoT)

Manisha Malik, Kamaldeep and Maitreyee Dutta

Abstract With the introduction of smart cities, transportation, industries, and even smart health care, the Internet of Things (IoT) has risen as a domain of great potential, impact and development with Gartner forecasting about 20.4 billion Internet-connected smart devices connected by 2020. However, most of these IoT devices can be easily compromised. Quintessentially, these IoT devices have limited computation, memory, and processing capacity, and thus they are more susceptible to attacks than conventional devices like computers, tablets, or smartphones. In this paper, we present and review significant security issues in IoT. We review and analyze security protocols and challenges with respect to standard IoT protocol architecture. More importantly, we discuss, how certificateless cryptography (CL-PKC) is being used to solve many security problems in IoT. Furthermore, we tabulate such proposals with an emphasis on open research problems in this domain.

Keywords Internet of Things · Certificateless public key cryptography · IoT security

1 Introduction

The rapid growth in miniaturization, the advancement in communication technologies, and the number of Internet-connected devices have contributed to pioneering advancement in our society. These have enabled the shift from the real world to the digital world. In other words, the way in which we communicate with each other and the environment has changed drastically. The Internet of Things (IoT), emerging as a combination of technologies from WSN to RFID, enables the sensing, actuation, and

M. Malik (✉) · Kamaldeep · M. Dutta
CSE Department, NITTTR, Chandigarh, India
e-mail: manisha.cse@nitttrchd.ac.in

Kamaldeep
e-mail: kamal.cse@nitttrchd.ac.in

M. Dutta
e-mail: d_maitreyee@yahoo.co.in

M. Dutta et al. (eds.), *Proceedings of International Conference on IoT Inclusive Life (ICIIL 2019), NITTTR Chandigarh, India*, Lecture Notes in Networks and Systems 116,
https://doi.org/10.1007/978-981-15-3020-3_5

communication of electronic devices over the Internet. These electronic devices can be any wearable, appliance or any development and storage platforms like servers. In this way, the IoT enables the transition of cities to smart cities, homes to smart homes and electrical grids to smart grids. The research firm, Gartner predicted that approximately 20.4 billion Internet-connected devices will exist by 2020 [1]. The huge expansion of the IoT can only be realized with the development of standard protocols and mechanisms to reduce the heterogeneity in this domain. Apart from the heterogeneity and integration issues in IoT, the security challenge must also be kept in mind. Although the IoT can facilitate the digitization of devices itself, the security of such devices is still a key challenge. In this context, the certificateless cryptography based on the PKC concept has the potential to offer solution to the impeding security problem in the IoT. It is a variant of identity-based cryptography (IBC) without the key escrow problem. This type of cryptography also completely removes the need for certificates for authenticating the public key. In this paper, we will review and analyze some of the CL-PKC-based research solutions proposed for security in IoT. The main contributions of the paper are: Identification and analysis of different approaches of integration of certificateless cryptography with IoT to ensure security. Survey of existing certificateless based research solutions to ensure robust security mechanisms in IoT. Evaluation and comparison of the surveyed works.

2 Standard IoT Architecture and Security Protocols

The realization of IoT comes from the amalgamation of diverse communication technologies which include WSN, RFC, Bluetooth, NFC, GSM, and IPv6. In particular, IPv6 and WSN are becoming key enablers as research and standardization bodies often term IoT as WSN connected with the Internet. The WSN which once supported isolated sensing and actuating environments are now being integrated with the Internet infrastructure, a process that is also accentuated by the development of 5G technology. From the beginning of this process, research and standardization efforts by the IEEE and IETF have immense significance. A major breakthrough in this regard is the development of the 6LoWPAN adaptation layer by the IETF in 2007. This layer bridges the IPv6 communication to the constrained communication environment of WSN, forming a standardized protocol stack as shown in Figure. In Fig. 1, we present standard communication protocols and respective security protocols and challenges in the context of standard protocol stack by IETF.

3 PKC-Based Key Management in IoT

It is irrefutable that security is essential to realize a secure IoT and cryptography does play a major role to achieve it. Cryptography is the study of cryptosystems. A cryptosystem is a suite of cryptographic primitives (or algorithms) required to implement

IoT Layer	IoT Protocol	IoT Security Protocol	Security Challenges	
Application Layer	CoAP	DTLS	• Large DTLS handshake messages • Not well suited for CoAP proxies • Unable to support multicast communications	Cross Layer Security Challenges (Key Management, Intrusion Detection, Fault Tolerance)
Transport Layer	UDP			
Network & Routing Layer	IPv6, RPL	IPSec, RPL Security	• Lack of threat models for RPL according to different applications • Lack of mechanisms to prevent internal attacks • Node authentication and key retrieval	
Adaptation Layer	6LoWPAN	-	• No specific security mechanism • Lack of support for packet fragmentation attacks	
Data Link Layer	IEEE 802.15.4	IEEE 802.15.4 Security	• Security is optional • Does not specify any keying model • Management of initialization vector values • Insecure acknowledgement messages • No support for group keying	
Physical Layer				

Fig. 1 Standard layers and protocol stack for the IoT

a particular security service. Cryptographic primitives consist of hash functions, algorithms on encryption and decryption (symmetric/asymmetric), authentication, and digital signatures. Crytpoprimtives by themselves are not useful but must be integrated together to build a cryptosystem. For example, the well-known Transport Layer Security (TLS) protocol is a great example of a cryptosystem that combines these primitives to achieve strong confidentiality, integrity, and authentication. A major component of any cryptosystem is the key which is used by the cryptographic algorithm to transform plaintext into ciphertext. On the basis of the type of key involved, there are two types of cryptosystems: Symmetric Key Cryptography (SKC) and Asymmetric Key or Public Key Cryptography (PKC). In SKC, a single secret key or symmetric key is shared by the communicating parties to perform the operations of encryption and decryption. Prior to 1970, all cryptosystems employed symmetric key cryptography but even today it has high relevance and is used extensively in various cryptosystems. Examples of symmetric key algorithms like Advanced Encryption Standard (AES), Data Encryption Standard (DES), etc. The application of symmetric cryptography is the encryption of data/traffic. The advantage of this type of cryptography is the less processing power to run symmetric algorithms. However, one of the major challenges that hinder the deployment of symmetric key cryptography is key establishment. Due to sharing the same key between the sender and receiver, the secure distribution of this key among communicating parties is a big challenge. Besides, the number of keys to enable communication between n parties is of the order of n^2.

This issue was resolved in 1970 with the introduction of PKC. PKC is a type of relatively new cryptography that lays its foundation on mathematical problems that are difficult to solve. In PKC, two different keys, i.e., a public key that is shared with everyone and a secret private key are used to encrypt and decrypt the data.

Both these keys are mathematically related but it is computationally infeasible to extract the private key from the respective public key. This type of cryptography is generally used for authentication, key exchange, and digital signatures. However, there is one issue with public key cryptography—how to ascertain the authenticity of the claimed public key, i.e., the respective public key which the receiver is claiming is actually his and not spoofed by any malicious user. To accomplish this, public key infrastructure (PKI) consisting of a third party is implemented to build trust between parties and also ensure the authenticity of public keys. Kerchoff's principle states that the security of any cryptosystem should lie in its keys and everything else including the algorithm should be considered public knowledge. In cryptosystem, this management of keys is known as key management which includes the generation, secure exchange, storage, usage, and replacement of keys. It also includes key servers, user procedures, cryptographic protocols and their design. On the basis of the cryptographic primitive family involved, key management schemes are of two types—pre-distribution of symmetric key and asymmetric key schemes.

In the scheme of pre-distribution of the symmetric key, the communicating parties begin by sharing common parameters. These pre-shared parameters can be the symmetric key itself or some dump bytes loaded into the node before its operation. Asymmetric Key Schemes, on the other hand, are based on expensive mathematical operations, they are never used for the direct encryption of a block of data. However, they are used to encrypt small block particularly, for the encryption of secret keys for distribution. These schemes are called asymmetric key schemes. The characteristics of both schemes are depicted in Fig. 2.

Among all asymmetric key schemes, public key infrastructure (PKI) is the most commonly used approach to ensure the authentication of public keys. Since its inception, PKI has been the backbone of Internet security to deliver the basic elements of

Key Management Scheme	Characteristics
Symmetric Pre-Distribution Scheme	• Assume that the communicating parties initially share some common credentials. These pre-shared credentials can be the symmetric key itself or some random bytes flashed into the node before its deployment. They may even deploy a server or a key distribution center to distribute the keys to nodes. • Provide low computation overhead which is suitable for constrained nodes, but, they have their own disadvantages, like high memory for storing keys, low scalability, high communication overhead and vulnerability to node capture attacks.
Asymmetric Key Scheme	• Employ asymmetric algorithms and are widely deployed in the conventional Internet. Although asymmetric key schemes have low memory requirements, high scalability and resilience to attacks, they employ computationally intensive operations which increase the energy consumption and computation cost in the context of IoT.

Fig. 2 Characteristics of key management schemes

security in communications using authentication and encryption. Due to the proliferation of heterogeneous devices in IoT, it makes a strong solution for securing the IoT. The purpose of PKI is to manage keys and certificates. PKI is a type of key management system that employs certificates to provide authentication and public keys for encryption. So, it is clear from the above discussion that symmetric encryption is fast but the symmetric key exchange is not scalable and asymmetric encryption is expensive but asymmetric key schemes are scalable. Most protocols in the conventional Internet use hybrid schemes that are using asymmetric cryptography for symmetric key exchange and using symmetric cryptography for encryption of data. The same convention has also been applied to security protocols in IoT like DTLS, IPsec, and RPLSecurity. However, conventional symmetric encryption schemes like AES, DES have large source code and require a number of processing cycles, a lot of research is needed to formulate lighter symmetric ciphers. Besides, the applicability of asymmetric schemes for key exchange in the IoT has major inconveniences like the computation cost and energy consumption. Besides, the distribution of public keys in an authenticated manner, i.e., using PKI or public key authority is another research challenge for IoT.

4 Cl-PKC

Designed for succinct public key management, Certificateless Public Key Cryptography (CL-PKC) is a variant of IBC removing the key escrow problem(KEP). This type of cryptography also completely removes the need for certificates for authenticating the public key. CL-PKC was first discussed by the authors in [2] in the year 2003. CL-PKC introduces a special arrangement of public/private keys wherein the private key is partially produced by a trusted third party called the Key Generation Center (KGC). Such a key is termed as the Partial Private Key (PPK). As illustrated in Fig. 3, the process of public/private key generation in CL-PKC is as follows:

- Initially, the KGC calculates the PPK from the identifier of the device and a master key (MK).
- The device then joins the PPK with its own secret in order to calculate its own private key.
- The device calculates its public key by combining its own secret with the public parameters of KGC.

It is clear from the process that CL-PKC is free from the KEP of IBC as the responsibility of the device's private key is withdrawn from the KGC. Also, public keys do not rely solely on their identities. Due to their high efficiency in bandwidth consumption, storage needs and computation costs, CL-PKC schemes are particularly useful in IoT to ensure security services like signatures [9], data encryption [11], data aggregation, access control, key agreement and proxy decryption among others. The focus of our paper, however, is only on key agreement protocols based on CL-PKC for IoT.

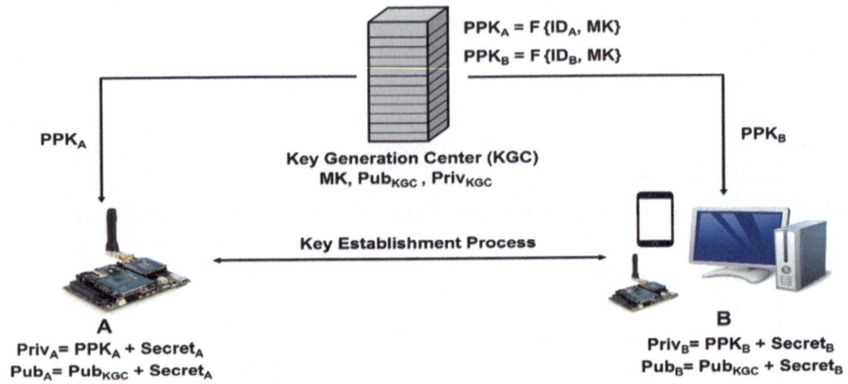

Fig. 3 Certificateless Schemes

5 Survey of Security Protocols Based on CL-PKC in IoT

This section reviews CL-PKC-based security solutions in IoT. These solutions fulfill either one or more security requirements to ensure a robust and secure IoT network.

The authors in [3] employed bilinear pairings to design a certificateless public key encryption (CL-PKE) scheme, which is secure in the random oracle model assuming that factorization of numbers and bilinear Diffie–Hellman problem are intractable. Their scheme used only one pairing and two scalar multiplication operations while encrypting and one pairing operation while decrypting making it suitable for resource-constrained IoT networks. Additionally, using this CL-PKE scheme, the authors even proposed a protocol to ensure the confidentiality and integrity of information in the Smart Car scenario. Likewise, the authors in [4] proposed a mutual authentication scheme based on CL-PKC to protect against replay and node impersonation attacks.

The authors in [5] used the certificateless signcryption scheme to ensure access control in industrial IoT networks. The access control scheme has four phases of initialization, registration, authentication, and revocation. In this scheme, the service provider node acts as the KGCs. The advantage of this scheme is that it shifts the computational node from sensor nodes to the gateway.

Various CL-PKC proposals have been intended for WSN conditions, anyway, just a couple of them fit the attributes of IoT systems and applications. In 2015, Seo et al. [6] proposed a CL-PKC-based key management for dynamic WSN, which can likewise be connected to IoT systems. This scheme used ECC with 160-piece keys and guaranteed validation through a signcryption scheme. The base station has the KDC and doles out a novel identifier, an individual key and a certificateless open/private key pair, to every node. This scheme supported key refreshing and renouncement, and guaranteed in reverse and forward secrecy.

Table 1 CL-PKC-based security solutions in IoT

Cryptographic primitive	Relevant references
Authenticated key management	[4, 7, 6, 3]
Non-repudiation signature	[10, 9]
Access control	[5]
Confidentiality and integrity	[3]
Data privacy	[8, 11]

With regard to the IoT, a novel ECC-based signcryption scheme (ECKSS) for MIKEY was proposed in [7]. In this work, the authors proposed two novel certificateless key transport mechanisms for MIKEY, with the two being PKI free and dependent on the elliptic curve Korean signcryption scheme for wrapping keys. This plan can be termed as certificateless, since it depends on restricting the open estimations of the discussing parties with the public keys created by a confided in key administration server.

In the domain of IoT data storage and privacy, the authors in [8] suggested a secure distributed data storage scheme based on blockchain and CL-PKC. The public ledger of blockchain advertises the public key of IoT device which is generated as per the CL-PKC rules. The drawback of advertising public key is overcome by deploying blockchain which spreads the public key to all entities in a network.

There have been multiple attempts to develop lightweight certificateless signature schemes for IoT devices. The authors in [9] presented computationally efficient certificateless signatures for IoT smart objects by replacing bilinear pairing operation with ECC point based operations. Similarly, the authors in [10] proposed certificateless signature scheme for IoT and demonstrated the KGC impersonation and public key replacement attacks on the conventional CL-PKC-based signature generation algorithm (Table 1).

As is evident from our discussion, though CL-PKC removes the computational overhead of traditional PKI, the cost of encrypting the public key is still significant. It is also important to note that solutions should also focus on ensuring the interoperability of CL-PKC-based IoT networks with the certificate-based outside network.

6 Conclusion

In this paper, we perform an analysis of IoT protocols and respective security challenges as per the IEEE standardized protocol stack. We focus on secure key bootstrapping particularly on certificateless PKC given its significance in terms of computational efficiency, scalability, and robustness against key escrow attacks. Our analysis is structured around various security services that CL-PKC aims to solve for IoT networks.

References

1. 20.4 billion iot devices by 2020, https://www.otarris.com/on-trend-20-4-billion-iotdevices-by-2020/. Accessed 3 Sept 2019
2. S.S. Al-Riyami, K.G. Paterson, Certificateless public key cryptography, in *Advances in Cryptology—ASIACRYPT 2003*, ed. by C.-S. Laih (Springer, Berlin 2003), pp. 452–473
3. R. Guo, Q. Wen, H. Shi, Z. Jin, H. Zhang, Certificateless public key encryption scheme with hybrid problems and its application to internet of things. Math. Prob. Eng. **2014** (2014)
4. D.Q. Bala, S. Maity, S.K. Jena, Mutual authentication for iot smart environment using certificate-less public key cryptography, in *2017 Third International Conference on Sensing, Signal Processing and Security (ICSSS)* (IEEE, 2017), pp. 29–34
5. F. Li, J. Hong, A.A. Omala, Efficient certificateless access control for industrial internet of things. Futur. Gener. Comput. Syst. **76**, 285–292 (2017)
6. S. Seo, J. Won, S. Sultana, E. Bertino, Effective key management in dynamic wireless sensor networks. IEEE Trans. Inf. Forensics Secur. **10**(2), 371–383 (2015)
7. K.T. Nguyen, N. Oualha, M. Laurent, Novel lightweight signcryption-based key distribution mechanisms for mikey, in *Information Security Theory and Practice*, ed. by S. Foresti, J. Lopez (Springer International Publishing, Cham, 2016), pp. 19–34
8. R. Li, T. Song, B. Mei, H. Li, X. Cheng, L. Sun, Blockchain for large-scale internet of things data storage and protection. IEEE Trans. Serv. Comput. 1 (2018)
9. K.-H. Yeh, C. Su, K.-K. R. Choo, W. Chiu, A novel certificateless signature scheme for smart objects in the internet-of-things. Sensors **17**(5) (2017), https://www.mdpi.com/1424-8220/17/5/1001
10. X. Jia, D. He, Q. Liu, K.-K.R. Choo, An efficient provably-secure certificateless signature scheme for internet-of-things deployment. Ad Hoc Netw. **71**, 78–87 (2018)
11. M. Ma, D. He, N. Kumar, K.-K.R. Choo, J. Chen, Certificateless searchable public key encryption scheme for industrial internet of things. IEEE Trans. Ind. Inf. **14**(2), 759–767 (2017)

Performance Evaluation of Advanced Machine Learning Algorithms for Network Intrusion Detection System

Sharfuddin Khan, E. Sivaraman and Prasad B. Honnavalli

Abstract In the past decade, there is a terrific growth on Internet and at the same time, we have seen an increase in malicious attacks on government, corporate, military, financial organizations. To overcome these attacks, Intrusion Detection Systems (IDSs) are developed and accepted by many institutions to keep a monitor on intrusion and additional harmful behavior. However, these IDSs still have some obstacles that are low detection accuracy, False Negatives (FN) and False Positives (FP). To overcome these problems, Machine Learning (ML) techniques are used which help in increasing the intrusion detection accuracy and greatly decreases the false negative rate and false positive rate. In this paper, we have considered five algorithms for evaluation, namely Decision Tree (D-tree), Random Forest (RF), Gradient Boosting (GB), AdaBoost (AB), Gaussian Naïve Bayes (GNB) on UNSW-NB15 dataset. And we found that Random Forest is the best classifier based on the following metrics detection accuracy, F1 score, and false positive rate.

Keywords Decision tree · Random forest · Gradient boosting · AdaBoost · Gaussian Naïve Bayes · Intrusion detection system · Machine learning · UNSW-NB15

1 Introduction

Presently, due to enormous growth in computer networks and applications, there are many new challenges that arise for research in the field of cybersecurity. Any activities that are capable to compromise the principles of computer systems can be referred to as attack/intrusion, e.g., confidentiality, integrity, availability [1]. The existing security solutions such as Firewall, Authentication, and Encryption which are first preferred for defense doesn't have the ability to give total safeguards for computer systems and computer networks. The firewall system that we have can be bypassed in modern attack environments and the firewall does not go in-depth to

S. Khan (✉) · E. Sivaraman · P. B. Honnavalli
Department of Computer Science and Engineering, PES University, Bangalore, India
e-mail: Sharfuddinkhan20@gmail.com

M. Dutta et al. (eds.), *Proceedings of International Conference on IoT Inclusive Life (ICIIL 2019), NITTTR Chandigarh, India*, Lecture Notes in Networks and Systems 116,
https://doi.org/10.1007/978-981-15-3020-3_6

analyze the network packets. Due to these drawbacks, IDSs security mechanisms are developed to attain far above the ground protection for network security infrastructure [2].

Network Intrusion Detection Systems (NIDS) is used to find out unsafe actions by monitoring network traffic flow. The intrusion detection technique can be divided into two forms. First, misuse/pattern-based detection which recognizes bad pattern by comparing received data on the network and the patterns maintained as a database of attacks such as malware. Second, anomaly-based detection in which we have a model of good traffic any deviation from this is considered an attack. Despite the deployment of newly developed intrusion detection systems, to fight against network intrusions, attackers still find new methods to deploy refined malicious code that are able to go around the IDSs and cause harm to resources. Machine Learning (ML) technology uses mathematical algorithms in the design of IDSs is a key feature in improving the performance of IDSs. ML technologies are used to train the IDS using a network security related dataset. The IDSs which are completely trained have the ability to detect malicious network activities and can also predict new attacks which are referred to as zero-day attacks.

The remaining contents of the paper are described in the following manner, in Sect. 2 literature survey, Sect. 3 describes the dataset used, in Sect. 4 a brief description of the proposed system, Sect. 5 has the description of performance evaluation metrics for IDS, and Sect. 6 presents results and discussion, finally Sect. 7 has a conclusion.

2 Literature Survey

Many works have been done on machine learning classifiers algorithm to weigh up the performance on several datasets of various intrusion detection systems that are as follows. In [3], the author proposed the process model, according to this one important key factor in increasing the detection rate for IDS is preprocessing as it has an impact on the accuracy of classifiers. Preprocessing coupled with artificial neural networks has given many successful detection rates. But this model is developed using the NSL-KDD which is a very old dataset that no longer reflects the current traffic scenario. In [4], the performance of the ML classifiers, namely SVM, Random Forest, K-means, Decision tree is evaluated when trained using the KDD99 and alternative network datasets. And the results indicated that classifiers trained with UNSW-NB15 dataset have a good f1 score and thus favoring UNSW-NB15 as a modern substitute to the benchmark datasets. However, only a small portion of the UNSW-NB15 is used for the research. And a large amount of this dataset is still uncharted. In [5], the author developed a model for the multi-level network detection method. These models have three steps: first, we need to create reliable rules to identify network abnormalities by understanding signatures from network traffic data; second, the predictive model is generated to decide each category of attack; third, integrate visual analysis tools to analyze and validate the identified intrusions. Using this model, the author has obtained near to 96 percent accuracy in identifying exact attack categories.

In [6], the author proposed a new approach that is used to alter the classification to decrease false positive rate in intrusion detection by applying the upgraded algorithm called as self-adaptive Bayesian algorithm. With the help of this technique, different categories of attacks in KDD99 datasets can be classified properly with far above the ground classification rates and reduced false positive rate in minimum response time. In [7], the proposed system, a new learning algorithm is presented by the combination of D-tree and NB algorithms for adaptive network intrusion detection. And they performed some test on the proposed algorithm against the existing learning algorithm by using the KDD99 dataset, according to the result obtained from the proposed system has given far above the ground detection rate and near to the ground false positive rate for different types of network intrusions.

In [8], the proposed system's author has used three different datasets which are given as follows: NSL-KDD, UNSW-NB15, and phishing to check the performance of six classifiers such as Random Forest, D-tree, SVM, ANN, KNN, and Naïve Bayes. By considering the results provided by the author, KNN and D-tree are the top performance classifiers. When testing time, detection accuracy and false positive rate metrics are considered. In [9], the author has used Random Forest classifier on the three types of NIDS, namely signature, anomaly, and hybrid network based IDSs, and the dataset used is KDD99. The results show the performance is high for the proposed misuse-based approach than the best KDD results; moreover, anomaly detection technique achieved far above the ground detection and false positive is near to the ground. In [10], the author has proposed a new layered intrusion detection system by fusing the different machine learning techniques to offer high-performance IDS for various kinds of attacks, however recent dataset has not been used for this system and compared with the old datasets. In [11], the authors have conducted research on different challenges that occur in the network intrusion detection mechanism using machine learning techniques and leaves a hint in finding the possible solutions.

3 Dataset

A. *Orientation of a NIDS dataset*

A network intrusion detection system can be referred to as relational data. NIDS takes a group of data attributes as inputs. Every data entry contains attributes that are in different types such as floating, integer, binary, and nominal [12]. The labels have two values for each entry of data either 0 or 1, where 0 represents normal, and 1 for abnormal.

B. *Problem with previous benchmark datasets KDDCUP99 and NSL-KDD*

Mainly there are three disadvantages: first, these datasets were generated more than a decade ago; second, these datasets do not reflect the modern network traffic scenario;

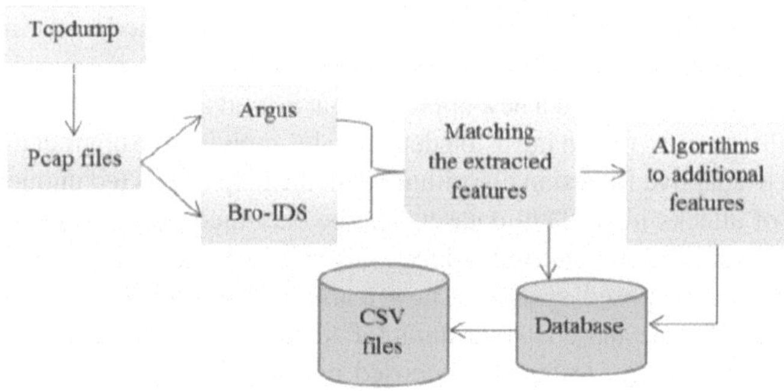

Fig. 1 Architecture for generating UNSW-NB15 dataset

third, this dataset does not completely correspond to the current low traces attack environment.

C. *The proposed UNSW-NB15 dataset*

The proposed dataset was created in the "University of New South Wales" in Australia for the development of this dataset a synthetic environment is created in the cyber range lab of the "Australian Centre for Cyber Security" (ACES). The unprocessed network packets were produced by using the ixia perfect storm tool which has the potential of generating combined traffic that contains both the current normal network traffic and abnormal traffic, i.e., attacks [13]. The TCP dump tool was used to generate Pcap files. The dataset created from this environment contains attacks that can be classified into nine forms, namely shell code, backdoors, analysis, denial of service, exploit, fuzzers, reconnaissance, generic, worms. The proposed dataset contains 49 features that are extracted tools, namely Bro-ids, Argus, further there are 12 algorithms that are developed as shown in Fig. 1 [14].

4 Methodology

The proposed methodology consists of different classifiers and principles that are explained in the following paragraphs. The implementation of our proposed system begins with the collection of UNSW-NB15 dataset which is the latest benchmark network dataset, next the features are extracted from this dataset by using the extra tree classifier, next the set has been divided into two parts, namely Training data and Test data in the ratio 80% and 20%, respectively, in the proposed system five different ML classifiers are dynamically trained using the training dataset as shown in Fig. 2, then the datasets are verified using the test data, the algorithm with the best accuracy will be dynamically selected for the prediction. Once we get the model with the best accuracy, we then pass the input data to this model and observe the results which are either normal or malicious data that is nothing but an intrusion.

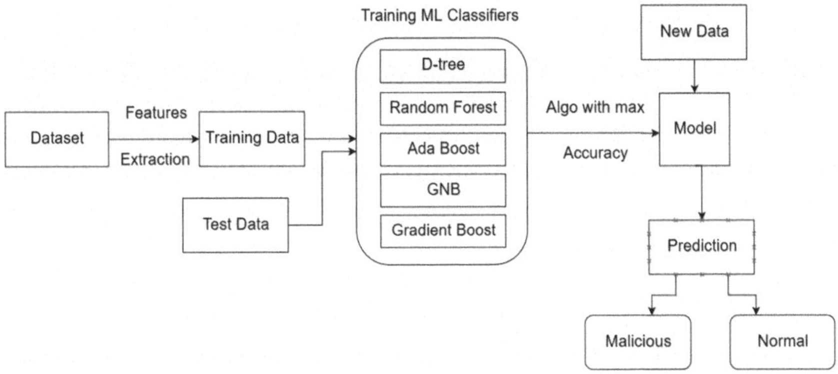

Fig. 2 Architecture of the proposed system

In the proposed system, five ML classifiers are used such as Decision tree, Random Forest, Gradient Boost, AdaBoost, Gaussian Naïve Bayes. And they are described below.

A. *Decision Tree*

Decision Tree is one of the most popular and powerful tools for prediction and classification. A D-Tree works similar to that of a flowchart that has a tree-like structure, in which each lane from the top (root) node to bottom (leaf) signifies a rule. In that path, each internal node indicates a test on attribute and every stem indicates a result of the test, further each bottom (leaf) node represents a class label which is referred to as attack [15].

B. *Random Forest Classifier*

As the name denotes it builds forest by combining the trees and makes the forest random. This forest is an ensemble of D-trees. There are two reasons to choose this algorithm, one, is it runs efficiently one large datasets, second this algorithm learns very fast [16].

C. *Gaussian Naïve Bayes Classifier*

GNB is a special type of algorithm, it is generally used when the features have continuous values, the prediction is based on the probability of each attribute which fit into each group that are utilized for prediction, in this classifier probability of each instance are calculated and whichever has the highest probability class value that is considered for prediction [17].

$$P\left(\frac{A}{B}\right) = \frac{P\left(\frac{B}{A}\right) * P(A)}{P(B)}$$

D. *AdaBoost Classifier*

AdaBoost is one among the most famous ML classifiers, implementing AdaBoost is simple, this classifier is broadly used for pattern detection problems like IDS, face recognition, AdaBoost has low susceptibility to overfitting than most machine learning algorithms, it also corrects the misclassification made by weak classifiers [18].

E. *Gradient Boost Classifier*

The Gradient Boost classifier converts the weak learner into strong learner, i.e., this ML classifier iteratively constructs a group of feeble D-tree learner using boosting technique, so that it can trim down the loss function of the gradient boost classifier and iteratively train decision tree, after each iteration, it compares the predicted labels with the true labels, and if any errors then in next iteration it will correct the previous mistakes [19].

5 Performance Metrics of IDS

There are many evaluation metrics through which different classifiers can be studied and analyzed. The confusion matrix in Fig. 3 acts as a base for calculating the various metrics such as accuracy, precision, false negative rate, recall, true positive rate, false positive rate, true negative rate for intrusion detection systems.

A. *Accuracy*: It is the most important metric to evaluate the classifier. It is the percentage of correct prediction of attack or normal traffic. It is determined by using the equation.

$$Accuracy = \frac{TP + TN}{TP + TN + FP + FN}$$

B. *False Positive Rate (FPR)*: It is determined using the formula.

$$FPR = \frac{FP}{TN + FP}$$

		Predicted class	
		Attack	*Normal*
Actual class	*Attack*	TP	FN
	Normal	FP	TN

Fig. 3 Example of confusion matrix for IDS

C. *Recall (R)*: It is the portion of instances that are correctly predicted as positive to the actual size of attack class, and is calculated using the equation.

$$R = \frac{TP}{TP + FN}$$

D. *Precision (P)*: It estimates the probability that a positive prediction is correct and it is calculated using the equation.

$$P = \frac{TP}{TP + FP}$$

E. *F1 score*: It is the gauge of the accuracy of a test, where an F1 score is good if it has the value near to 1, and worst when the value reaches near to 0, is determined using the equation.

$$F1 - score = \frac{2 * P * R}{P + R}$$

F. *Training time*: The time required to train a classifier to build a model in seconds.

6 Results and Discussions

Figure 4 shows that Random Forest has the best accuracy followed by Decision tree and AdaBoost, the Gradient Boost comes in fourth position in terms of accuracy and we found that Gaussian Naïve Bayes has the low accuracy among the five classifiers used. Table 1 represents the results of each classifier's performance trained using the UNSW-NB15 dataset. The detection accuracy of each machine learning classifier is

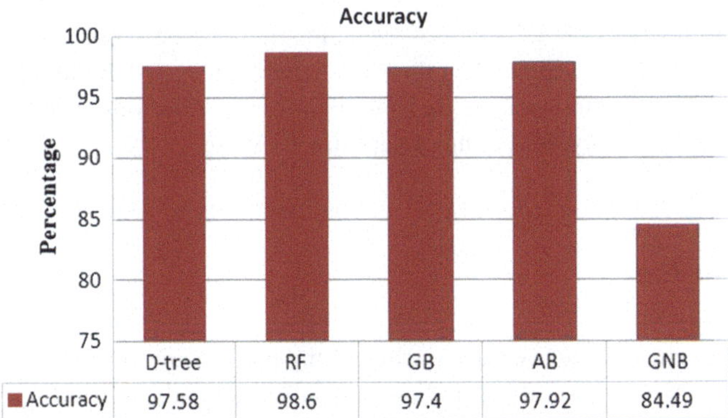

Fig. 4 Detection accuracy of classifiers on UNSW-NB15 dataset

Table 1 Performance results of classifiers

	Accuracy	Time (s)	F1 score
D-tree	97.85	00.75	0.972
Random Forest	98.60	06.97	0.983
Gradient boost	97.40	12.36	0.969
AdaBoost	97.92	21.22	0.976
Gaussian Naïve Bayes	84.49	00.23	0.823

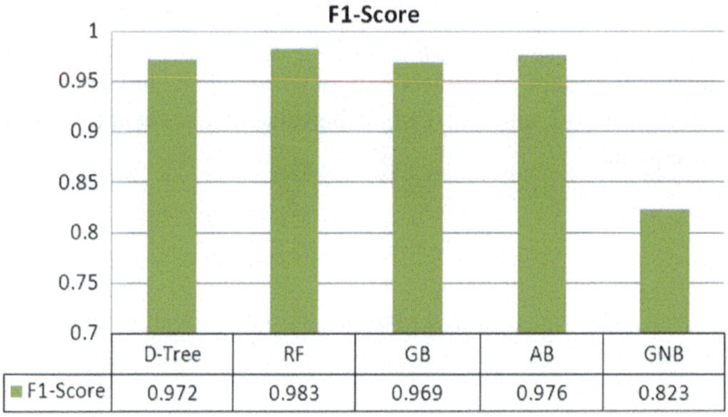

Fig. 5 F1 score of classifiers

high on the used dataset. Gaussian Naïve Bayes has a low accuracy of 84.14% but it has a good execution time of 0.97 (s). Gradient Boosting and AdaBoost have the best accuracy of 97.40% and 97.92%, respectively, but they have the worst execution time of 12.36 (s) and 21.22 (s), respectively. Decision tree has an accuracy of 97.58 percent and has a very good execution time 0.75 (s). Random Forest has the highest accuracy rate of 98.60% and F1 score 0.98% with a time of 6.97 (s) to build the prediction model.

Figure 5 illustrates the F1 score of each classifier, the F1 score is good if it has the value near to 1, and worst when the value reaches near to 0. Random Forest has the highest F1 score and Decision tree, Gradient Boost, AdaBoost also has good F1 score but Gaussian Naïve Bayes has a very low F1 score.

7 Conclusion

In the proposed research, we have evaluated the performance of machine learning classifiers, namely D-Tree, Random Forest, AdaBoost, Gradient Boosting, Gaussian Naïve Bayes, these algorithms are trained and tested using the UNSW-NB15 dataset. The best algorithm is recognized by using performance metrics such as F1 score, accuracy, FPR. From the results, it can be concluded that Random Forest beats the other algorithms for the given dataset. It has an accuracy of 98.6%. This work can be extended by considering the different algorithms and different attributes for intrusion detection.

References

1. R. Heady, G. Luger, A. Maccabe, M. Servilla, *The Architecture of a Network Level Intrusion Detection System* (Computer Science Department, University of New Mexico, New Mexico, Tech. rep., 1990)
2. M. Aydın, M. Ali, A. Halim Zaim, K. Gökhan Ceylan, A hybrid intrusion detection system design for computer network security. Comput. Electr. Eng. 517–526 (2009)
3. R. Thomas, D. Pavithram, A Survey of Intrusion Detection Models based on NSL-KDD Data Set (Dubai, UAE, 2018)
4. A. Divekar, M. Parekh, V. Savla, R. Mishra, M. Shirole, Benchmarking datasets for Anomaly-based network intrusion detection: KDD CUP 99 alternatives 2018, in *IEEE 3rd International Conference on Computing, Communication and Security, Kathmandu* (Nepal, 2018)
5. S.-Y. Ji, B.-K. Jeong, S. Choi, D.H. Jeong, A multi level intrusion detection method for abnormal network behaviors. J. Netw. Comput. Appl. **62**, 9–17 (2016)
6. D.Md. Farid, M.Z. Rahman, Anomaly network intrusion detection based on improved self adaptive bayesian algorithm. J. Comput. **5**(1) (2010)
7. D.Md. Farid, N. Harbi, M.Z. Rahman, Combining naive bayes and decision tree for adaptive intrusion detection. Int. J. Netw. Secur. Appl. **2**(2), 12–25 (2010)
8. M.F. Suleiman, B. Issac, Performance comparison of intrusion detection machine learning classifiers on benchmark and new datasets, in *28th International Conference on Computer Theory and Applications (ICCTA 2018)* (2018)
9. J. Zhang, M. Zulkernine, A. Haque, Random-Forests-Based Network Intrusion Detection Systems (IEEE, 2008)
10. U. Cavusoglu, A New Hybrid Approach for Intrusion Detection Using Machine Learning Methods (Springer, Berlin, 2019)
11. N. Sultana, N. Chilamkurti, W. Peng, R. Alhadad, Survey on SDN Based Network Intrusion Detection System Using Machine Learning Approaches (Springer, Berlin, 2018)
12. P. Gogoi et al. Packet and flow based network intrusion dataset, in *Contemporary Computing* (Springer, Berlin, 2012), pp. 322–334
13. UNSW-NB15, https://www.unsw.adfa.edu.au/unsw-canberra-cyber/cybersecurity/ADFA-NB15-Datasets/
14. N. Moustafa, J. Slay, UNSW-NB15: a comprehensive data set for network intrusion detection systems (UNSW-NB15 network data set), in *Military Communications and Information Systems Conference (MilCIS)* (IEEE, 2015), pp. 1–6
15. F. Tsai, Y.F. Hsu, C.Y. Lin, W.Y. Lin, Intrusion detection by machine learning: a review. Exp. Syst. Appl. **36**(10), 11994–12000 (2009)
16. P. Aggarwal, S.K. Sharma, An empirical comparison of classifiers to analyze intrusion detection, in *2015 Fifth International Conference on Advanced Computing Communication Technologies (ACCT)* (IEEE, 2015), pp. 446–450
17. M.C. Belavagi, B. Muniyal, Performance evaluation of supervised machine learning algorithms for intrusion detection. Procedia Comput. Sci. **89**, 117–123 (2016)
18. J. Zhang, M. Zulkernine, A. Haque Random-forests-based network intrusion detection systems. IEEE Trans. Syst. Man Cybern.—Part C: Appl. Rev. **38**(5) (2008)
19. M. Kulariya, P. Saraf, R. Ranjan, G.P. Gupta, Performance analysis of network intrusion detection schemes using apache spark, in *International Conference on Communication and Signal Processing* (India, 2016)

Low-Light Visible and Infrared Image Fusion in NSST Domain

Apoorav Maulik Sharma, Ayush Dogra, Bhawna Goyal, Renu Vig and Sunil Agrawal

Abstract Multisensor images are captured to facilitate the human visual system under different environmental circumstances. Nowadays, images are captured in a wide range of electromagnetic spectrum. The focus of this manuscript will be on images captured from two sensors, viz. Visible and Infrared. The images captured from these sensors are integrated efficiently to enhance the information content of fused images and the performance is evaluated with the help of various metrics available. In this manuscript, fusion methods based on four benchmark techniques are presented on a challenging dataset and evaluated on the basis of various evaluation metrics and also visually and according to the results obtained this can be established successfully that the Shearlet-based methods are most efficient in terms of subjective as well as objective quality.

1 Introduction

Nowadays, images are captured in almost every range of electromagnetic spectrum. Images captured in different domains of the spectrum reveal different kinds of information, for instance, visible images have the information which is normally visible to the human perception system. However, the infrared images reveal the information which is not obvious to human perception as the infrared sensors capture the heat radiations emanated by the objects present in the scene. Image fusion techniques can be applied to these images captured to merge the complimentary available information from two images into one single frame. When done efficiently, this not only helps in increasing the information content of the image but also increases other vital features of an image like contrast, detail, and luminance. The fusion of multisensor images finds its applications in fields like medical diagnosis, civil and military

A. M. Sharma (✉) · B. Goyal · R. Vig · S. Agrawal
UIET, Panjab University, Chandigarh, India
e-mail: apoorav1207@gmail.com

A. Dogra
Center for Biomedical Engineering, IIT Ropar, Punjab, India

M. Dutta et al. (eds.), *Proceedings of International Conference on IoT Inclusive Life (ICIIL 2019), NITTTR Chandigarh, India*, Lecture Notes in Networks and Systems 116,
https://doi.org/10.1007/978-981-15-3020-3_7

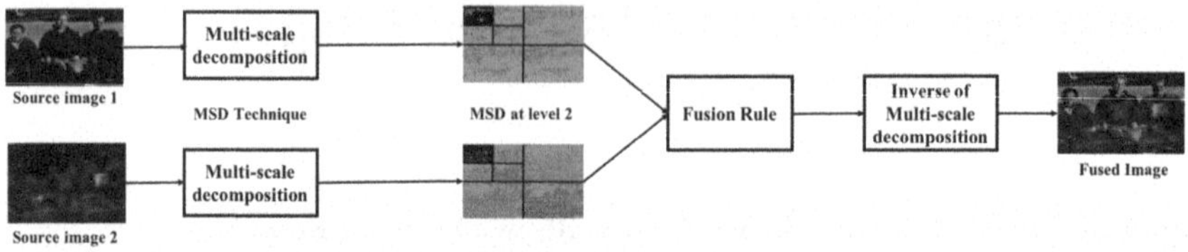

Fig. 1 Multi-scale decomposition scheme [2]

surveillance, concealed weapon detection. Moreover, multi-sensor image integration can be applied to images captured in low light. Firstly, the images captured with the help of visible light sensors have very low luminance and contrast along with very low detail [1]. On the other hand, images captured with the help of IR sensors have better contrast, luminance, and detail but they lack details obvious to the human eye. Secondly, images captured in low light contain Gaussian noise. So, the goal of image fusion is to reduce the noise in the final fused image and to compensate for the shortcomings of each other [2].

The general fusion process, shown in Fig. 1, requires decomposition of the input (source) images into finer and relatively coarser levels. This process is called multi-scale decomposition (MSD). This process is carried out either in the spatial domain or transform domain and it is done in order to separate the images into their base (smoother or finer or low-pass) and detail (coarser or high-pass) equivalents. This is an important step in this regard as the technique chosen determines not only the quality of decomposition but also the performance of the fusion methodology employed. After the decomposition, detail and base layers are either fused directly using some appropriate fusion rule or they are enhanced prior to final fusion to improve the evaluation results as well as the visual quality of the final fused image [2, 3].

Starting with MSD, there are numerous ways available to decompose the image in both the transform domain as well as the spatial domain. Some examples in transform domain include Discrete Cosine Transform, Wavelets and its shift-invariant versions like Shift-Invariant Wavelet Transform (SIWT), Dual-Tree Complex Wavelet Transform (DTCWT), Curvelets, Contourlets, Shearlets, etc. These transforms evolved in a written order to achieve properties like shift-invariance, better computation speed, etc. Examples in the spatial domain include filters like Average filter, Gaussian filter, Bilateral filter, Joint/Cross-Bilateral filter, Guided filter, etc. These filters also evolved in the same order to acquire better smoothing operation while preserving the edges and contours or preserving high-pass information.

Going further, the base and the detail layers can be enhanced using some saliency detection tools or some other visibility improving technique. There are wide varieties of these techniques which are explored by researchers. Finally, the base and detail layers can be merged using some appropriate fusion rule. There are some most popularly used fusion rules such as Choose max, Choose min, Average fusion rule, Weighted average fusion rule, Fusion rule based on activity measurement, etc.

The fusion rules are also called coefficient selecting methods as they are used to identify coefficients for the fused image out of multiple source images [2, 3]. A typical MSD-based image integration technique is shown in Fig. 1.

2 Related Work

Many state-of-the-art techniques have been designed over the years to efficiently fuse the images obtained from multiple sensors, one such technique is the Ratio of Low-Pass pyramid (ROLP) technique. The roots of the ROLP method lie in the Laplacian pyramid or Difference of Low-Pass (DOLP) pyramid technique proposed by Burt and Adelson [4]. The DOLP method is a multi-resolution decomposition method in which each layer of the pyramid is obtained by subtracting the successively Gaussian-filtered versions of the image from the expanded versions of low resolution (down-sampled) image from next stage. In ROLP, division is used instead of subtraction. The ROLP method considers the human visual system which is sensitive to changes in contrast to local parts. Due to this, the ROLP method is also known as the "Contrast Pyramid" decomposition technique. This method is useful for the purpose of detection and recognition. The methodology for fusion using this method is described in the following section [5]. The third technique will be discussed is Singular Value Decomposition (SVD) based image fusion. The image can be considered as an array of pixel values with each row and each column representing a dimension. Originally, the goal of SVD is to reduce the dimensions of the input data so that it can be represented sparsely. Suppose A is an image which can be mathematically represented with the help of SVD as [6, 7]

$$A(m \times n) = U(m \times r)S(r \times r)V' \tag{1}$$

where r is the rank of the matrix, U and V are matrices of right and left singularity, respectively, and S is a diagonal matrix in which the eigenvalues are placed from highest to lowest order of magnitude. Clearly, in this type of decomposition, the highest eigenvalue represents the base layer and the other subsequent eigenvalues represent the detail layers [6]. The fourth and most important technique is the Shearlet-based image fusion technique. The Shearlet transform is like the wavelet transform. The only difference is the exploration in multiple directions. Hence it can be said that Shearlet transform has improved directionality. The Wavelet transform lacks directionality due to its association with only two parameters, viz., scaling and translation. The Shearlet transform introduces one more parameter in the scene, i.e., the shear parameter. This shear parameter is used to obtain different directions of operations of the Wavelet operator. The Shearlet transform of a signal gives out base layer and multiple details calculated in different directions [7].

3 Image Fusion Methodology

To achieve the goal of fusion each input image is broken down into a set of base layer and detail layers using discussed methods. In the first and second cases of DOLP and ROLP, the first layer or the base layer is obtained by taking the difference/ratio of the source image (say G_0) to the expanded version of blurred and down-sampled G_0 (say G_1). The blur is obtained by applying the Gaussian low-pass filter to the input images. The subsequent detail layers are obtained by further blurring G_1 by changing the Gaussian filter kernel characteristics and again down-sampling blurred G_1 to obtain say G_2 then again taking the ratio of G_1 to the expanded version of G_2 and so on. So, each level (say L) can be obtained as

$$L_i = G_i \big/ expand(G_i + 1) \tag{2}$$

where G_i is the blurred and down-sampled version of the original source image. After decomposing the source images according to the procedure discussed above they are fused using Choose max fusion rule in which for every position of the pixel the maximum value is chosen out of multiple source images. This fusion rule is most popular because after the total fusion process the image obtained is better in contrast and luminance and hence it is better visually. But, this is not the case all the times so care must be taken in choosing an appropriate fusion rule. After applying the fusion rule, the reconstruction is done by multiplying (reverse of division as done during decomposition) as

$$G_i = L_i * expand(G_i + 1) \tag{3}$$

It can be noted that this equation is a direct consequence of Eq. (2) [5]. In the third case of SVD, the method of singular value decomposition is applied to each source image, due to which each source image is decomposed according to Eq. (1). So basically, three matrices are obtained for each source image after SVD decomposition. As discussed earlier, the base layer and the subsequent detail layers can be obtained by matrix multiplication of singularity matrices and the relevant eigenvalues. In SVD decomposition, the source images are decomposed in wavelet-like fashion in which one base layer is obtained and three details layers, viz. horizontal detail, vertical detail and diagonal detail are obtained. These layers are fused separately using choose max fusion rule and then these fused layers are reconstructed to make a fused image. The fused image can be obtained by simple matrix multiplication shown in the equation. It can be noted here that the matrices U and V in the equation are orthogonal matrices hence they facilitate sparse representation of the signal under consideration [6].

In the fourth and final case, source images are decomposed using Shearlet transform [8]. As described earlier, the Shearlet transform decomposes the image and provide details in multiple directions. So, multiple subbands are obtained for each source image. These subbands are then fused separately as shown in Fig. 2. For the experiments done the base layer is fused using choose max fusion rule and the rest

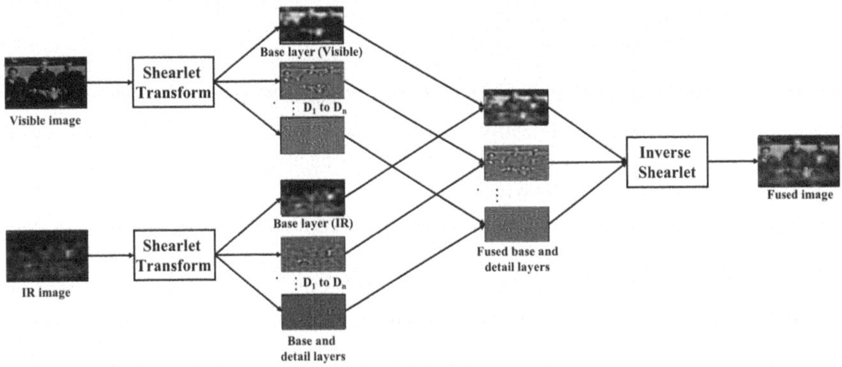

Fig. 2 IR and visible image fusion using Shearlet transform

of the subbands are fused using average fusion rule. This fusion rule tends to average the pixel values to obtain the final pixel value for fused image. After applying the fusion rule, the image is reconstructed back by applying Inverse of the transform. In this context, it is called Shearlet reconstruction [9–12].

To compare the fusion results, there are various mathematical tools available in the form of metrics. These metrics are divided into two main categories.

1. Classical or reference-based metrics.
2. Gradient or non-reference-based metrics.

As the name suggests, the classical or reference-based metrics require a reference image to evaluate the fusion results and non-reference-based metrics require no reference image to evaluate the fusion results. Nowadays, non-reference-based metrics are more popular and give a clearer picture of the performance of the fusion process. The examples of reference-based metrics are Peak Signal-to-Noise Ratio (PSNR), Standard Deviation (SD), Mutual Information (MI), Structural Similarity Index Measure (SSIM), Cross-Correlation, Percentage Fit Error (PFE), etc. [13] and the examples of non-reference-based metrics are metrics proposed by Xydeas and Petrovick [14] that are Q_{ABF}, L_{ABF} and N_{ABF}. Where Q_{ABF} is the fusion rate. Its value lies between 0 and 1. The value 1 depicts a 100% fusion rate. L_{ABF} is the loss of information and N_{ABF} is the amount of artifacts. Although, there is a proven disparity, which sometimes exists between these objective results and the visual quality of the image still these non-reference-based metrics are the most important metrics which are used in all recent works regarding image fusion to evaluate the quality of the fused image.

4 Experiments and Results

For the experimentation, source images used are captured in very low-light conditions, as shown in Fig. 3. These source images lack proper luminance, contrast and hence lack proper visibility. Furthermore, these source images are noisy which makes them even more challenging to work with.

The visible image contains features that are normally visible to the human eye. The IR image contains concealed weapon information. When fused together they not only fix the low visibility problem but also helps in concealed weapon detection which is a major application of fusing this dataset. The source images are registered bitmap images of size 200 * 256 with a predefined 64 * 3 colormap. The same colormap is also used for the fused image. All the experiments are performed on MATLAB R2015b on a 64-bit processor. The fusion process is carried out according to the techniques discussed above. Although, there is a technique for fusion, based on Cross Bilateral Filter (CBF) [15] which provides the best fusion result for this dataset these particular techniques are chosen because DOLP is the most primitive state-of-the-art technique which was originally proposed for compact coding of images. ROLP was first introduced for the purpose of fusion only [5]. SVD-based technique is very unique among various other image fusion techniques regarding its dynamics and the Shearlet is chosen because it is the latest state-of-the-art transform which is not explored much in this field but holds the potential to produce excellent results both objectively and subjectively (Table 1).

Fig. 3 Visible image and infrared image

Table 1 Fusion rate, loss of information and amount of artifact values for image fusion using SVD, ROLP, and shearlet transform

Technique	Q_{ABF}	L_{ABF}	N_{ABF}
DOLP	0.7792	0.2193	0.0015
ROLP	0.6309	0.0500	0.3191
SVD	0.6129	0.3864	6.4e–4
Shearlet	0.7364	0.2636	7.03e–5

Fig. 4 Image fusion results from left to right using DOLP, ROLP, SVD, and NSST

The evaluation of results is done on two bases viz. objective and subjective. According to the objective fusion results, evaluation is done on the basis of metrics while on the subjective scale the evaluation is done visually. According to evaluation metrics, it is evident that the DOLP pyramid-based technique performs better than both ROLP, SVD, and Shearlet-based technique. The Shearlet-based fusion technique is very efficient in removing the artifacts (Fig. 4).

The details can be obtained in many more directions. Fusing the individual low-frequency and high-frequency information layers is a much more effective process in the case of Shearlets because of the availability of detail information from multiple dimensions. In DOLP and ROLP, the levels of decomposition can be adjusted and it is observed that best results are obtained at decomposition level 1. The reason behind is that a lot of information is thrown away during the down-sampling operation employed in multi-resolution techniques like DOLP and ROLP. As the level of decomposition is increased more down-sampling operations are applied to the image due to which image quality further degrades. The same is the case for SVD.

In the case of SVD also information is lost due to down-sampling operations. Hence, the level of luminance of the fused image is very low. The down-sampling operation is applied to the image to make it adjustable to changing kernels of the filter applied. In the experiments performed, we used a non-sub-subsamplesampled version of Shearlet transform so that no information is lost in the down-sampling operation. It is important to note here that in DOLP and ROLP technique the loss of information is very low. It's 5% but the amount of artifacts is very high which is also evident in visual inspection of fusion result. In the other two techniques, the loss of information is high and amount of artifacts is low.

5 Conclusions and Future Directions

There are many techniques which are explored in the domain of image fusion. While some techniques perform well on the objective evaluation scale, on the other hand, some techniques provide very good visual results. There are very few techniques that prove better on both evaluation scales. This conclusion is drawn from the analysis of the results that Shearlet transform holds the potential to provide better evaluation results while enhancing the visual quality of the overall fused image.

References

1. J. Ma, Y. Ma, C. Li, Infrared and visible image fusion methods and applications: a survey. Inf. Fusion **45**, 153–178 (2019). https://doi.org/10.1016/j.inffus.2018.02.004
2. S. Li, X. Kang, L. Fang, J. Hu, H. Yin, Pixel-level image fusion: a survey of the state of the art. Inf. Fusion **33**, 100–112 (2017). https://doi.org/10.1016/j.inffus.2016.05.004
3. A. Dogra, B. Goyal, S. Agrawal, From multi-scale decomposition to non-multiscale decomposition methods: a comprehensive survey of image fusion techniques and its applications. IEEE Access **5**, 16040–16067 (2017). https://doi.org/10.1109/ACCESS.2017.2735865
4. H. Ghassemian, A review of remote sensing image fusion methods. Inf. Fusion **32**, 75–89 (2016). https://doi.org/10.1016/j.inffus.2016.03.003
5. P.J. Burt, E.H. Adelson, The laplacian pyramid as a compact image code. Readings Comput. Vis. Elsevier **31**, 671–679 (1989). https://doi.org/10.1016/B978-0-08-051581-6.50065-9
6. A. Toet, A morphological pyramidal image decomposition". Pattern Recognit. Lett. **9**, 255–261 (1989). https://doi.org/10.1016/0167-8655(89)90004-4
7. V.P.S. Naidu, Image fusion technique using multi-resolution singular value decomposition. Def. Sci. J. **61**, 479 (2011), http://dx.doi.org/10/gfkwnz
8. H. Andrews, C.L.I.I.I. Patterson, Singular value decomposition (SVD) image coding. IEEE Trans. Commun. **24**(4), 425–432 (1976), http://dx.doi.org/10.1109/TCOM.1976.1093309
9. Q.W. Lim, Nonseparable shearlet transform. IEEE Trans. Image Process. **22**(5), 2056–2065 (2013), http://dx.doi.org/10.1109/TIP.2013.2244223
10. S. Häuser, G. Steidl, Fast Finite Shearlet Transform (2012), https://arxiv.org/pdf/1202.1773.pdf
11. G. Kutyniok, D. Labate, Shearlets: Multiscale Analysis for Multivariate Data (Springer Science and Business Media, Berlin, 2012), http://dx.doi.org/10.1007/978-0-8176-8316-0
12. X. Liu, Y. Zhou, J. Wang, Image fusion based on shearlet transform and regional features. AEU—Int. J. Electron. Commun. **68**(6), 471–477 (2014). https://doi.org/10.1016/j.aeue.2013.12.003
13. P. Jagalingam, A.V. Hegde, A review of quality metrics for fused image. Aquat. Procedia **4**, 133–142 (2015). https://doi.org/10.1016/j.aqpro.2015.02.019
14. C.S. Xydeas, V. Petrović, Objective image fusion performance measure. Electron. Lett. **36**(4), 308 (2000), http://dx.doi.org/10.1049/el:20000267
15. B.K. Shreyamsha Kumar, Image fusion based on pixel significance using cross bilateral filter. Signal, Image Video Process. **9**(5), 1193–1204 (2015). https://doi.org/10.1007/s11760-013-0556-9

Automated Sleep Stage Classification Based on Multiple Channels of Electroencephalographic Signals Using Machine Learning Algorithm

Santosh Kumar Satapathy and D. Loganathan

Abstract Sleep quality assessment is the preliminary approach in sleep scoring which becomes to identify sleep quality and diagnose sleep-related diseases. Currently, Automatic Sleep Stage Classification (ASSC) methods have obtained by most of the researchers for sleep scoring during sleep tests because it gives comfortable and error-free recordings from different electrode positions without the intervention of clinicians. The major objective of this study is to analyze some existing contributions done by researchers and also in this study we have obtained all electrode positional values of electroencephlogram (EEG) signal from patients. The electrode placed in the patient brain as per 10–20 electrode placement guidelines. Here we have considered EEG signals from the ISRUC-SLEEP public sleep data repository for this experimental work. This proposed work mainly focused on discriminating between wake and sleep. In this study, we have extracted some more features which become easier to identify the different sleep stages. A total of 34 numbers of features from the frequency domain were extracted. The main challenging of this research paper is to discriminate between which channel of EEG signals gives the best sleep score and which classification methods are more suitable for detecting the sleep abnormality. In this research work, basically we investigated which channels are and which classification algorithm is to be more appropriate identifying the different stages of disorder in sleep patterns from particular categories of subjects (healthy and Unhealthy) with various age classes. Study results that out of 4 EEG channels, F4-A1 placed electrode give good accuracy with all base classifiers such as SVM (96.6%), DT (95.8%), and KNN (95.6%). With ensemble classifiers, it resulted in 97.1% accuracy to classify in between the wake stage and sleep stage.

Keywords Sleep scoring · EEG · Machine learning classifier

S. K. Satapathy (✉) · D. Loganathan
Pondicherry Engineering College (PEC), Puducherry, India
e-mail: santosh.satapathy@pec.edu

M. Dutta et al. (eds.), *Proceedings of International Conference on IoT Inclusive Life (ICIIL 2019), NITTTR Chandigarh, India*, Lecture Notes in Networks and Systems 116,
https://doi.org/10.1007/978-981-15-3020-3_8

1 Introduction

1.1 Background

Sleep has an important role in human life. Its impacts indirectly associated with our health. Proper keep up sleep hours make our body strength both physically and mentally [1]. It is unquestionable that, if any type of disturbances toward sleep, it may impact on the physical and mental condition of the human body. Its long-term continuity may leads to sleep disorder and its impact on various aspects of human life in different contexts like deteriorating effects on our healthy body and safety and its consequences toward monetary load on both the personal and communal levels. Now in large scale, the sleep labs and research labs comprehensively conducted to analyze human sleep for different purposes. As per the report from the Centre of Disease and Control Studies, two-thirds of the population of the world affected with sleep-related diseases and its ratio is more in the case of adults daily. As per the survey from different research groups, it has found that more than a third of American adults affected with sleep-related diseases because of a lack of sleep daily [2]. Sleep disorder may happen in any age group of human lives and its consequences also different in related deteriorate quality of life of the human body. The most common sleep-related diseases like obstructive sleep apnea, sleep-walking, night-terror, hypersomnia, insomnia, circadian rhythm, narcolepsy, and parasomnias occurred due to abnormality in stages of sleep. Nowadays sleep-related complaints are giving upmost priority to identify the sleep disorder stages and its associated side medical problems and make find a proper clinical diagnosis against different types of sleep diseases.

Generally, diagnosis for sleep disorders is traditionally performed in sleep laboratories, where sleep-disordered subject's physiological signals considered and it was diagnosed through the manual manner by sleep experts. Traditionally sleep disorder disturbances diagnosed as per recommended R&K guidelines [3] and after since from the year 2005, as per American Academy Sleep Medicine (AASM) rules [4], sleep disorder diagnosis going on. Concerning the diagnosis of the disorder, the first step is to analyze irregularities of sleep. As per assessment the most important part of the sleep study is that collection electrophysiological signals from subjects during sleep hour [5, 6]. First of all to know about the irregularity of sleep through the PSG study or sleep test. Generally, in the PSG study, several electrophysiological signals collected from the patient. In this sleep study, we have followed the AASM manual's guidelines for proper diagnosis of different exceptions happened during bedtime recorded signals from sleep disease subjects. During this sleep disorder investigation, the most essential step is to scrutiny the hypnogram of gathered brain signals from subjects, thorough that we are trying to identify sleep transition between different stages [7]. In this investigation, the complete process as per the AASM manual regarding sleep stage classification (SSC), as per the AASM criterion, the total sleep hours divided into three stages, which are wake, non-rapid eye movement (NREM), rapid eye movement (REM). Further, the NREM stage divided into three stages as per AASM standards such as N1, N2, and N3. The recorded information varies with different stages throughout the whole overnight with different periods. For observation regarding abnormality, we have split the whole recordings into different time frames, called epochs. Here we have worked on 30 s epochs. All these processing is done through a super specialty hospital, where sleep specialists review the PSG recordings and detect the sleep patterns [8]. According AASM standrads, here we have considered sleep patterns during sleep stage scoring and the behaviour of sleep patterns described in Table 1.

Generally for proper diagnosis of sleep disorders, physicians or sleep specialists recommended strongly AASM guidelines and EEG channel. Here we have obtained the noninvasive testing technique and with the help of that number of electrodes has fixed on the scalp of subjects. Those recorded signals are used for detecting the irregularities on sleep stages. As per the AASM manual, we have recorded different positional electrode placement values and it has managed with AASM 10–20 electrode system [9].

EEG signals are specifically widely accepted in the field of neurology and though we have extracted the EEG signals from the brain, so that it has strongly recommended sleep applications. Generally, most of the researchers have adopted both EEG single-channel and multichannel during sleep data analysis. With regards to channel selection during recording from subjects are also one of the important step and as per the different collected views, its quite difficult because in this situation the subject must fixed by so many electrodes and which alternatively create unnecessary disturbances during recording, which ultimately hamper to reach the best classification accuracy in between wake and sleep stages [10]. Other limitations also observed during sleep tests that sleep scoring from a large number of recorded epochs manually

Table 1 Summary of EEG sleep patterns as per AASM standards

Phase	Classification criteria
Wakefulness (W)	It is the state before decline to asleep. In this state, the brain signal is very rapidly changed. Here alpha waves (8–13 Hz) are most significant
Stage 1 (N1)	It is the first stage of sleep, in this state the subject is in the drowsy mood. During this state, generally, the activity of the brain slows down. Comprising of theta waves (4–8 Hz) is most suitable for this category
Stage 2 (N2)	Generally here we found sleep spindles and sometimes also K-complexes occurred (11–15 Hz). Normally in this state, the eye movement of subjects is stopping
Stage 3 (N3)	It is the deep sleep state and most of the epochs being high voltage and low-frequency activity. Here around 20–50% of epochs are delta waves. The range of frequency occurred (2–4 Hz)
Rapid Eye Movement (REM)	Here, the subjects' eye remains closed but it moves rapidly. Beta waves (>13 Hz) are more significant for this state

through sleep experts or physicians being some human error related to monitoring. To overcome the challenges raised during manual clinical settings to detect the irregularities of patterns of sleep, for that reason most of the researchers recommended automatic observation procedure to test sleep disorder. Through this method more advantages for sleep experts for proper examination of collected brain signals and it is helpful for speed up toward the diagnosis of different types of diseases in related sleep deformity. It becomes easier for clinical staff for monitoring sleep scores from patient's very easier manner which is more comfortable for proper diagnosis in connection to different sleep diseases. Another disadvantage related to PSG test is a lot of wire connectivity to patient's body during recording the sleep score but it may fail to give correct score because of patient disturbances in related to more wire connectivity in body, which becomes more uncomfortable for patients but nowadays, it may also lower by using single-channel recording of fixed electrodes on scalp and it becomes comfortable for patients with fewer cablings attached in body [11]. Some common approaches selected by researchers for analyzing disruption during sleep time: (1) Waveform recognition [12] (2) Spectral frequency bands [13] (3) Decision Trees [12] (4) Hidden Markov Models [14] (5) Fuzzy Systems [15]. As per recent observation that neural networks have been widely used in the field of diagnosis sleep stages using EEG signal [16, 17].

1.2 Contribution and Paper Organization

Broadly for this type of study, we have considered the EEG signals from different public repository datasets. Earlier so many researchers have used EEG datasets from Physio Net using the Sleep-EDF database and ISRUC-SLEEP dataset. The recorded signals have been visualized by sleep experts through different frequency sub-bands. Categorizing these EEG signals sub-bands based on the characteristics differences in several frequencies found at each level of sleep stages. For proper analysis of sleep stages behavior from sleep rhythm, we need to divide the EEG signal into different sub-bands based upon the different frequency levels such as delta sub-band (0.5–4 Hz) frequency, theta sub-band (4–8 Hz) and alpha sub-band (8–13 Hz) frequency. Beta in between 13 and 35 Hz and gamma waves in between of greater than 30 Hz. In this study, we have reduced the noise factor through Butterworth bandpass filters. In addition to that different other discriminating evaluation features including Spectral Power, Wave Index, Activity, Mobility, Complexity Parameters, Entropy, and Skewness and Kurtosis are measured from each frequency ranges of different frequency bands. Furthermore extracted features are fed to a supervised learning classifier for identifying the sleep state and confirm that the recorded signals belong to either wake state or sleep state, as per our proposed research objective.

This research paper has been planned as follows: Sect. 2 briefly describes related work already done through different researchers. Section 3 explains the complete description of the recommended methodology. Feature extraction complete description in Sect. 4. Section 5 describes the adopted feature selection techniques. In Sect. 6

explains both experimental results and discussions about the different comparative results. At last, we specified the conclusion in Sect. 7.

2 Related Work

Sharma et al. have proposed a single channel of EEG based sleep stage classification. Here the author has obtained time–frequency features from recorded signals from subjects. The model has reached the overall accuracy from 91.5% to 98.3% for five state classifications [18].

Z. H. Ge et al. has obtained multichannel for detecting the abnormalities in different sleep transition states. In this work, the author has focused on five state classification and considered power spectral density features for classifying the states of sleep. The working model based on a multilayer feed-forward neural network and has achieved 90% accuracy toward identifying irregularities [19].

In [20], the research work was carried out considering both categories of subjects such as healthy subjects and effected with sleep diseases and also the author has considered the EEG signal for this study and he has employed the neural network classifier for classifying and achieved 91% accuracy.

Tim and Stefan have designed a work on the classification of sleep stages based on different physiological signals such as EEG, ECG, EOG, and EMG. Here the author has used a decision tree classifier and has achieved the result of classification accuracy as 94.5% as subject to discrimination in between effected with effect and not effected with sleep disorder [21].

In [22], the author has accepted a single channel of EEG for their experimental work; here the author has obtained multimodal signal segmentation. In this work, the author has extracted nonlinear features from the input channel. The extracted features are fed into the HMM classifier. In this research work, two different sources of datasets have used such as sleep-EDF and sleep-EDF Expanded.

In [23] the authors have used the sleep-EDF datasets for their experimental work here extracted both linear and nonlinear features and for classification process, they have obtained two classification techniques such as SVM and artificial neural network and have achieved 90.30% and 89.93% accuracy, respectively.

Kaare B. Mikkelsen et al. have obtained the EEG signal from placed ear electrodes positions. In this study, the authors have made a comparison between manual investigations through sleep experts and automatically generated sleep scores. It was observed that more good accuracy was found from the automatic evaluation process and it has reached an overall accuracy of 87–93% for different sleep state's classification processes [24].

M. B. Uddin et al. have obtained adults dataset for the sleep study, where they have considered respiratory and oximetry signals fetched from subjects in both single and multiple channel-wise. He has found that both the input signals are more sufficient to detect sleep apnea. In this scenario, the only binary decision-making process was more suitable but he has observed that for multiple signals multiple classification techniques were more appropriate for best decision-making in a subject to sleep disease diagnosis [25].

In [26], the authors have considered different gender subjects for sleep disorder identification, where they have made a comparison with different time frame epochs like 1 s sub-epoch versus 5 s sub-epoch. In this research work, the author has considered epochs in consecutive and non-consecutive manners with different periods. It was observed that 91% accuracy achieved for 1 s sub-epoch and 70% accuracy has reached for 5 s epoch toward the detection of abnormality.

Agarwal et al. [27] considered five-channel PSG data and computed features as amplitude, energy, dominant rhythm, spindles. For classification of sleep states, here the author has used the K-means clustering algorithm.

Berthomier et al. [28] have taken single-channel EEG and applied a logical based neuro-fuzzy system for classification pf sleep states.

Karkovska et al. [29] extracted PSG signals consisting of signals from EEG, EOG, and EMG. They have been implemented quadratic discriminant analysis for the analysis of sleep irregularity.

Ronzhina et al. [30] considered single-channel EEG, extracted power spectral density and attained ANN for classification.

In [31] Different types of entropy-based features have used for distinguishing between different sleep phases.

3 Methodology and Procedure

3.1 Proposed Method

This study proposed an active procedure that could be smoothly implemented in a hardware implementation, which is being easier for the clinical system to categorize between wake stages and sleep stages. With the help of this proposed method, the physicians and sleep experts easily identify the different sleep-related diseases. First of all, the EEG signal records collected from the Medicine Department of the Hospital of Coimbra University (CHUC). After that those recorded signals is filtered and break down into different frequency sub-band. In addition, we have extracted temporal, frequency, time-domain features. Finally, we are following a supervised learning approach for classifying the sleep stages from selected feature vector. The complete work layout of the classification methodology is shown in Fig. 1.

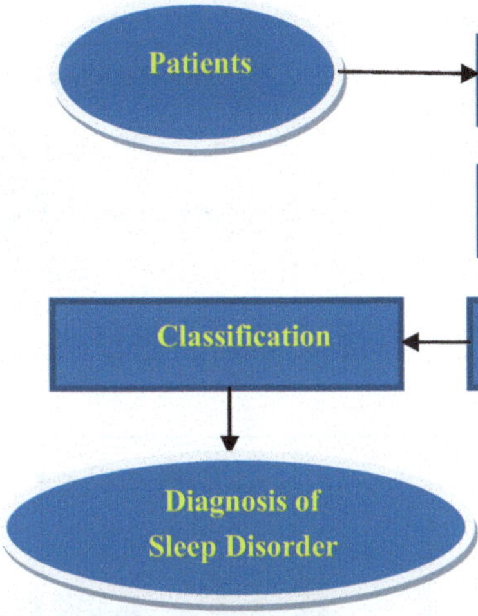

Fig. 1 Classification methodology

3.2 EEG Dataset

The records used for this experiment are collected from ISRUC-SLEEP dataset. This data recorded from the sleep lab of the Coimbra University (CHUC), which is publicly available and this recorded data is open to access for those who are working his/her research in sleep subjects. This dataset recorded information from adults group including healthy subjects and subjects with sleep disorders under medication [32]. In this study, we have considered one patient and their multiple channels recording to be considered. Each signal was sampled at 200 Hz and was segmented epochs into 30 s data as per the AASM manual guidelines. The data recordings are visualized and examined by two sleep experts. In [32], brief information has to be mentioned. Here, we have only employed one subject and only the EEG signals of patients are recorded and are used in the experiments of this study. In this work, we have extracted four EEG channel such as F3-A2, C3-A2, O1-A2, F4-A1 from the patient. We also considered its corresponding hypnograms files in which their corresponding sleep patterns are managed. This pattern numbered the different sleep stages as wakefulness (W–0), non-rapid eye movements (NREM1-1, NREM2-2, and NREM3-3), and rapid eye movements (REM-5). For objective clarification, Table 2 shows 30 s epochs samples of the EEG signals which were to be considered as inputs and further experiments in this study.

Table 2 Input signals from various EEG channels

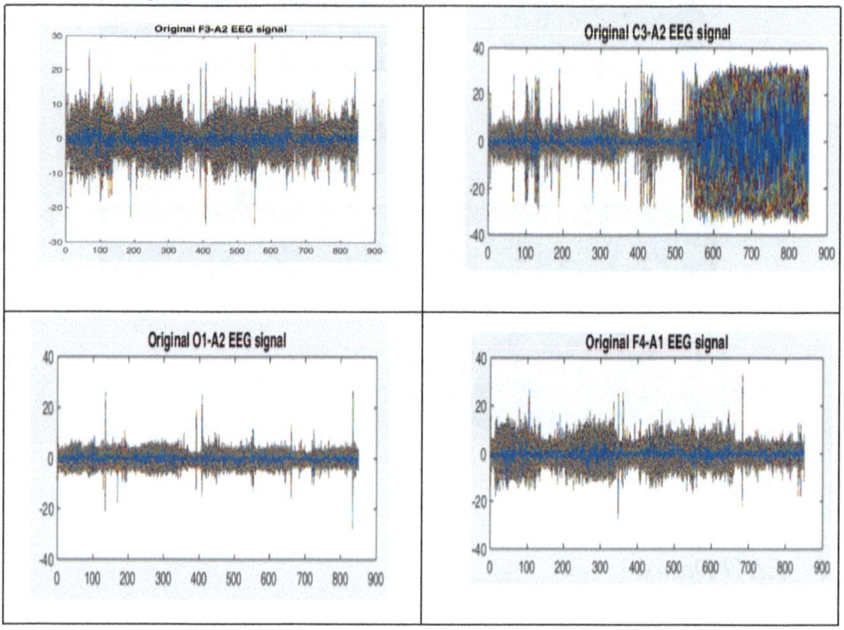

4 Feature Extraction

In this study, we have extracted the features from single-channel EEG; here we have recorded F3-A2 electrode positional value from subjects during sleep in the night time period. From a single channel, we have extracted 34 discriminant features have extracted at each sub-bands frequency for the classification process. The extracted features from different input channels are mentioned in Table 3.

Table 3 Features extracted from different channels

Label	Description
F1,F3,F5,F7,F9 F2,F4,F6,F8,F10	Relative Spectral Power (RSP) with different frequency sub-bands
F11,F12,F13	Wave Index for different frequency sub-bands
F14, F15,F16,F17,F18	Centre Frequency (fc) for various frequency length
F19,F20,F21,F22,F23	Bandwidth (fσ) for different frequency range patterns
F24,F25,F26,F27,F28	Spectral value at centre frequency (sfc) in distinct frequency sub-bands
F29,F30,F31	Hjorth Parameter
F32,F33,F34	Entropy

5 Feature Selection

In this step, we have focused that how best features are fed into classifiers. Machine learning techniques generally require a suitable number of features to predict the intended outputs correctly. Using a large number of properties, it may hamper the proper prediction and ultimately it has effected to accuracy level and led to poor performances in many situations. Therefore, we are trying to avoid such a situation as much as possible with the help of feature selection techniques.

Here we have selected the effective features from extracted features, which are more useful with regards to help for best diagnosing the sleep irregularities [33]. For identifying the suitable combination of features for classification activity we have applied a feature selection algorithm is and how it has best to predict the abnormality regarding sleep states. As per our review of some existing research work in this area, most of the authors have prescribed principal component analysis (PCA) and sequential selection methods [34]. The selected features used for classifying the sleep stages are mentioned in Table 4.

Table 4 Selected features combinations for classification

Channel	Best feature combination	Classifier
F3-A2	F1, F2, F5, F8, F12, F14, F3, F4, F9, F10, F16, F25, F26, F27, F30, F31, F34 (19 Features)	SVM
		DT
		KNN
		ENSEMBLE
C3-A2	F1, F2, F5, F6, F10, F11, F14, F15, F3, F4, F8, F9, F12, F13, F20, F21, F23 (17 Features)	SVM
		DT
		KNN
		ENSEMBLE
O1-A2	F1, F2, F9, F10, F23, F28, F3, F8, F14, F22, F31, F32 (12 Features)	SVM
		DT
		KNN
		ENSEMBLE
F4-A1	F5, F11, F12, F14, F15, F16, F27 (7 Features)	SVM
		DT
		KNN
		ENSEMBLE

6 Results and Discussion

The proposed study was followed by the automatic sleep stage classification and considered multiple channels of EEG signals. Here we have collected the values of the recorded channels from the ISRUC-SLEEP database. In this study, we have to deal with one subject, from that subject we have recorded data from multiple channels using 10–20 electrode placement system like F3-A2, C3-A2, O1-A2 and F4-A1. Each channel recorded information regarding wake epochs and sleep epochs from each subject. In this research work, we have tried to discriminate in between which electrode data gives the best accuracy to classifying in between the wake stage and sleep stage. Here we have obtained 34 features from each channel and that features we have input to classifiers for further classifying in between sleep stages. To further compare the performances of classification in between wakefulness versus sleep epochs for that reason we have used three base classifiers like SVM, DT, KNN and ENSEMBLE (Bagging method) used in sleep scoring. In this study, we have compared both channel recordings and classifier methodology is related to decide that which channel data and which classification algorithm gives the best accuracy in order to discriminate easily between wake and sleep stage. To evaluate the performance of our proposed experiment, 34 sets of features were extracted from multiple channels of EEG signal and each channel carries 850 epochs with 6000 continuous data samples individually and evaluates the performances in terms of accuracy in order to classify between wake stages and sleep stages. In this study, we have obtained cross-validation techniques that are applied to operation both on the training dataset and the test dataset portion into the selected classifiers. In order to evaluate the common performances for all classes, three metrics are used here. The equations of the above mentioned performance metrics are shown below.

Accuracy (ACC) $= (\mathrm{TP} + \mathrm{TN}) / (\mathrm{TP} + \mathrm{FP} + \mathrm{TN} + \mathrm{FN})$ [35]

Sensitivity (Sen) $= \mathrm{TP} / (\mathrm{TP} + \mathrm{FN})$ [36]

Specificity (Spc) $= \mathrm{TN} / (\mathrm{TP} + \mathrm{FN})$ [37]

From Tables 5, 6, 7 and 8 we have mentioned different classification performances for different channels of EEG signals. From Table 5, we have found the performances for the F3-A2 channel recorded data from the subject. Finally, as per computation, we have found that ensemble classifier accuracy is said to be best in order to classify between wakes and sleep stages. From Table 6, it was observed that for the C3-A2 channel, the SVM classifier gives the best accuracy regarding sleep stage classification. From Table 7, for O1-A2 channel, it has shown that the ensemble classifier

Table 5 Comparisons of results of F3-A2 channel through different classifier

Classifier metric	SVM (%)	DT (%)	KNN (%)	Ensemble (%)
Accuracy	94.8	92.9	94.7	95.3
Sensitivity	96.4	94.4	96.1	96.4
Specificity	91.0	89.5	91.4	92.6

Table 6 Comparisons of results of C3-A2 channel through different classifier

Classifier metric	SVM (%)	DT (%)	KNN (%)	Ensemble (%)
Accuracy	91.6	88.8	88.1	91.6
Sensitivity	97.2	94.9	92.9	97.9
Specificity	78.0	74.8	77.5	77.1

Table 7 Comparisons of results of O1-A2 channel through different classifiers

Classifier metric	SVM (%)	DT (%)	KNN (%)	Ensemble (%)
Accuracy	96.4	94.7	92.9	96.5
Sensitivity	98.3	96.1	96.6	98.3
Specificity	92.2	91.4	84.4	92.6

Table 8 Comparisons of results of F4-A1 channel through different classifiers

Classifier metric	SVM (%)	DT (%)	KNN (%)	Ensemble (%)
Accuracy	96.5	95.7	95.6	97.0
Sensitivity	98.1	96.6	97.8	98.4
Specificity	93.0	93.7	90.6	93.7

obtained the best accuracy incomparable to the other three base classifier in order to classify different sleep stages. As per Table 8, it has observed that for channel F4-A1, the ensemble classifier gives the best accuracy in order to classify between sleep stages. Finally, we have observed that F4-A1 channel recorded information from subject gives the best accuracy in order to best classification in between wake stage and sleep stage and in case of classification algorithm, overall for all 4 channels as base classifier SVM is found to be a suitable method and it has been found that for all 4 channels with SVM base classifier we have achieved 90% of average accuracy. In this study, we have also obtained an ensemble classifier approach, where we have noticed that for all channels, we have achieved 91% of average accuracy in order to classify in between the wake stage and sleep stage of the subject. We have observed that by using the ensemble bagging method we achieved better accuracy for the classification of each sleep stage. As per our study here out of 4 channels, F4-A2 electrode positional data is best suitable for diagnosing. In Table 9, we have represented the confusion matrix of this experimental work.

This proposed study is also subject to some limitations. Firstly we have not focused the further classification sleep stages as N1, N2, and N3. We are still requiring more novel representative features and analysis mechanisms for addressing several

Table 9 Confusion matrix representation of individual input channel

F3-A2		Wake	Sleep	C3-A2		Wake	Sleep
SVM	Wake	112	21	SVM	Wake	203	55
	Sleep	15	602		Sleep	16	576
		Wake	Sleep			Wake	Sleep
DT	Wake	105	28	DT	Wake	193	65
	Sleep	10	607		Sleep	30	562
		Wake	Sleep			Wake	Sleep
KNN	Wake	103	30	KNN	Wake	200	58
	Sleep	9	608		Sleep	43	549
		Wake	Sleep			Wake	Sleep
ENSEMBLE	Wake	236	22	ENSEMBLE	Wake	236	22
	Sleep	23	569		Sleep	23	569
O1-A2		Wake	Sleep	F4-A1		Wake	Sleep
SVM	Wake	238	20	SVM	Wake	240	18
	Sleep	10	582		Sleep	11	581
		Wake	Sleep			Wake	Sleep
DT	Wake	236	22	DT	Wake	242	16
	Sleep	23	569		Sleep	20	572
		Wake	Sleep			Wake	Sleep
KNN	Wake	218	40	KNN	Wake	234	24
	Sleep	20	572		Sleep	13	579
		Wake	Sleep			Wake	Sleep
ENSEMBLE	Wake	239	19	ENSEMBLE	Wake	242	16
	Sleep	10	582		Sleep	9	583

problems regarding classification. Secondly, our proposed method is unaware of how efficiently discrimination between healthy subject and sleep disorder affected subject. In the proposed method, there are still many issues to be explored in the future. In these studies, we have only considered only 4 channels of EEG signal. In a future study, we will try to include number channels of EEG signal and also add other signals like ECG, EOG, EMG signals.

7 Conclusion

This research work proposed a unique method in the subject of an investigation on abnormality and identifies different sleep stages from multiple channels of EEG signal. In this study, we have tried a comparison between different electrode placed channels' records collected from the subject. Here we have used 34 features in different classifiers and measuring for best discrimination to classify among different sleep stages. According to the result of different classification algorithms given from Tables 5, 6, 7 and 8, it was observed from the results mentioned in Tables 5, 6, 7, and 8, incomparable to base classifiers the performances are best in ensemble classification techniques. We have observed that our ensemble bagging method achieved 91% average accuracy which is better than other classifiers used in this experiment for the same input signals. Here in this experiment we have considered the advantages of multiple classifiers obtained for this research work for analysis and compare the various performances of classification accuracy in sleep stage irregularities that happened during sleep time and also we have found that which channel is more effective in terms to best prediction to sleep disorder.

In our method, we used four channels of EEG signal for experimentation, in the future we can apply this method for other channels of EEG signals and also add signals like EOG, EMG, and ECG to improve classification accuracy.

References

1. T.L. Skaer, D.A. Sclar, Economic implications of sleep disorders. Pharmacoeconomics. **28**(11), 1015–1023 (2010)
2. Y. Liu, A.G. Wheaton, D.P. Chapman, T.J. Cunningham, H. Lu, J.B. Croft, Prevalence of Healthy Sleep Duration Among Adults–United States (2014)
3. A. Rechtschaffen, Kales, A Manual of Standardized Terminology, Techniques and Scoring Systems for Sleep Stages of Human Subjects. U.G.P.Office (Washington DC Public Health Service, 1968)
4. A.L.C.C. Iber, S. Ancoli-Israel, S.F. Quan, *The AASM Manual for the Scoring of Sleep and Associated Events: Rules, Terminology and Technical Specification* (American Academy of Sleep Medicine, Westchester, USA, 2005)
5. F. Mendonça, S.S. Mostafa, F. Morga do-Dias, J.L. Navarro Mesa, G. Juliá-Serdá, A.G. Ravelo-García, A portable wireless device based on oximetry for sleep apnea detection. Computing 1–17 (2018)
6. Y.-Y. Cheung, B.-C. Tai, G. Loo et al., Screening for obstructive sleep apnea in the assessment of coronary risk. Am. J. Cardiol. **119**(7), 996–1002 (2017)
7. E.A. Nofzinger, Neuroimaging and sleep medicine. Sleep Med. Rev. **9**(3), 157–172, https://doi.org/10.1016/j.smrv.2004.07.003
8. S. Khalighi, T. Sousa, J.M. Santos, U. Nunes, ISRUC-sleep: a comprehensive public dataset for sleep researchers. Comput. Methods Prog. Biomed. **124,**180–192
9. R.B. Berry, Fundamentals of Sleep Medicine (Elsevier Saunders, Philadelphia, 2012)
10. G. Zhu, Y. Li, P. Wen, Analysis and classification of sleep stages based on difference visibility graphs from a single channel EEG signal. IEEE J. Biomed. Health Inform. **99**, 1 (2014)
11. S-F Liang, C.-E. Kuo, Y.-H. Hu, Y.-H. Pan, Y.-H. Wang, Automatic stage scoring of single-channel sleeps EEG by using multiscale entropy and autoregressive models. IEEE Trans. Instrum. Meas. **61**(6), 1649–1657 (2012)
12. M. Hanaoka, M. Kobayashi, H. Yamazaki, Automatic sleep stage scoring based on waveform recognition method and decision-tree learning. Syst. Comput. Jpn. **33**(11), 1–13 (2002)
13. J.C. Principe, S.K. Gala, T.G. Chang, Sleep staging automaton based on the theory of evidence. IEEE Trans. Biomed. Eng. **36**(5), 503–509 (1989)
14. A. Flexer, G. Gruber, G. Dorffner, A reliable probabilistic sleep stager based on a single EEG signal. Artif. Intell. Med. **33**(3),199–207 (2005)

15. P. Piñero, P. Garcia, L. Arco, A. Álvarez, M.M. García, R. Bonal, Sleep stage classification using fuzzy sets and machine learning techniques. Neuro Comput. **58–60**, 1137–1143 (2004)
16. R.K. Sinha, Artificial neural network and wavelet based automated detection of sleep spindles, REM sleep and wake states. J. Med. Syst. **32**(4), 291–299 (2008)
17. R.K. Sinha, EEG power spectrum and neural network based sleep-hypnogram analysis for a model of heat stress. J. Clin. Monitor. Comput. **22**(4), 261–268, 299 (2008)
18. M. Sharma, D. Goyal, P.V. Achuth, U.R. Acharya, An accurate sleep stages classification system using a new class of optimally time-frequency localized three-band wavelet filter
19. Z.H. Ge, Y.F. Sun, Sleep stages classification using neural networks with multi-channel neural data, in *Lecture Notes in Computer Science* (Springer, Berlin, 2015), pp. 306–316
20. L. Derong, P. Zhongyu, S.R. Lloyd, A neural network method for detection of obstructive sleep apnea and narcolepsy based on pupil size and eeg. IEEE Trans. Neural Netw. **19**(2), 308–318 (2008)
21. T. Schluter, S. Conrad, An approach for automatic sleep stage scoring and apnea-hypopnea detection, in *2010 IEEE 10th International Conference on Data Mining (ICDM)* (2010), pp. 1007–1012
22. H.N. Yoon, S.H. Hwang, J.W. Choi, Y.J. Lee, D.U. Jeong, K.S. Park, Slow-wave sleep estimation for healthy subjects and osa patients using rr intervals. IEEE J. Biomed. Health Inf. (2017)
23. B.A. Savareh, A. Bashiri, A. Behmanesh, G.H. Meftahi, B. Hatef, Performance comparison of machine learning techniques in sleep scoring based on wavelet features and neighboring component analysis. PeerJ. **2018**(7), 1–23 (2018)
24. K.B. Mikkelsen, D.B. Villadsen, M. Otto, P. Kidmose, Automatic sleep staging using ear-EEG. Biomed. Eng. Online **16**(1), 1–15 (2017)
25. M.B. Uddin, C.M. Chow, S.W. Su, Classification methods to detect sleep apnea in adults based on respiratory and oximetry signals: a systematic review. Physiol. Meas. **39**(3) (2018)
26. E. Malaekah, D. Cvetkovic, Automatic sleep stage detection using consecutive and non-consecutive approach for elderly and young healthy subject, in *ISSNIP Biosignals Biorobotics Conference BRC* (2014)
27. R. Agarwal, J. Gotman, Computer-assisted sleep staging. IEEE Trans. Biomed. Eng. **48**(12), 1412 (2001)
28. C. Berthomier, X. Drouot, M. Herman-Stoca, J. Berthomier Prado, D. Bokar-Thire et al, Automatic analysis of single channel sleep EEG: validation in healthy individuals. Sleep **30**(11), 1587–1595 (2007)
29. D.H. Wolpert, The lack of a priori distinctions between learning algorithms. Neural Comput. **8**(7), 1341–1390 (1996)
30. M. Ronzhina, O. Janouek, J. Kolrov, M. Novkov, P. Honzk, I. Provaznk, Sleep scoring using artificial neural networks. Sleep Med. Rev. **16**(3), 251–263 (2012)
31. L. Fraiwan, K. Lweesy, N. Khasawneh, H. Wenz, H. Dickhaus, Automated sleep stage identification system based on time–frequency analysis of a single EEG channel and random forest classifier. Comput. Methods Prog. Biomed. **108**(1), 10–19 (2012)
32. K. Sirvan, T. Sousa, J.M. Santos, U. Nunes, Comput. Methods Prog. Biomed. **124,** 180–192 (2016)
33. T. Lan, Feature Extraction Feature Selection and Dimensionality Reduction Techniques for Brain Computer Interface. Ph.D. Thesis, Oregon Health and Science University (Portland, OR, USA, 2011)
34. P. Khatwani, A. Tiwari, A survey on different noise removal techniques of EEG signals. Int. J. Adv. Res. Comput. Commun. Eng. **2**, 1091–1095 (2013)
35. B. Sen, M. Peker, A. Çavuşoğlu; F.V. Çelebi, A comparative study on classification of sleep stage based on EEG signals using feature selection and classification algorithms. J. Med. Syst. **38**, 1–21 (2014). [CrossRef] [PubMed]

36. A.R. Hassan, M.I.H. Bhuiyan, Computer-aided sleep staging using complete ensemble empirical mode decomposition with adaptive noise and bootstrap aggregating. Biomed. Signal Process. Control **24**, 1–10 (2016). [CrossRef]
37. L. Fraiwana, K. Lweesy, N. Khasawnehc, H. Wenzd, H. Dickhause, Automated sleep stage identification system based on time–frequency analysis of a single EEG channel and random forest classifier. Comput. Method Prog. Biomed. **108**(2013), 10–17 (2011)

Ontology-Driven Traffic Scenario Modeling for Situational Assessment and Decision-Making at Expressway Toll Plaza

Annu Mor and Mukesh Kumar

Abstract Intelligent Transportation System (ITS) are a combination of technological sub-systems used to enhance safety as well as performance of surface transportation.A vital element for success of ITS is interoperability among connected road elements.In this study,Toll Plaza ontology is designed to minimize traffic congestion by gate switching for stop–go scenario.Decision-making rules are designed with the help of Indian traffic regulation guidelines along with exogenous variables such as weather conditions and holiday.The proposed approach can be used for further situational assessment to design the existing infrastructure more efficiently for the intelligent transportation operations in smart cities.The proposed framework is subjected to minimize the infrastructure cost,operational cost,day-to-day prediction of de-congestion time for priority vehicles and segment travel time to improve social life.

Keywords Intelligent transportation system ontology · Gate switching · Indian traffic regulations guidelines · De-congestion time

1 Introduction

In the Globe, India has the second-highest surface transportation network with 4.24 million kilometers of road segment length, about 65% of freight and 80% passenger traffic is carried by the roads [1]. The road traffic density is the highest in the world and numbers of vehicles are growing exponentially. The existing road infrastructures are not capable of handling a large number of vehicles.

Advanced Traffic Management Systems (ATMSs) are umbrella applications and technologies designed with the purpose of enhancing existing infrastructure and

A. Mor (✉) · M. Kumar
University Institute of Engineering and Technology, Panjab University, Chandigarh, India
e-mail: annu_mor@pu.ac.in

M. Kumar
e-mail: mukesh_rai9@pu.ac.in

M. Dutta et al. (eds.), *Proceedings of International Conference on IoT Inclusive Life (ICIIL 2019), NITTTR Chandigarh, India*, Lecture Notes in Networks and Systems 116,
https://doi.org/10.1007/978-981-15-3020-3_9

safety in Intelligent Transportation System. These systems allow many control strategies such as management and monitoring of various elements of road network [2]. Due to the dynamic nature of surface transportation, the use of ontologies is vital for ensuring interoperability among physical entities and different applications included in Advanced Traffic Management Systems (ATMSs) [3].

Ontology is a powerful tool to model traffic scenarios which consists of set of concepts, relationships between concepts with properties and set of rules that support to perform automatically reasoning and draw conclusions based upon inference [4]. Ontologies are designed to reduce operational cost, use-ability of knowledge and reasoning among different domains especially allow real-time capabilities [5]. Knowledge base representation as well as reasoning (KR & R) is essential for decision-making.

Traffic congestion today is not limited to big metropolitan areas, but also is observed even in medium-sized cities and highways Toll Plaza. Intelligent Transportation System (ITS) contributes less at Toll Plazas [6]. Toll plaza entity still needs to be explored for traffic flow prediction, which helps to design effective traffic guidance strategy [7]. Design of toll plaza include number of toll lanes, merging patterns and service provided in toll lane [8].

Under normal operation, dynamically changing toll plaza capacity might be unnecessary. However,during non-recurring congestion on freeway corridor due to construction, catastrophic failure (heavy rain, earthquake) or festival events, there may be a sudden rise in demand for alternate routes [9]. This sudden rise in demand, coupled with limited government resources, may necessitate dynamically determining toll plaza operations.

Ontology-based Toll Plaza congestion avoidance along with exogenous factors (weather conditions and holiday) in Indian road network context seems missing in literature. Therefore, it is necessary to design the toll plaza ontology encoded with decision-making rules for dynamically capacity estimation based on real-time demand. In this study, Ontology-based decision-rules are designed for different toll plaza scenarios using traffic regulation guidelines for Indian transportation networking.

To estimate the congestion cost by estimation of congestion class and provide a better direction for decision-making, a country like India, as growing economy, is still in the planning phases. Cities and highways are still under construction not as the case like developed countries. This characteristic helps to reduce the accident rate, reduce road construction, operational cost and enhance traffic capacity. The paper is organized as follows: Sect. 2 covers the role of ontology in ITS, Sect. 3 propose framework with situation assessment, lastly the conclusion and future work part.

2 Role of Ontology in ITS

Previous works showed that ontology plays an important role in intelligent transportation system operations in different traffic scenarios such as automated vehicle, driver assistance system, ontology for traffic management and VANET ontology. In this section, ontology-based ITS operations are described:

Freddy et al. [10] designed STAR_CITY model based on integration of heterogeneous historical, real-time and quasi stream sensor data using ontology. W3T (Time Ontology) and space (W3C Geo Ontology) are designed including traffic related social media feeds. Quasi stream data includes road works, weather conditions, events and city maps. The model predicted severity of congestion.

Pierfrancesco et al. [11] suggested novel smart city ontology named as knowledge Model 4 city (KM4city). The public administration data sets such as population, accidents, votes, energy consumption, museum restaurant and weather status are used in ontology designing. KM4city works on reconciliation algorithm to remove inconsistency and incompleteness for data integration. Ontology-based knowledge base is accessed via SPARQL queries along with SlIK link discovering algorithm. Performance metric used are Precision, Recall and F-measure.

Achilleas et al. [12] have given a frame work known as OsMoSyS for smart city planning and management. OsMoSyS is based on web interface using WOB(Web Ontology Browser) for RDF data through force-directed graphs and multi-pane interfaces. It is mainly based on JavaScript and HTML5.

Gregor et al. [13] proposed a structured ontology for Intelligent Transportation System including 14 applications. For semantic interoperability, semantic clustering algorithm is used for information retrieval. The heterogeneous interoperability SOA (Service Oriented Architecture) and SOAP (Simple Object Access Protocol) act as communication protocol between services. For dimensional reduction, hierarchical clustering and ultra-metric tree were applied.

3 Propose Framework

The proposed framework consists of 5 layers as shown in Fig. 1. The following layers are Sensor Layer, Data Processing Layer, Ontology Layer, Agent Layer and Application Layer.

3.1 Sensor Layer

In sensor layer, different sensors are used such as Loop Detector sensor, Magnetic Tape sensor and weather sensor for traffic flow estimation and weather conditions nearby to city where traffic data is being considered. Toll Plaza sensor data contains

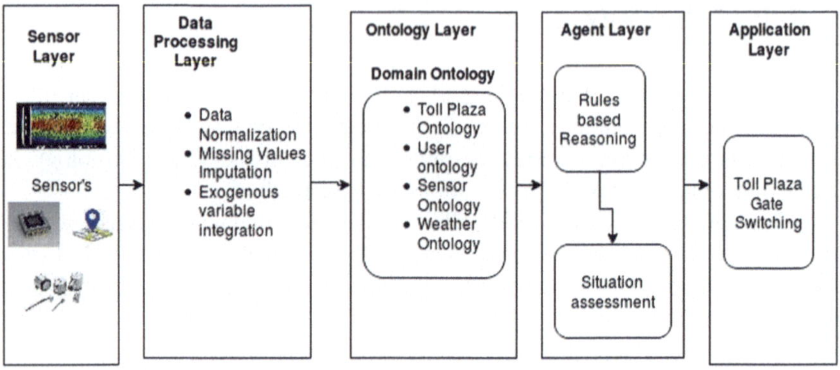

Fig. 1 Framework for Toll Plaza Gate switching

features that are vehicle types, mode of payment, vehicle entry time and vehicle's details.

3.2 Data Processing Layer

In data processing layer, Firstly data is normalized to reduce the influence of highly dimensional attribute. The missing values are imputation by MEDIUM.The traffic flow data are integrated with exogenous factors such as temperature, humidity and visibility. The weather data is collected from Internet.

3.3 Ontology Layer

The ontology layer is a graphical representation of different entities such as road network, vehicle type, toll plaza entities and weather conditions with object and data properties. Different sensors are combined into one framework with the help of ontology for information sharing from sensor-to-sensor. For avoiding the inconsistency in knowledge representation toll plaza, user agent, weather and event ontology are designed. The concepts are connected with object and data properties along with axioms for inference purpose useful in query results.

3.3.1 Taxonomy for Toll Plaza Domain Ontology (TPDO)

Intelligent Transportation System is considered as a subclass of Cyber-Physical Systems (CPS) due to connection among physical systems such as road types vehicle types and road segments (such as Bridge, Tunnel and Toll Plaza) as shown in Fig. 2.

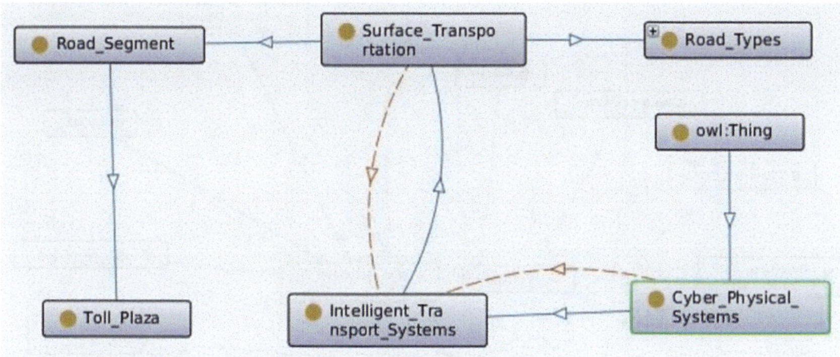

Fig. 2 Taxonomy used for Toll Plaza ontology

The Toll Plaza ontology consists of classes: road type, vehicle type, toll plaza properties such as capacity, density, service time, headway time, queue length and sensor class. The taxonomy for toll plaza ontology and vehicle types with relationship is shown in Fig. 3.

Classes are connected with object and data properties with individual instances as presented in Fig. 5. The ontology was designed in OWL-RDF language using the Protégé tool 5.0 using Pellet Reasoner [14] to check the consistency (Fig. 4).

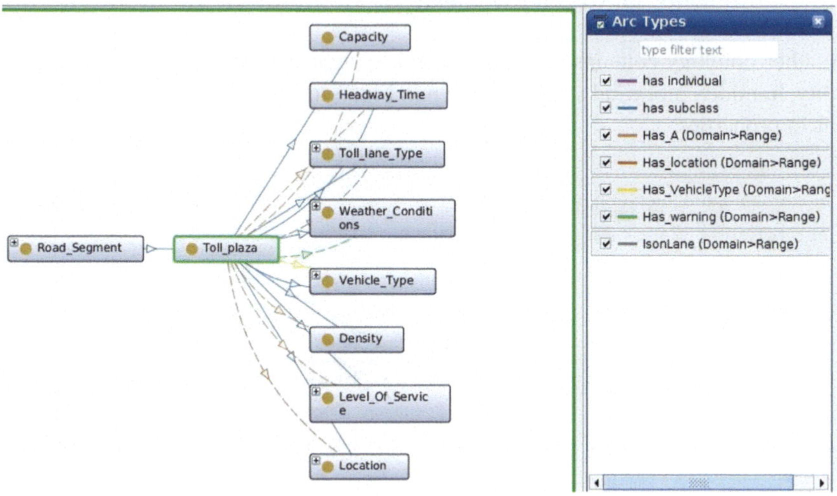

Fig. 3 Taxonomy for Toll Plaza with relationship among concepts

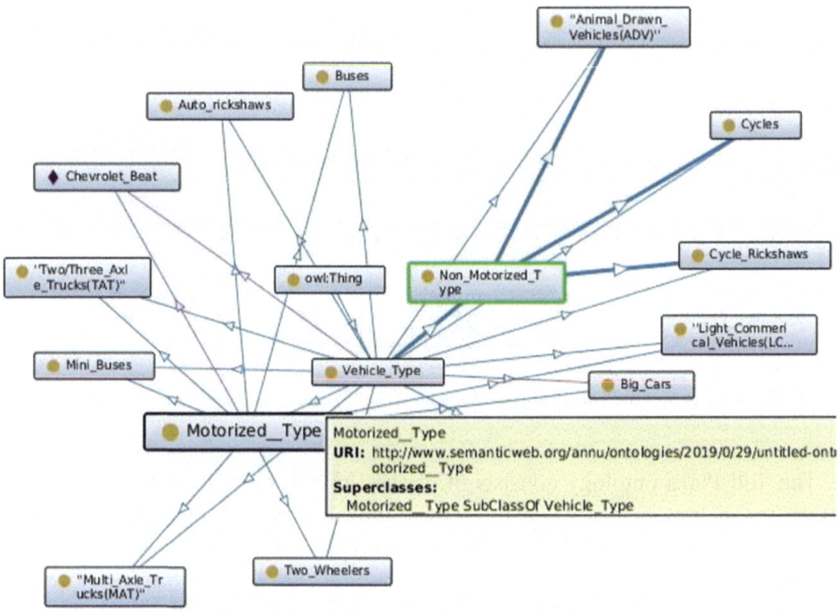

Fig. 4 Taxonomy along with relationship to characterize vehicle class

3.3.2 Taxonomy for Weather Ontology

Weather ontology contains concepts related to environmental factors such as visibility, humidity and wind speed as shown in Fig. 5 to gather weather attributes with object properties.

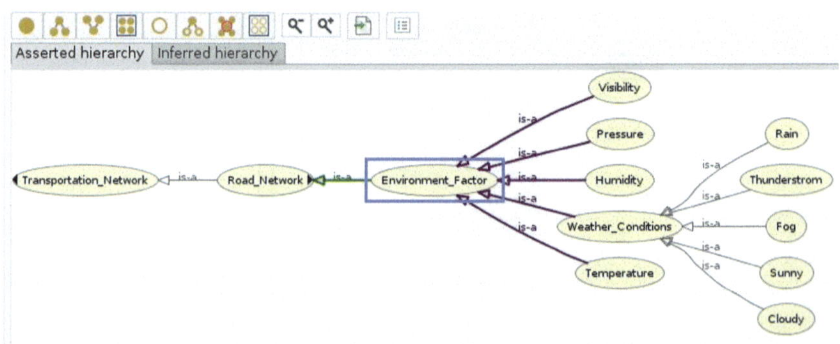

Fig. 5 Taxonomy for environment factors

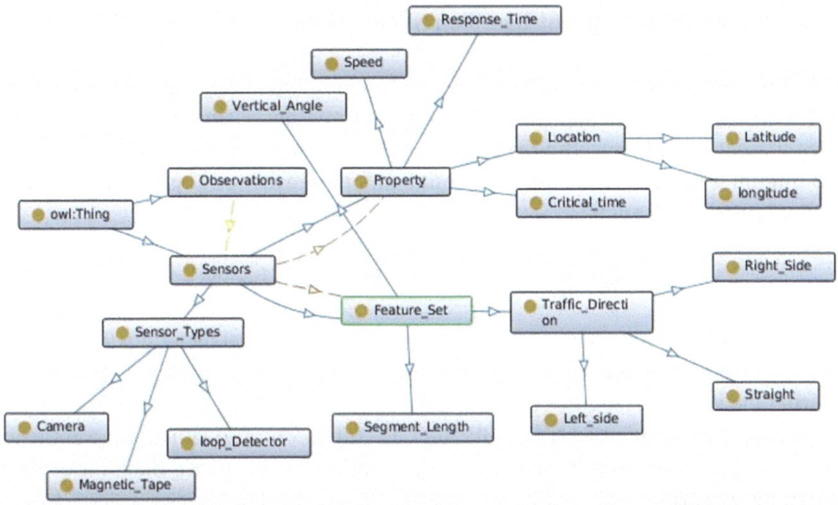

Fig. 6 Taxonomy for sensor with relationship among concepts

3.3.3 Taxonomy for Sensor Ontology

For integration and sharing heterogeneous data information from different sensors is a tedious work in Intelligent Transportation System. To access knowledge from sensors, ontology model provides common vocabulary.The sensor ontology has many attributes such as sensor device, location and observation property shown in Fig. 6.

3.3.4 Taxonomy for User Ontology

The agent layer is used by multi-agent such as priority vehicle agent (driver and passenger), road agent (Toll plaza authorities) and environment agent, to access the information from ontology-based knowledge base.

3.4 Agent Layer

In agent layer, different actions are taken using knowledge stored in ontology through queries. SPARQL [14] is used as query access language in ontology. Environment agent is used to indicate the minimum and maximum range for weather conditions such as rain, fog, humidity and visibility.

3.4.1 Situation Assessment Based on SWRL Rules

Toll Plaza ontology is encoded with Semantic Web Rule Language (SWRL) rules [15]. Different rules are applied to gate switching by increasing the number of toll lanes, based on Indian traffic regulation guidelines rules. Each toll plaza contains a minimum 4 lanes which act as constraints in ontology. It can be defined as exact cardinal restrictions such as

$$\text{Toll\_ Plaza contain minimum 4 Toll\_ Lanes.}$$

In this study, gate switching principal scenarios are focused including the weather conditions and guidelines. Toll plaza follows a rule First In First Out—FIFO considering the segment length access as well as queue length act as a reference for the congestion state estimating a "STOP and GO" situation. The situation assessment related to gate switching as shown in Fig. 7 shows the sequence diagram of gate switching scenario.

To accomplish the queue length at particular instance for particular segment, user agent queries the toll plaza ontology to fetch the data about the queue length and travel time in congested Toll lane. The gate switching is done based on traffic flow including the exogenous variables. Toll Plaza ontology contains the concepts related to each Toll Lane such as number of vehicles, travel time, service time, Location coordinated (Latitude & Longitude) and maximum traffic density in particular lane.

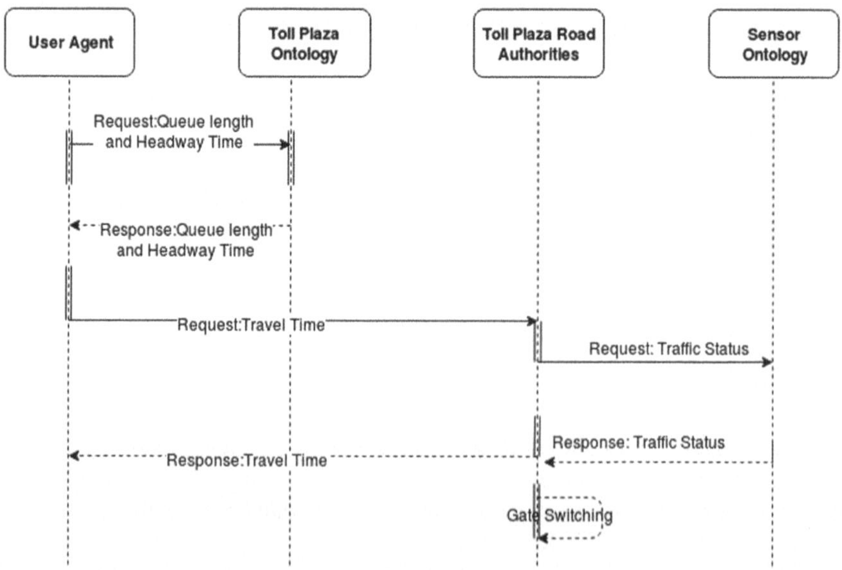

Fig. 7 Sequence diagram of Gate switching at Toll Plaza using SWRL

Toll plaza capacity and segment travel times correlate with each other [8]. The number of toll lanes can be dynamically adjusted by analyzing historical data patterns. Toll plaza is designed with three general gates with one ETC (Electronic Toll Collection) gate as basically used in Indian Road network. In case, traveler agent (say ambulance) makes queries to sensor ontology regarding congested state during peak hours to know up to date traffic status and travel time (decongestion time). The travel time [16] is calculated as given below:

$$Td + [L - (Ls - Lt)]/Vc + Ls/Vs \tag{1}$$

where,

Td Travel Time (s),
L Distance from entity generation to booth gate,
Ls Distance from merging point to booth gate,
Lt Traffic Length,
Vc Average speed and
Vs Average speed in waiting queue

The travel time is computed, the value is compared with threshold value in particular toll lane segment. If predicted travel time is greater than or equal to threshold value then, it is assumed that toll lane is congested and gate switching rule get fired as described below: These rules are applied when traffic status varies from one instance of Expressway to another instance of Expressway (when to convert 4 lanes to 6 lanes, 6 lanes to 8 lanes and so on), by estimation of Level of Service (LOS). As mentioned in Indian traffic guidelines, LOS_B act as threshold value for gate switching. Rule for gate switching get triggers as:

Toll_Lane (?TL,6) ^ has_Traffic_volume (?TL,?TV)^ swrlb:lessThan (?TV,450) ^has_traffic_status (?TS,Slow)^has_LevelOfService (?TL,"LOS_B")^swrlb:lessThan (?TV,1 5600)^has_Speed_Limit (?TL,?S)^swrlb:greaterThan (?S,10)^has_Head way_Time_in_sec (?TL,?HT)^swrlb:lessThan (?HT,8)- > UseGateSwitching (?TL,8).

Additional environmental factors also affected the traffic flow such as weather conditions (rain, fog, cloudy), temperature, humidity, visibility and wind speed which are used as data properties in weather ontology, connected to toll plaza ontology using the property *Has weather condition* SWRL can be applied to know the impact of external factors on traffic volume following rule can be triggered:

Toll_Lane (?TL,4)^has_Traffic_Status (?TS,Saturated")^has_weekday (?wd,"Monday")^h s_Time_duration (?Td,9a.m))^has_Speed_Limit(?S,20)^has_Headway_Time_in_sec (?HT,10) has weather_condition (?TL,"Rain")- > SlowTraffic_Status (?TL,"Low").

3.5 *Application Layer*

The application layer is designed for smart toll plaza gate switching for dynamic capacity estimation by increasing the number of toll lanes in peak-hour or non-peak hour with the impact of exogenous factors.

4 Conclusion and Future Work

In this study, an ontology-based framework designed for dynamic capacity estimation at toll plaza through gate switching. The framework performs different actions in automatic way based on decision-rules.

The proposed framework is efficient to enhance the static information using the traffic regulations, by implementing the generic decision-making rules including exogenous factors such as weather conditions and holiday. Ontology-based situational assessment using decision-making rules is based on real circumstances. Toll Plaza authorities can use the same knowledge base for designing or planning purpose. The proposed framework gives higher priority to toll lanes that are congested as compared to others.

Further, the framework is extended to add the multi-media properties to use decision-making rules for traffic regulations based scenarios. The traffic environment is dynamic which can't be always in true and false state, how to handle the uncertainties, major missing part in OWL language needs to be addressed. From the technical view, enhance the framework to support action engineering for testing in real-time application.

References

1. M.S. Bains, S.S. Arkatkar, K.S. Anbumani, S. Subramaniam, Optimizing and modeling tollway operations using microsimulation: case study sanand Toll Plaza, Ahmedabad, Gujarat. India. Transp. Res. Rec. **2615**(1), 43–54 (2017)
2. A. Bezuglov, G. Comert, Short-term freeway traffic parameter prediction: application of grey system theory models. Expert Syst. Appl. **62**, 284–292 (2016)
3. F. Lécué, R. Tucker, V. Bicer, P. Tommasi, S. Tallevi-Diotallevi, M. Sbodio, Predicting severity of road traffic congestion using semantic web technologies, in *European semantic web conference* (2014), pp. 611–627
4. M. Katsumi, M. Fox, Ontologies for transportation research: a survey. Transp. Res. Part C: Emerg. Technol. **89**, 53–82 (2018)
5. N. Anand, M. Yang, J.H.R. Van Duin, L. Tavasszy, GenCLOn: an ontology for city logistics. Expert Syst. Appl. **39**(15), 11944–11960 (2012)
6. P. Chakroborty, R. Gill, P. Chakraborty, Analysing queueing at toll plazas using a coupled, multiple-queue, queueing system model: application to toll plaza design. Transp. Plan. Technol. **39**(7), 675–692 (2016)

7. S.K.S. Fan, C.J. Su, H.T. Nien, P.F. Tsai, C.Y. Cheng, Using machine learning and big data approaches to predict travel time based on historical and real-time data from Taiwan electronic toll collection. Soft. Comput. **22**(17), 5707–5718 (2018)
8. Y. Zhang, W. Zhang, D. Zhang, F. Liu, S. Liu, A redesigned back Toll Plaza with new merge pattern, in *2018 17th International Symposium on Distributed Computing and Applications for Business Engineering and Science (DCABES)* (IEEE, 2018), pp. 272–275
9. C. Kim, D.K. Kim, S.Y. Kho, S. Kang, K. Chung, Dynamically determining the toll plaza capacity by monitoring approaching traffic conditions in real-time. Appl. Sci. **6**(3), 87 (2016)
10. F. Lécué, S. Tallevi-Diotallevi, J. Hayes, R. Tucker, V. Bicer, M. Sbodio, P. Tommasi, Smart traffic analytics in the semantic web with STAR-CITY: Scenarios, system and lessons learned in Dublin City. Web Semant.: Sci., Serv. Agents World Wide Web **27**, 26–33 (2014)
11. P. Bellini, M. Benigni, R. Billero, P. Nesi, N. Rauch, Km4City ontology building vs data harvesting and cleaning for smart-city services. J. Vis. Lang. Comput. **25**(6), 827–839 (2014)
12. A. Psyllidis, OSMoSys: a web interface for graph-based rdf data visualization and ontology browsing, in *International Conference on Web Engineering* (2015), pp. 679–682
13. D. Gregor, S. Toral, T. Ariza, F. Barrero, R. Gregor, J. Rodas, M. Arzamendia, A methodology for structured ontology construction applied to intelligent transportation systems. Comput. Stand. Interfaces **47,** 108–119 (2016)
14. Pellet, http://clarkparsia.com/pellet/. Accessed 20 Aug 2019
15. SPARQL, http://sparql.org/. Accessed 20 Aug 2019
16. I. Horrocks, P.F. Patel-Schneider, H. Boley, S. Tabet, B. Grosof, M. Dean, SWRL: a semantic web rule language combining OWL and Rule ML, http://www.w3.org/Submission/SWRL/ ,last. Accessed 18 Aug 2019
17. Y. Wang, J. Ning, S. Li, Research on merging pattern after toll based on simulation, in *2017 3rd International Conference on Economics, Social Science, Arts, Education and Management Engineering (ESSAEME 2017)* (Atlantis Press, 2017)

IoT Sensing, Monitoring, Networking and Routing

Congestion Aware and Stability Based Routing with Cross-Layer Optimization for 6LowPAN

Anita Sethi, Sandip Vijay and Vishal Kumar

Abstract **Zero Touch** network requires no human involvement other than high level implementation and declarative independent intents. Machine and human interaction learning help machines to achieve the human objective efficiently and enhances the security domain. At the different places of the network, computation and intelligence decisions with cloud environment will enhance the performance of the network. Scalability and heterogeneity in IoT framework are important parameters for performance of 6LoWPAN network. Protocol stack of NGN is a key part of protocol for supporting limited processing capabilities, low memory and limited power constrained power supply devices to Internet. Heavy network traffic causes congestion in the network which degrades the network performance and influences the QoS aspects goodput, throughput, E2E delay, jitter, Energy consumption, reliability and latency. Numerous congestion control heuristics are discussed based on traffic modeling, queue management and hybrid algorithms. This paper represents the queue management in the heterogeneous ad hoc network. Impact of minimum and maximum value of queue size on goodput and with respect of packet size is discussed in this paper. Throughput, delay and Jitter for different objective functions are compared and burst rate in case of 6LoWPAN is represented. For IoT application requirements, a novel traffic modeling based on congestion for future work is summarized.

Keywords Traffic modeling · 6LoWPAN · Jitter · E2E delay · Goodput · Throughput

1 Introduction

In smart city projects, environment of the place, traffic monitoring and management and end users with their appliances plays the most important role. Sensors on the road can easily detect the traffic jam, damaged roadways, polluted environment and

A. Sethi (✉) · V. Kumar
ICFAI University, Dehradun, India
e-mail: seep4g@gmail.com

S. Vijay
Shivalik Engineering College Dehradun, Dehradun, India

M. Dutta et al. (eds.), *Proceedings of International Conference on IoT Inclusive Life (ICIIL 2019), NITTTR Chandigarh, India*, Lecture Notes in Networks and Systems 116,
https://doi.org/10.1007/978-981-15-3020-3_10

rerouting based on traffic condition with the help of GPS enabled equipment are important tasks performed by vehicle traffic monitoring system. Smart light systems at the streetlight with sensors to detect the human or vehicles displacement can automatically ON/OFF when activity is detected in the area help to save energy as well as money and enhance the security. Early detection of abnormal pollution at some places in the city, or fire and water level detection can alert the people to take some safe and secure action. For healthy environment, public places like trash bins or restrooms equipped with sensors for cleanliness detection and informing the authority to take the right action can help in clean city projects [1]. Roadways maintenance and bridge safety can be improved by building sensors into the infrastructure and alerts and automatic maintenance can be initiated. Aggregation of data, processing, data mining and analysis of real time conditions with the help of integration of services can help in recovery process of stolen devices like bikes.

Distributed sensors throughout the home form home network, data is collected by gateway and cloud hosted network access the data, analyzes it and transmit it to the end user having portable device forms the Smart Home architecture. In different locations of home, cameras can be placed to monitor the home activity, in case of detecting abnormal behavior like carbon and smoke detection alert can be transmitted to mobile. Home appliances can be connected, and quantity of products can be detected to take decisions, like fridge, oven and electric appliances. Energy management can handle like light in a room, temperature of the room and daytime can be adjusted while away from the home [2]. Injured and sick people using IoT sensors can be analyzed to observe numerous aspects of body by doctor without getting hospitalized and instructed (Fig. 1).

2 State of Art

Based on the traffic control technique, traffic rate is handled originated from the source to reduce the congestion by minimization of number of packets injected into network. Three mechanisms are used in Congestion detection and avoidance CODA proposed by Wang et al. [3]: past and present both channel loading conditions are used for congestion detection and status of buffer occupancy at each receiver. On congestion detection, backpressure message is broadcasted to every node in the route. Each node detects the network traffic on receiving the backpressure message and detects the congestion condition and propagates the message in network. Based on the theoretical value of throughput, source node regulates the rate. In case if source event rate is lesser than fraction of channel throughput, it is regulated by source itself otherwise closed-loop control is triggered and regulated by sink node. Real time experimentation with TinyOS shows that average energy with lesser penalty in CODA is reduced by using open-loop congestion control and without congestion control method.

Base on hop-by-hop communication and cross-layer optimization heuristic, congestion control technique based on priority levels (PCCP) is proposed by Tang et al.

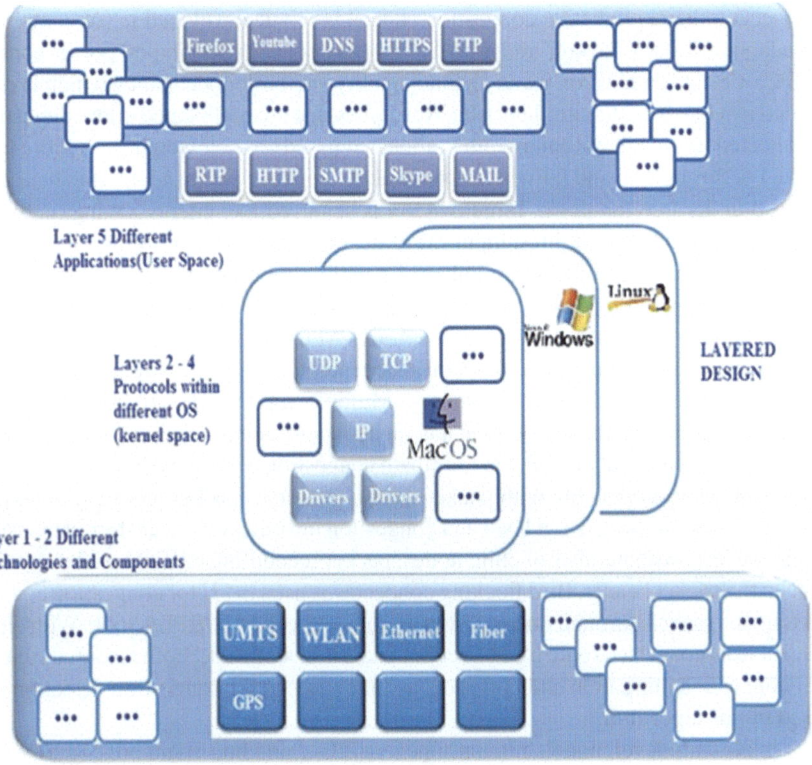

Fig. 1 Layered architecture for IoT framework

[4]. Priority based rate adjustment is activated on implicit congestion notification generated by intelligent congestion detection. At the MAC layer, packet service time and packet inter-arrival time is used to detect congestion and informed by piggy-backed in the header to rest of the nodes. Priority index and congestion degree are used to adjust the rate in both single and multipath routing scenarios. High link utilization, energy consumption improvement, packet delay and loss minimization are achieved by using PCCP as compared to congestion control and fairness (CCF).

Reliability and energy saving using five levels of congestion region is suggested in event to sink reliable transport protocol for WSN (ESRT) [5]. Optimal operation region, congestion with low reliability, congestion with high reliability, no congestion with high reliability and no congestion with low reliability are five regions of congestion with objective to stay in the optimal operation region. Buffer occupancy reflects the congestion level, for example, buffer overflow represents the congestion and sink node activates the bit for congestion in header of the succeeding packets. On detection of reliability level higher than threshold, sink node requires to save energy by decreasing the reporting rate otherwise to achieve the desired reliability level, sink adjusts reporting rate.

Decentralized predictive congestion control heuristic developed in [6] comprises of adaptive CSMA backoff and adaptive flow works with distributed power control (DPC), detects congestion using channel quality and buffer occupancy which is estimated by channel estimator heuristic. Resources that are reallocated after congestion is detected like available bandwidth is allocated based on their weights to ensure fairness. For throughput and fairness improvement, dynamic weight adaptation heuristic is used under tree topology. Targeted QoS is achieved like energy saving, network efficiency and increased throughput. Modification according to original priority of data packets, waiting time and number of neighboring nodes is done in cross-layer active predictive congestion control heuristic (CL-APCC). Based on the queuing theory and congestion avoidance concept, CL-APCC predicts the input and output rates of node for next period and operates depending on data flow trends, node rates, network condition and buffer occupancy. Multiple applications are hosted by each node with priorities and every node has many parents simultaneously and forwards data to single parent at one time in multipath congestion control mechanism. Ratio of average packet service rate with packet scheduling rate is packet service ratio that is used to detect the congestion level and congestion information is transferred by using piggybacking with number of child nodes, packet scheduling rate and packet service rate in the packet header. Bird flocking concept was used in [7] for congestion control to transfer packets through uncongested region in 6LoWPAN/RPL/CoAP. With election of least congested route for delivery of packets to sink node, both resource control for congestion mitigation and buffer occupancy strategy for congestion detection are used in this algorithm.

Link quality level, node's remaining energy, E_{TX} and hop count are combined to quantify and form hybrid routing metrics, e.g., lexical and additive manner based on system and user requirement [8]. To ensure application's requirement, lexical combination heuristic is preferred, while additive heuristic supports a flexible manner to form hybrid routing metrics using a metric weight pair. At MAC layer, link performance is measured using R-metric which is combined with packet loss due to MAC contention and E_{TX} with Q-metric which provide load distribution function as a performance parameter. In PER_{-HOP}, E_{TX} metrics distribute E_{TX} value to each node along a path from source to destination instead of additive E_{TX} metric. This performs well in large scale networks as compared to OF_0, E_{TX}. Performance metrics for 6LoWPAN are divided as node metric contains node state and parameters like hop count and energy level and link metric latency, throughput, reliability and link color for the security purpose [5]. Multipath routing optimization technique called M-RPL which decreases the packet loss and resolves the network congestion by using dynamic adaptive routing is proposed [9]. Node density for M-RPL is 20 nodes which combines ETX and transmitted packets at a node metrics by selecting the shortest path dynamically to reduce E2E delay, congestion and packet lost to enhance the network performance.

3 Effective Channel Capacity

Actual channel capacity of IEEE standard decides the buffer loss probability model, for example, maximum of 250 kbps data rate is supported by the 2.4 GHz IEEE 802.15.4 standard. No of active nodes, collision and overhead of ACK packet transmission and channel access heuristics decide the effective data rate which is smaller than 256 kbps. Nearest path to the destination is elected based on hop count limit by OF_0. Packets generated by application layer are transmitted to MAC layer through network layer and sicslowpan layer and queued in buffer and transmitted to $null_{rdc}$ layer using NET_{STK_RDO} start procedure for packet transmission on radio. If CHANNEL is busy in receiving a packet or received a packet and ACK is waiting for transmission, $TX_{collision}$ signal is transmitted else nullrdc calls $N_{ETSTKradiotransmit}$ procedure. $MAC_{ACKWAIT}$ is the duration till $null_{rdc}$ waits for receiving ACK packet which is followed by $ACK_{DETECTAFTERWAIT}$ and transmits TX_{OK} in case of success to CSMA layer otherwise TX_{NOACK} is transmitted by nullrdc layer. In case of ACK packet, CSMA layer dequeues the packets and prepares the next packet; otherwise in case of collision detection and failure of ACK packet ,it waits for random backoff time[time, time + 2BEXtime], where time is equal to 1/channelcheckrate. After completion of backoff, interval packet is retransmitted and MAC_{MAX} FrameRetries for retransmission attempt packet is dropped by CSMA layer. Turnaround time is required by MAC layer to wait and then send ACK packet. In case of successful transmission of a packet, effective data rate is

$EDR_{max} = N/T_{NOCOL}$, where data packet length is N and real time required to transmit 127 bytes of data length.

$$T_{NOCOL} = T_{data} + \text{turnaroundtime} + T_{ack} + T_{WAIT}$$

Where data packet time and acknowledgment time waiting time is represented by T_{data}, T_{ack} and T_{WAIT}, respectively. In real time communication, collision occurs due to which nodes retransmit the collided packets and data rate decreases with an increase in the collision probability. Extra time is required for retransmission of collided data packets, ADR, actual data rate with probability of collision is determined as

$$ADR = N\big/ (1 - P_{collision})\, T_{nocoll} + P_{collToll}$$

Calculation of T_{coll} is $T_{data} + MAC_{ACKwaitduration} + T_{backoff} + T_{nocoll}$

With 5% collision probability.

Active nodes present in the transmission range of node entered in the $back_{off}$ period can utilize this duration for their data packets transmission which is the proper channel utilization. Actual channel capacity depends on network circumstances like collision probability, no of active nodes, idle wireless channel time utilization rate. Effective channel capacity estimation in beaconless mode is measured with collision less and random $back_{off}$ time.

Based on the distance using hop count objective function OF_0 elects the nearest route to the destination node. With minimum rank value, OF_0 elects its preferred node and increases the hop count value by a strictly positive normalized scalar value to obtain mean value of hop count. OF_0 does not reflect the link condition and heavy packet drop loss occurrence in the wireless connection or at nodes present in the route [10]. Expected number of successful transmissions ETX_{OF} of packet on a route is another parameter to observe the performance of network. With minimum number of transmissions required to transmit the packet to destination, ETX_{OF} adds two values, link value of neighbor node and route value to the HELLO message in the heuristic. Every node in the route elects a minimum value of ETX_{OF} with least packet channel loss, which represents the congestion level in the network but not the location of congestion occurrence [8]. AVG_{DEL} is used with the objective to minimize the delay from source to destination which is estimated as cumulative sum of connection-by-connection delay along the route to the destination. This parameter is compared with ETX_{OF} in terms of delay over a 50node density network [11]. Remaining energy-based performance parameter uses minimum value between path cost and node's energy [12].

Optimized Link state heuristic is performed with multipoint relay selection [13], link state advertisements and neighbor discovery process. 1-hop neighbor is elected by every router which is closer to it with bidirectional communication using HELLO msg and status of the link is transmitted sporadically. For efficient communication without delay, a set of linked nodes is recognized which helps in easy maintenance of route connectivity. Link information with optimized path election is transmitted in topology control; TC message is specified in LSA. Periodically, information of active nodes in the network convergence and accurate topology map is updated and broadcasted to neighbor node (Table 1).

Table 1 Different attributes of optimized link state heuristic

Link constant	Attributes
Link metric constants	Minimum = 1 and Maximum = 16776960
Willingness constants	Willingness_Always = 15, Willingness_Default = 7 and Willingness_Never = 0
Time constant	Time granularity C = 1/1024 s, represents time interval
LINK_METRIC (2 Octets)	Kind and direction bits 0–3, exponent (b) 4–7; Mantissa (a) 8–15;
Link Metric TLV types and directions	*0–Incoming link metric;* *1–Outgoing link metric;* *2–Incoming neighbor metric;* *3–Outgoing neighbor metric;*
MPR (1 Octet)	*Broadcasting, routing and broadcast + Routing*
NBR_ADDR_TYPE (1 Octet)	*Source, routable and routable + Source*

Link metric value contains the link direction which is kept in router records used for route maintenance. In a message, 12 bits are used for link metrics value which is in compressed form divided into two fields of 8-bits and 4-bits. Route metrics are formed by these link metric values up to a limit of 256 with exact format of compressed representation. A modified form of link metric with 12 bits is divided into two fields of 8-bit mantissa and 4-bit exponent. Compressed form of link metric is obtained as

1. Select the minimum value of b such that v + 256 <= 2(b + 9)
2. A = (v − 256(2b − 1))/(2b)−1 with rounding nearest integer.

4 Network Performance

IEEE 802.15.4 LR-WPANs designed for data rates of 250 kbs or lower are low power, low throughput communication networks which are used for creating wireless sensor networks WSNs and IoT applications with minimum complexity. Ultra-wideband (UWB), Chirp Spread spectrum (CSS) and Direct sequence spread spectrum different modulation techniques at PHY layer and ALOHA and CSMA-CA at MAC layer can be defined in LR-WPANs. WirelessHART, SNAP, MiWi, ISA100.11a, Zigbee are basis of IEEE 802.15.4; each of which further extends the standard by developing the upper layers with 6LoWPAN. Existence of human in a room is obtained by wireless sensor placed in different rooms of an apartment in a building and information is transmitted to controller. Controller nodes are used to decide the adjustment of the lighting conditions in different rooms, while collecting the information received from sensor nodes periodically. They transmit control packets to different devices in the rooms to set their properties or turn them on and off accordingly. All nodes use the IEEE 802.15.4 narrowband model to communicate. With controller at the center different paths are set up according to star topology. At random initialization time, 10-byte udp packet is transmitted to controller by all sensors each second. Same specifications are also used by controller to reply the packets to devices. Default values of radio parameters and background noise power of −110 dBm are used. Each sensor is sending one packet per second to the controller that transmits 8 packets per second to randomly selected lamp. ACKs are transmitted in response.

5 Result Analysis

Energy consumption of controller is greatest as compared to sensors and devices which consumes approximately same amount of energy. Controller consumes 25% more energy as compared to sensors and devices while transmitting eight times more packets due to energy consumption was dominated by the idle radio state. Difference between energy consumption between sensors and devices is very small due to data

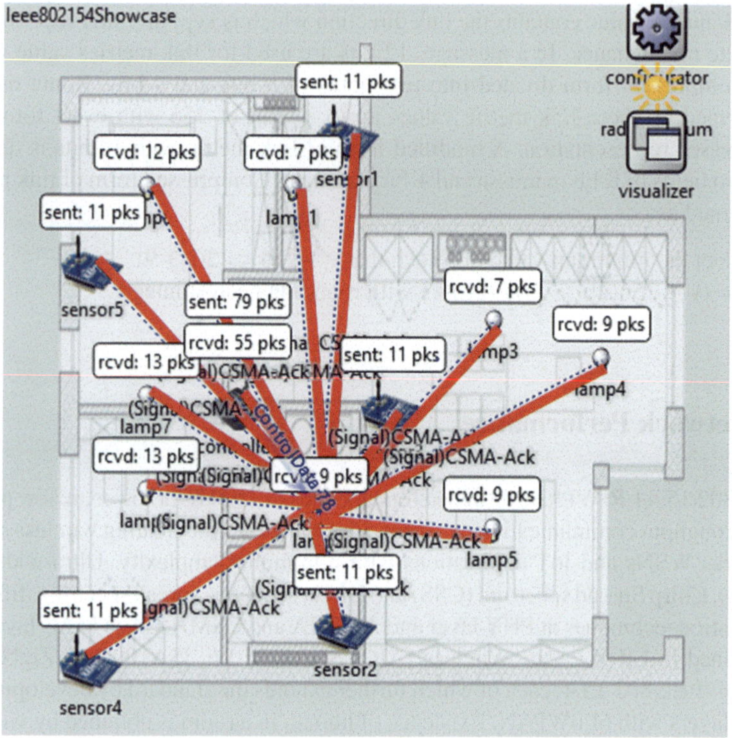

Fig. 2 Transmission of packets in smart home scenario

transmission by sensors which requires 1.7 ms, while devices transmit only ACKs signals in 0.7 ms (Fig. 2).

The controller transmitted in about 8% of the time, the sensors about 1%. Also, all transmissions were received by all nodes in the network at the PHY level, thus they shared the energy consumption due to reception (Fig. 3).

Poor connectivity, misconfiguration and congestion degrades the network performance by excessive packet drops. Visualization of packet drops can provide information of location of packet drops and easily solution can be searched which helps in debugging and analysis. Wireless network's poor connectivity causes source node to drop unacknowledged packets after the retry limits exceed and queue overflow due to congestion in network increases the packet drop in router bottleneck. Causes of packet drop in the network can be one or many from interface down, network address resolution failed, unroutable packets, retry limit exceed and queue overflow. In the HetNets, between source and destination, different devices of different configuration are connected, the bottleneck arises when speed network devices are used in the network in case of video streaming as an example source and destination are connected through high speed router and low speed switch with Ethernet cable. DropTailQueue

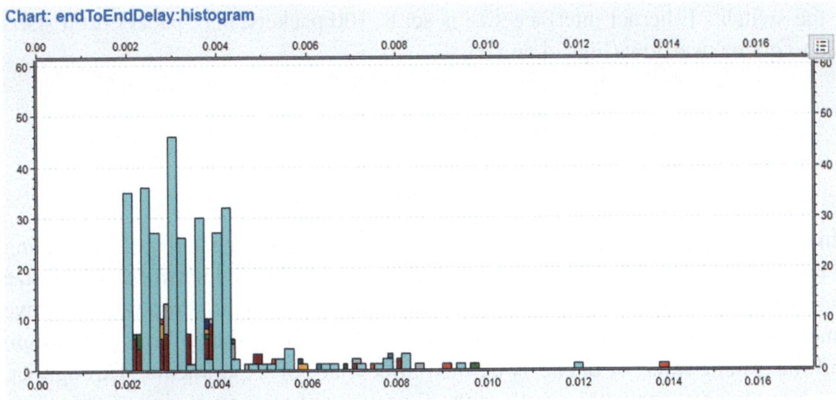

Fig. 3 End-to-End delay

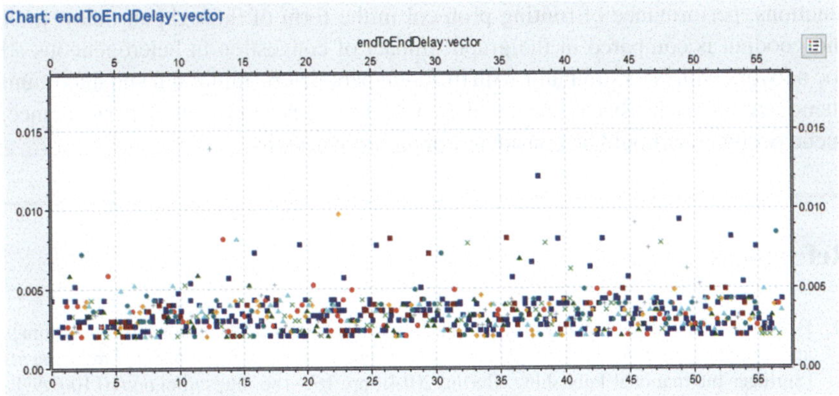

Fig. 4 End-to-End delay

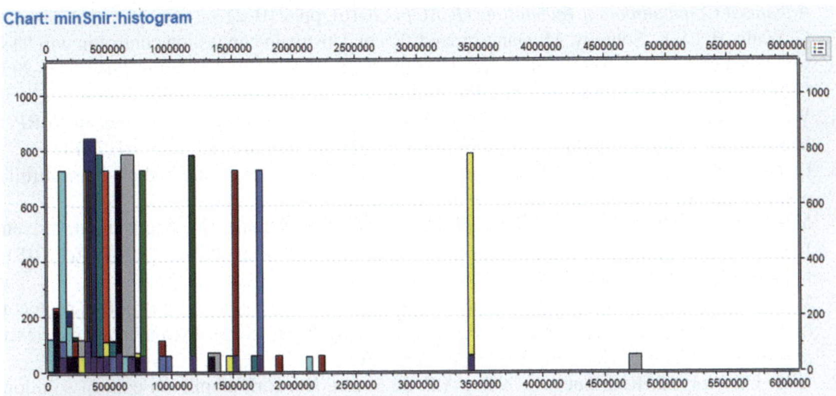

Fig. 5 SNIR

at the switch's Ethernet interface size is set to 100 packets, here packet drop starts due to queue overflow (Figs. 4 and 5).

6 Conclusion

This paper focused on buffer occupancy for resource scheduling and traffic modeling in routing protocol. When congestion occurs in the IoT based network, queue size should be adjusted according to the traffic for smooth operation. Proposed objective function is simulated using NS3 simulator with different queue sizes. Minimum and maximum value of queue is defined after a no. of experiments. Throughput, goodput are represented by graph with minimum and maximum size of the queue and different packet size. Packets received per unit time in case of 6LoWPAN are observed using wireshark and burst rate is also observed. With different objective functions, performance of routing protocol in the form of throughput, delay, jitter and goodput is compared in the graph. Impact of congestion in heterogeneous ad hoc network with NS3 simulator with different parameters: minimum and maximum queue size and packet size is presented. In order to enhance the network performance, queue occupancy should be considered in protocol designs.

References

1. F. Bonomi, R. Milito, P. Natarajan, J. Zhu, Fog computing: a platform for internet of things and analytics, in *Big Data and Internet of Things: A Roadmap for Smart Environments* (Springer International Publishing, Berlin, 2014), pp. 169–186, https://doi.org/10.1007/978-3-319-05029-4_7
2. W. Lee, K. Nam, H. Roh, S. Kim, A gateway-based fog computing architecture for wireless sensors and actuator networks, in *Proceedings of the 2016 18th International Conference on Advanced Communication Technology (ICACT)* (2016), pp. 210–213
3. C. Wang, B. Li, K. Sohraby, M. Daneshmand, Y. Hu, Upstream congestion control in wireless sensor networks through cross-layer optimization. IEEE J. Sel. Areas Commun. **25**(4), 786–795 (2007)
4. W. Tang, Z. Wei, Z. Zhang, B. Zhang, Analysis and optimization strategy of multipath RPL based on the COOJA simulator. Int. J. Comput. Sci. Issues (IJCSI) **11**(5), 27–30 (2014)
5. O. Gnawali, P. Levis, The ETX objective function for RPL. Internet draft: draft-gnawaliroll-etxof-00 (2010)
6. Y. Simmhan, P. Ravindra, S. Chaturvedi, M. Hegde, R. Ballamajalu, Towards a data-driven IoT software architecture for smart city utilities. Softw. Pract. Exp. **48**(7), 1390–1416 (2018). https://doi.org/10.1002/spe.2580
7. W. Xiao, J. Liu, N. Jiang, H. Shi, An optimization of the object function for routing protocol of low-power and lossy networks, in *Proceedings of 2nd International Conference on Systems and Informatics (ICSAI)* (IEEE, 2014), pp. 515–519
8. L.M. Oliveira, J.J. Rodrigues, C. Neto, A.F. de Sousa, Network admission control solution for 6LoWPAN networks, in *2013 Seventh International Conference on Innovative Mobile and Internet Services in Ubiquitous Computing (IMIS)* (IEEE, 2013), pp. 472–477
9. O. Gnawali, P. Levis, The ETX objective function for RPL. Internet draft: draft-gnawaliroll-etxof-00 [10] Gonizzi P, Monica R, Ferrari G (2013) Design and evaluation of a delay efficient

RPL routing metric, in *Proceedings of 9th International Wireless Communications and Mobile Computing Conference (IWCMC)* (IEEE, 2010), pp. 1573–1577
10. M. Sherburne, R. Marchany, J. Tront, Implementing moving target ipv6 defense to secure 6lowpan in the internet of things and smart grid, in *Proceedings of the 9th Annual Cyber and Information Security Research Conference* (ACM, 2014), pp. 37– 40 https://doi.org/10.1145/2602087.2602107
11. P. Gowthaman, R. Chakravarthi, Survey on various congestion detection and control protocols in wireless sensor networks. Int. J. Adv. Comput. Eng. Commun. Technol. (IJACECT) **2**(4), 15–19 (2013)
12. P.O. Kamgueu, E. Nataf, T.D. Ndié, O. Festor, Energy-based routing metric for RPL (2013). [Research report] RR-8208, INRIA, 14
13. J.-P. Vasseur, M. Kim, K. Pister, N. Dejean, D. Barthel, Routing metrics used for path calculation in low-power and lossy networks. RFC 6551 (2012)

A Survey on Handover in LTE Heterogeneous Networks

Manoj and Sanjeev Kumar

Abstract The Long-Term Evolution technology called 4G is a recent expertise that was presented by 3GPP for smooth system migration. The main issues in LTE networks are controlling overhead and overlapping in network controlling circuit. Signal Traffic is increased by a large number of connections. This will also increase the consumption of energy in the network. In cellular mobility management, handover is the major component of LTE. Handover is basically a process or mechanism that helps to move one call from one base station to another. Handover delay and complexity are being reduced by the use of Hard Handover. This work presents a review of Handover techniques under LTE to optimize its performance. This work summarizes various challenges that came under the handover system and the solutions to handle it. It presented various algorithms related to handover that improved the system performance by handling handover failure. The design of a successful system under handover is being done by the use of suitable handoff parameters and its optimization setting. It presents various handover techniques to improve the throughput of the network so that system performance can be enhanced.

Keywords LTE · Handover in LTE · Handoff management · Handover procedure, etc.

1 Introduction

In recent years, the rate of growth in telecommunication media has increased rapidly. In order to fulfill the requirement of co-operative and domestic needs, the network has been spread worldwide. The growth in mobile communication can be seen in the explosive number of users and changing trends to communicate which is likely to continue in the future. To meet with the requirements of high speed and expanding lifestyles, telecommunication came up with a solution named as LTE. The major development in the mobile communication system started in 1990. There was an

Manoj (✉) · S. Kumar
Department of CSE, Guru Jambheshwar University of Science & Technology, Hisar, India
e-mail: manojkoslia91@gmail.com

M. Dutta et al. (eds.), *Proceedings of International Conference on IoT Inclusive Life (ICIIL 2019), NITTTR Chandigarh, India*, Lecture Notes in Networks and Systems 116,
https://doi.org/10.1007/978-981-15-3020-3_11

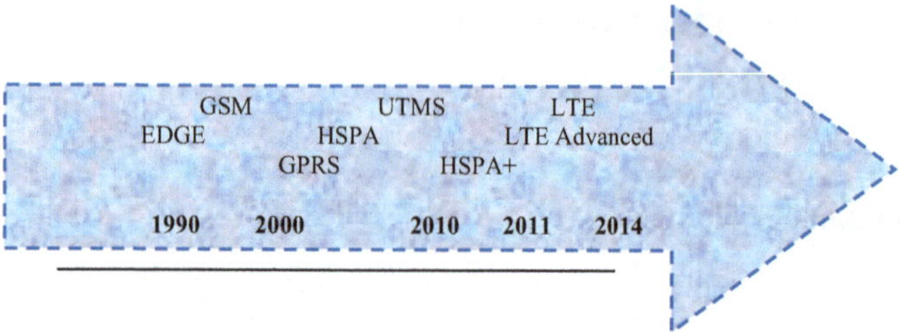

Fig. 1 Evolution of 3GPP family technology

enormous change in communication technology from 1990 to present which can be seen in the following flow chart diagram as shown in Fig. 1. The system starts with circuit technology in the 1980s known as 1G technology. It was used for voice communication. After development in digital systems, 2G was launched in the 1990s. Then technology starts growing rapidly in terms of speed and advancement. LTE starts from 2011 and now used as an advanced version of LTE with large bandwidth. With the evolution of 3GPP technology, subscribers are able to get an enhanced and effective rate of speed and its availability [1].

With the extension of cell phones, the request for high information rate and QoS increments quickly. In this way, 3GPP has indicated new models for versatile correspondence on GSM (Global System for portable correspondence)/EDGE and Universal Mobile Telecommunications System (UMTS).

In the correspondence framework, the structure with high information rate participates an imperative part in day by day era. For providing a large number of users with high information speed at a lower cost, we need to implement techniques like LTE. Figure 2 shows the network architecture of the LTE system in which it consists of one Mobile Management Entity (MME) and two gateways called serving and path

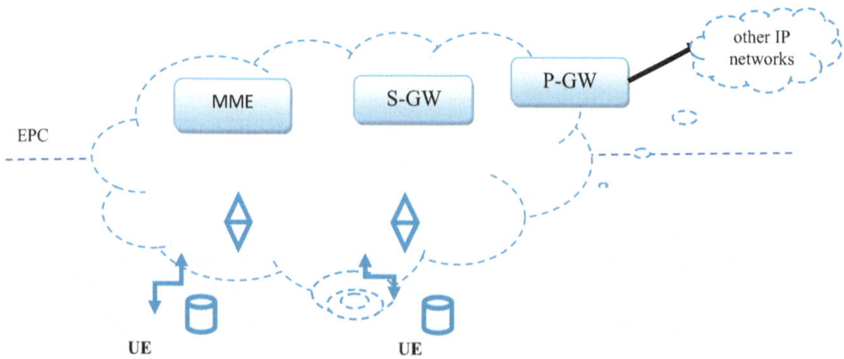

Fig. 2 LTE network architecture

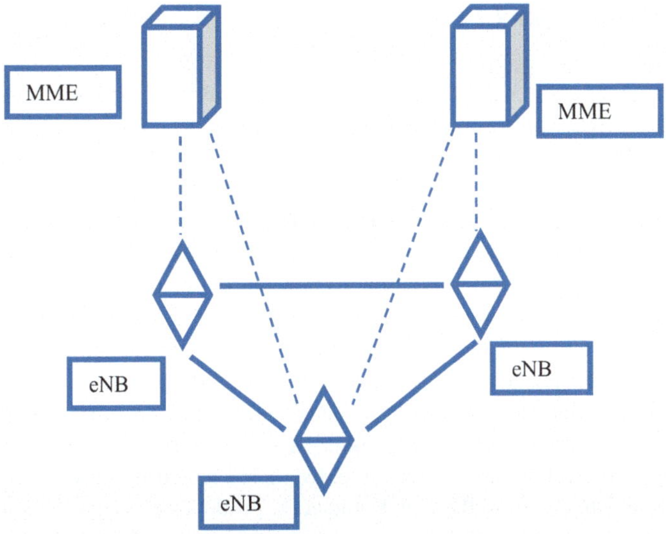

Fig. 3 Network architecture extracted from 3GPP

gateway. The IP address is provided by P-GW and the local anchor is handled by S-GW. The network access and mobility are controlled by MME in the system.

There are basically four abnormal state spaces: E-UTRAN, User Equipment (UE), Evolved Packet Core Network and Services. The initial three states are utilized in Internet Protocol Connectivity Layer to make IP based accessibility. It is exceptionally enhanced and offers all administrations over IP and furthermore expels the prerequisite of circuit-exchanged nodes and interfaces present in the past 3GPP frameworks. UEs, more often than not alluded to as a portable terminal, interrelate with eNodeBs by means of LTE–Uu interface. The LTE network architecture is shown in Fig. 2 and its network architecture extracted from 3GPP is shown in Fig. 3. LTE gives effortlessness in engineering when contrasted with past frameworks that advanced this design while in transit to less complex and viable level framework.

The principle point of LTE is to give streamlining to bundle exchanging administrations, which is essential for advanced throughput and high information rates and furthermore change in parcel conveyance delay. There is likewise thought of enhancement between systems administration and other stages like distinctive access systems. Advanced UMTS Terrestrial Radio Access Network is utilized at base station level that comprises of wise base stations called developed Node B (eNodeB). E-UTRAN relates to a straightforward matrix of eNodeBs sorted out by the X2 interface.

1.1 Motivation for LTE

The LTE is created by 3GPP as the innovation to deal with the interest of UMTS: an innovation that would give a powerful and practical remote access offered by other accessible advancements. It ought to likewise bolster the exponential development of broadband needs of versatile clients because of administration and system frameworks assembly. The arrangement of the versatile broadband ought not exclusively be restricted to home or work environment, however all over the place. The universal arrangement of portable broadband requires seeing from the two clients' and administrators' viewpoint. For clients, the worry is on the arrangement of high downlink information rate that will empower continuous client administrations like video spilling, internet gaming, and versatile TV. Notwithstanding the high downlink information rate, openness of a wide scope of cell phones, security, cost of administration and comfort are additionally significant. The worry of administrators, then again, ranges through issues, for example, expanded data transmission , movement from existing framework to the new framework, productive usage of remote range, and arrangement of higher ability to empower arrangement of new administrations. Facilitating these worries makes it imperative to institutionalize different parts of the LTE framework.

1.2 Article Organization

This article is organized as follows: Sect. 2 discusses surveys done earlier on Handover Mechanism on LTE. Section 2 describes the literature survey presented related to handover in LTE system. Various authors presented their work related to the handover system in LTE. Section 3 describes the handover in the LTE system, their techniques, types, etc. The handover procedure is also presented in this section. Section 4 describes the conclusion of this work.

2 Literature Survey

This section provides a literature survey related to the LTE handover system and provides various approaches related to them. Extensive work on Optimization of Handover parameters is presented by Harja et al. [1] who proposed a technique to evaluate the technique of handover. The parameters for evaluation are based on eNodeB with 43 dBm transmit power, used three cells per eNodeB, bandwidth value was 5 MHz, kept distance between nodes 2 km and speed was taken as it could have 46/20 s. The result showed that this algorithm was much better in terms of optimization efficiency as compared to other algorithms.

A review on schemes of Handover in LTE was presented by Li et al. [2] who presented handover schemes in heterogeneous LTE systems. So as to improve the system speed little cells, for example, femtocells, picocells, and microcells were used in advanced LTE. Cells are normally introduced at hotspots and conveyed with the large-scale cell in order to improve vitality proficiency and information rates. It has seen that the handover issue in full-scale femto HetNets could really compare to that in large-scale cell systems. Thusly, handover significantly affects the presentation of large-scale femto HetNets. Here, a portion of the handover systems was introduced, for example, load-balance-related handover plots, a vitality productive handover plan, and handover schemes in large-scale femto HetNets.

Management for group Handover in Mobile Terminals was presented by Hwang et al. [3], who exhibited an ongoing framework for gathering handover the board for V2x in Moving cell based LTE-propelled framework. The author proposed the gathering handover the board for versatile terminals appended to the moving cell dependent on Advanced Long-Term Evolution. Another was gathering the board conspire where it was conceivable to derive the handover calls for gathering individuals in the moving cell. He additionally depicted the substances of our proposed moving cell engineering, the convention heaps of control plane and client plane for the gathering the executives. Results demonstrated that the gathering handover plan can improve the vehicular client experience, and in this way can conceivably carry huge advantages to future remote correspondence frameworks.

A two advance handover scheme was presented by Park et al. [4]. It has seen that handover (HO) execution in heterogeneous systems arrangements was not as productive as in unadulterated large-scale organizations. It likewise demonstrated that rapid client hardware (UE) had influenced a lot higher handover disappointment (HOF) rate than low-speed UE. This made HO execution enhancement and commonplace issues. The author gave an answer that improves HO execution to HOF rate and ping-pong simultaneously. It comprised of early HO readiness and ping-pong evasion. The early HO readiness helps to relive HOF issues with sped up transmission of HO command message.

A two-step handover in LTE was presented by Jouilis et al. [5], who proposed a system coding based methodology for streamlining transfer speed limit during delicate handover in LTE-A framework. It has seen that the handover strategy turned out to be increasingly fragile over high portability speed and for ongoing administrations. Thus, it was viewed as that a roadway handover situation, wherein, a clogged objective node, as far as asset squares (RB), prompted a bundle misfortune case. To enhance asset designation during handover choices, a Network Coding Approach was presented. Results demonstrated that it improved the QoS and also reduced traffic. Additionally, this methodology limited the quantity of utilized RBs in the objective node during the handover choice. Subsequently, more assets were saved for imminent clogged.

A distributed handover scheme under data packet forwarding was proposed by Ko et al. [6], who displayed an improved and circulated information bundle sending plan in long haul development (LTE)/LTE-progressed (LTE-A) systems advanced node Bs (eNBs) to lessen the flagging overhead and the postponement happened in the information way exchanging plan which is a benchmark handover plot in LTE/LTEA systems. Likewise, a low unpredictability esteem emphasis calculation was displayed to unravel the optimality condition of MDP in a progressively successful way. Genuine follow-driven assessment results exhibited that the proposed plan decided the ideal length of the sending tie adaptively to applications' nature of administration (QoS) necessities.

A review on applications of fuzzy logic for LTE was presented by Mudassir et al. [7], who completed a study on fluffy rationale applications in remote and versatile correspondence for LTE. The point was to feature the standards of fluffy rationale applications in the territory of channel estimation, channel evening out, handover the board, and QoS of the board. Moreover, in LTE-progressed heterogeneous systems, fluffy rationale applications in the zone of interference of the board was additionally spotlighted. It has been seen that fluffy rationale based calculation simultaneously expanded the understandability and diminished the intricacy.

An enhanced scheme under handover in LTE was proposed by Lee et al. [8], who proposed an improved LTE handover plan utilizing NFV for LTE handover defer decrease. So as to help constant applications in versatile conditions, another handover innovation for low inactivity was required. Along these lines, the LTE handover technique was broke down dependent on the 3GPP standard and postpone the execution of LTE handover advancing NFV. According to proposed arrangements, a decrease in information cost and low inertness could be accomplished.

A method under optimization for Unnecessary Handover reduction was proposed by Li et al. [9] who presented an advanced technique for lessening superfluous handover in the LTE framework. So as to maintain a strategic distance from cycle and intricacy sort of handover disappointments, a streamlined handover procedure was presented, and expounded the handover flagging streams in favor of UE in subtleties, and utilized TTCN as reenactment device, the RRC layer as the recreation object and different layers as reproduction condition to mimic handover process. Through the recreation, consistency and rightness of the planned flagging procedure could be estimated. Thus, it was a pragmatic answer to stay away from pointless handover.

A self-organizing scheme under 4G was presented by Boujelben et al. [10] who presented a book about self-sorting out the scheme under LTE. It had seen that savvy means have been planned in 3GPP LTE-propelled standard to fundamentally upgrade inclusion and limit. Gigantic execution additions could be accomplished through expanding nodes thickness with the presentation of little cells. Recreation results had demonstrated that the exhibition of the given methodology through the assessment of some presentation parameters, for example, normal throughput and cell load. The acquired outcomes demonstrated the productivity of our novel way to deal with equalization of the heap between the various sorts of cells and to offer almost a similar throughput to all UEs paying little mind to their position.

A handover mechanism for self-organized network was presented by Isa et al. [11], who proposed a self-sorting out system-based handover instrument for LTE systems. The author proposed the gathering handover of the board for versatile terminals appended to the moving cell dependent on LTE-advanced. Another was gathering the board conspire where it was conceivable to derive the handover calls for gathering individuals in the moving cell. Specifically, the LTE system demonstrates wonderful enhancement in throughput and reduced system delay.

A handover performance improvement mechanism was presented by Park et al. [12], who proposed a thought by utilizing numerous handover arrangements to improve handover execution in LTE systems. A streamlined handover procedure was presented, and expounded the handover flagging streams in favor of UE in subtleties, and utilized TTCN as a reenactment device, the RRC layer as the recreation object and different layers as reproduction condition to mimic handover process. Because of the extraordinary handover execution gains, it was generally used in LTE frameworks, particularly in thick little cell systems and heterogeneous systems.

An adjustment method for handover improvement was presented by Sheu et al. [13] who presented a dynamic adjustment scheme in handover thresholds (DAHT) for off-loading enhanced node base station (eNB) in LTE (Long-Term Evaluation) small-cell networks. The proposed DAHT could distinguish whether user equipment (UE) was in static or mobile status from the strength of reference signal receiving power (RSRP). It was then determined that a handover factor (HF) with which the source eNB could increase or decrease the handover thresholds of RSRP. The purpose of dynamic adjustment in handover thresholds was to advance or delay the handover of a UE. Through the advanced and delayed handover, the load of an eNB could be relieved and its system performance could be improved. To demonstrate the proposed DAHT, it was implemented on a Linux-based LTE small-cell platform and then the load s measured on eNB and the residence time of UE to prove that the proposed DAHT can significantly improve the system performance.

Saeed et al. [14] displayed an upgraded fluffy rationale enhancement procedure dependent on client portability for LTE handover. It was assessed regarding the normal number of handover, framework throughput, and framework postpone dependent on client gear (UE) speed. The fundamental favorable position of this work was that UE speed is considered in the handover procedure. The proposed advancement procedure was successfully chosen for compelling Time-To-Trigger (TTT) in view of the UE speed, and it was helpful to locate the ideal handover edge (HOM) required for the handover procedure. The proposed calculation accomplishes an altogether improve in the handover execution when contrasted and the standard LTE and self-enhancement procedure. Results demonstrated that the proposed method adequately improves organize execution (limit the handover, boost the framework throughput, and limit the framework delay. We have thoroughly analyzed and summarized the summary from literature done in Table 1.

3 Handover in LTE

Mobility is a basic part of portable cell correspondence frameworks since it offers clear advantages to the end clients: low defer administrations, for example, voice or ongoing video associations can be kept up while moving even in rapid trains. Portability at fast is a test, and LTE as long haul advancement has guaranteed more

Table 1 Summary from literature survey

S.N.	Author's name	Techniques	Parameters	Conclusions
1.	T. Sheu, J. Sie	A scheme for dynamic adjustment in handover thresholds	Discussed about load status and residence time	• Evaluated that DAHT can significantly improve the system performance • Measured the load between the source and target node. It is observed that DAHT can significantly balance the loads between the source and target node
2.	S. Harja, Hendrawan	Parameters optimization and evaluation under handover	Considered throughput and optimization ratio	• Evaluated the handover performance parameters • Comparison between RSRP and RSRQ method was shown and RSRP proved better
3.	M. Saeed, H. Kamal	Fuzzy logic enhancement for optimization in handover	Average handovers and throughput	• Proposed an improved performance technique than standard LTE and other optimization techniques • Improved network performance in terms of handover minimization and maximization of throughput
4.	Y. Li, B. Cao, C. Wang	Schemes of handover in heterogeneous LTE networks	Spectrum efficiency	• Presented a technique for uninterrupted and quality service of mobile users • Presented an energy efficient and load balancing for handover

(continued)

Table 1 (continued)

S.N.	Author's name	Techniques	Parameters	Conclusions
5.	A. Mudassir, S. Akhtar, H. Kamel, N. Javaid	Used the fuzzy logic for channel estimation, QOS service	Computational time	• Proposed an application of LTE-advanced handover fuzzy logic application • It increased the understandability and reduced the complexity
6.	I. Jouili, K. Hassine, M. Frikha	Used the network coding method for soft handover	Packet loss ratio and no. of resource blocks	• Network coding approach (NCA) solution improves the quality of service (QoS) • Increased the number of available resource blocks (RB) in handover area offers a better service for RT users
7	Y. Hwang, J. Shin	Management of group handover for mobile terminals	Reduced burden	• Proposed management of group handover for mobile terminals • Described the entities for cells
8	I. Isa, M. Baba	Mechanisms for handover for self-organizing network	Hysteresis, throughput and time-to-trigger	• Presented a procedure for self-organizing handover scheme • Results showed that network performance was better and improvement in throughput

than previous advances to conquer this test. One of the primary objectives of the LTE radio system is to give quick and consistent handover starting with one cell then onto the next while all the while keeping system the executives basic. LTE innovation is intended to help portability for different versatile accelerates to 350 km/h or even up to 500 km/h. With the moving rate considerably higher, the handover will be progressively regular and quick. Handover is one of the key strategies for guaranteeing that the clients move unreservedly through the system while as yet being associated and being offered quality administrations. Since its prosperity rate is a key marker of client fulfillment, it is crucial that this technique occurs as quickly and as flawlessly as could reasonably be expected. In any case, the issue of giving consistent access turns out to be significantly increasingly significant in LTE since it utilizes hard handover (break-before-make).

Henceforth, streamlining the handover methodology to get the required presentation is viewed as a significant issue in LTE systems. Handover techniques are a key capacity of LTE eNBs. They are proposed to decrease interference time contrasted with the circuit-exchanged handover process in 2G systems. Handover inside an E-UTRAN. The system for when a UE is leaving a phone overseen by the eNB and entering a phone overseen by a second eNB. The essential objective of LTE or any remote structure for correspondence is to give snappy and reliable handover from one cell (a source cell) to another (a goal cell). The organization is to be kept up during the handover strategy, data move should not be deferred or should not be lost; for the most part, execution will be tainted fundamentally. LTE handover contributes an imperative occupation while moving a data beginning with one device then onto the following. In LTE, there are some predefined handover conditions for setting off the handover technique, moreover a couple of destinations concerning handover plan and streamlining.

3.1 Characteristics of Handover

Depending upon the required QoS, a consistent handover or a lossless handover is performed.

1. **Seamless Handover**
The seamless handover is the handover in which UE moves from one end of a cell to another end of another cell. Its main objective is to provide handover consistently. This type of information is commonly permissive however less tolerant of rescheduling, (for example voice administrations). Along these lines, consistent handover ought to limit the unpredictability and postponement albeit some SDUs may be lost.

2. **Lossless Handover**
In this handover, no information must be lost during this handover. This can be done by transmitting PDCP PDUs for which UE has not provided the acknowledgement back to the cell. In this handover, all deliveries which are in sequence be ensured by

data cells. Due to this, this method is very useful for services like download of files because it handles the loss of data due to the reaction of TCPs. It is also used in user planes and somewhere on the radio plane. The compression protocol for the header is reset because its data is not further provided from source end to destination.

3.2 *Techniques of Handover in LTE*

The techniques of handover in LTE are further divided into two parts, i.e., soft and hard handover. Soft handover is also known as Connect-Before-Break and hard handover is named as Break-Before-Connect, respectively. These are named after its working.

1. **Soft Handover, Connect-Before-Break**
In this handover, one link is always in connection mode before shifting to another link as shown in Fig. 4. These are used in the structure of WCDMA. In this, new links are joined to the structure before terminating another one. In this work, a central controller is provided to handle all activities related to them. Two or more cells can be connected at a time in this method. In softer handover, two or more cells are connected to a common physical structure. The one connection gets weaken due to another call in the process.

2. **Hard handover, Break-Before-Connect**
In this handover, only one link is in connection mode at a time as shown in Fig. 5. These are used in the structure of legacy systems in wireless communication. In this,

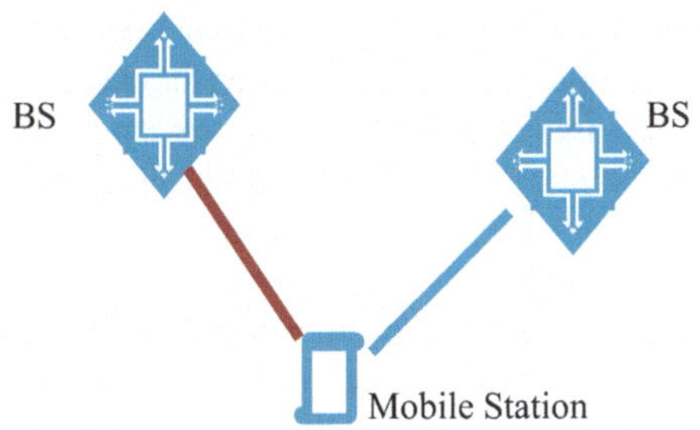

Fig. 4 Soft handover concept

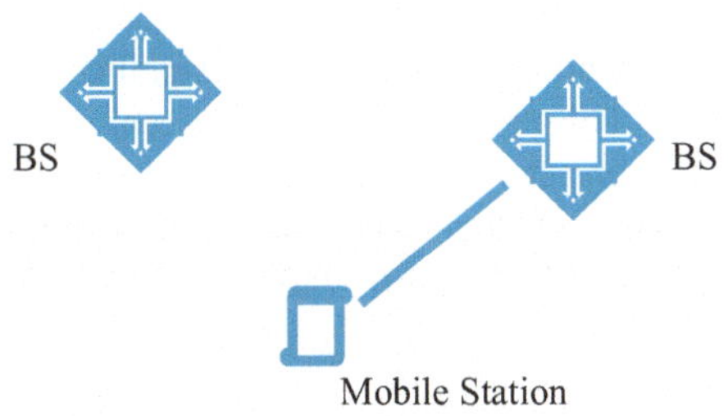

Fig. 5 Hard handover concept

Table 2 Techniques of Handover

Conventional techniques		
Handover techniques	Functions	Drawbacks
Hard handover	It is technique in which User has to break the existing connection to make a new connection	Tuning in non-smoother, delay gets increased, handover failure gets increased
Enhanced weighted performance	It helps to assign weights to parameters of handover	Poor convergence time, fluctuation problem
Fuzzy control handover techniques		
Fuzzy technique	Provides tuning to parameters, control speed variation	Fast convergence time, smooth tuning, reduce packet loss

new links are joined to the structure after terminating another one. In this work, a central controller is provided to handle all activities related to them. No two or more cells can be connected at a time in this method, only single-cell facility is provided in this handover scheme. In this handover, the controller has to break the previous connection for creating the new one.

The techniques of handover are presented in Table 2. It presented the methods and their functions with some drawbacks of each method used.

3.3 Procedure of Handover

The procedure of handover is explained as it contains many sections, namely preparation of handover in the first step, its execution in the next and completion of handover in the final step. In this work, the report of an event measurement is done by UE to source node eNB. It did not consider the evolved packet core for the handling of the plane. The signals are directly transferred between each node. The methodology appears in Fig. 6. There are a lot more examinations that are exhibited to accomplish upgrades and improvements in LTE handover, some with same and some are with various HO calculations and which take a various stages for various cases, however positively every one of them are done so as to get ideal handover rehearses that may deal with the exceptionally exact handover on cell limits of the LTE organize. This process may initiate or terminate in user terminal nodes or in radio systems.

Portability improvement is a significant viewpoint for the innovation in LTE since it will bolster versatility for different versatile accelerates to 350–500 km/h. With expanding the affecting velocity, the handover methodology will be progressively incessant and quick; accordingly, the handover execution will, in general, be increasingly significant particularly for constant administrations. Portability achievement

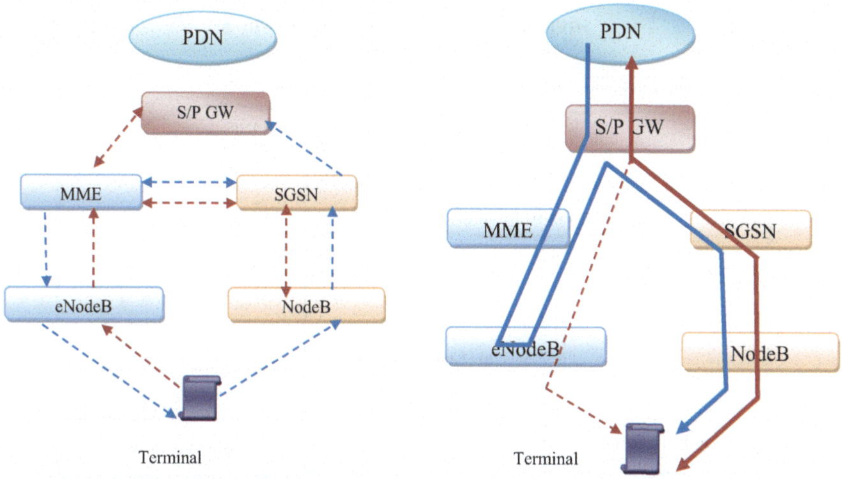

Fig. 6 Overview of handover in LTE

rate can be determined from 3GPP versatility KPI record, which can be detailed as demonstrated as follows:

In this procedure, all these phases are included for the completion of the handover. In the first phase, it prepares or selects the host node or eNB based on data analysis. The source node provides the request signal to the target eNB node. After receiving the request, it sends the acknowledgment back to the host node and provides the admission control in the process. After this, source eNB sends the command to UE for control. After this, it sends the signal to target and makes a connection for it. It sends the path switch request to switch gateway, after which acknowledgment is received back by S-GW. In the end, target resources are released to UE. The handover procedure is shown in Fig. 7.

1. **Minimize the Number of Handover Failures**
The call end because of handover ought to be maintained a strategic distance from other cell, and the strategy minimize number of failures when the mobiles start moving with one serving cell then onto the next by this scheme. This is a vital objective for handover structure and streamlining.

2. **Diminish Unnecessary Handovers**
It is constantly alluring to limit the quantity of handovers in light of the fact that over the top handovers increment the exchanging burden and decline the correspondence quality, and traffic limit of a framework. Moderating Ping-Pong impacts (in which the client more than once switches between nearby cells) and distinguishing the right objective cell can help maintaining a strategic distance from superfluous handovers.

3. **Diminish Handover that Initiated**
The handover system is dangerous in light of the fact that the call might be dropped because of the handover. The quantity of handover commencements will be altogether

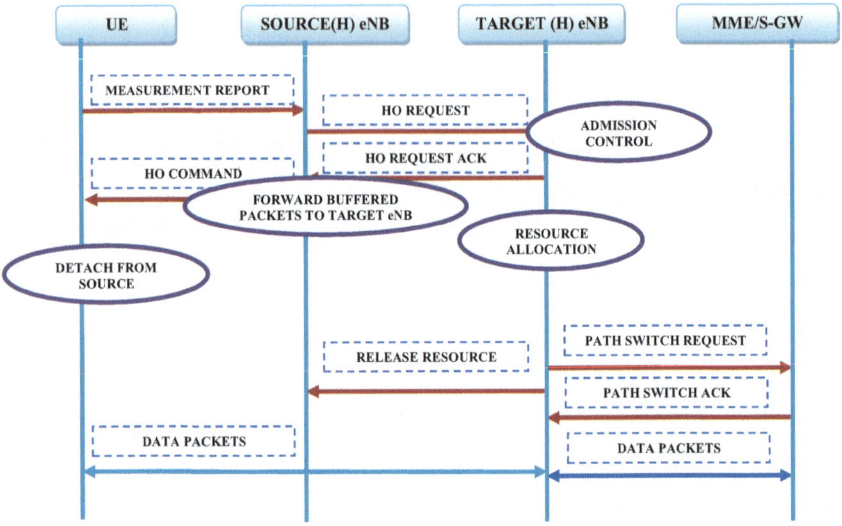

Fig. 7 LTE handover procedure

expanded if there are many Ping-Pong handovers or off base objective cell choice. Subsequently, it is significant for the administrator to limit the quantity of handovers to give decent support of their clients.

4. **Reduce Delay in Handover**
Handover ought to be quick with the goal that the client does not experience administration corruption or interference. This objective is progressively significant where there is an interference in the client plane.
5. **Improve Handover Connected Time**
Accomplishing this objective will be simpler if the handover is planned in a manner that drags out the measure of time that the UE is associated with the best cell. Henceforth, augmenting the all-out time that the client is associated with the best cell is a significant plan objective.

4 Conclusions

We know that LTE is a mobile standard that came into existence in 2008–09 after drawbacks in 3G and GSM. LTE is a packet switching technology that can provide data rates of 100 Mbps in the downlink. Handover plays an important role in cellular areas and so much of works have been done so far. The Handover in LTE provides a feature of multiple preparations at a time. This work summarizes various challenges that came under the handover system and the solutions to handle it. The authors presented various algorithms related to handover that improved the system performance by handling handover failure. Also the idea of handover for low-frequency and high-frequency systems is introduced . This work presented a review on the handover mechanism in the LTE system. In this paper, the handover problems within an LTE network in the vicinity of another LTE network was investigated. The techniques of handover are studied and also various schemes of handover to reduce handover failure rate and improving throughput are presented.

References

1. S.L. Harja, Hendrawan, Evaluation and optimization handover parameter based X2 in LTE network, in *International Conference on Wireless and Telematics*, pp. 175–180 (2017)
2. Y. Li, B. Cao, Handover schemes in heterogeneous LTE networks: challenges and opportunities. IEEE Wirel. Commun. 112–117 (2016)
3. Y. Hwang, J. Shin, Group handover management for V2x in moving cell based LTE-advanced system, in *IEEE ICTC*, pp. 1054–1057 (2015)
4. H.-S. Park, A.-S. Park, Two-step handover for LTE hetnet mobility enhancements, in *ICTC*, pp. 763–766 (2013)
5. H. Ko, G. Lee, D. Suh, An optimized and distributed data packet forwarding scheme in LTE/LTE-a networks. IEEE Trans. Veh. Technol. 01–11
6. A. Mudassir, S. Akhtar, A survey on fuzzy logic applications in wireless and mobile communication for LTE networks, in *International Conference on Complex, Intelligent, and Software Intensive Systems*, pp. 76–83 (2016)
7. C. Lee, S. Shin, Enhanced LTE handover scheme using NFV for LTE handover delay reduction, in *IEEE International Conference on Consumer Electronics-Asia*, pp. 01–02 (2016)
8. X.-W. Li, J. Wang, The optimized method of reducing unnecessary handover in LTE system, in *International Conference on Instrumentation, Measurement, Computer, Communication and Control*. (IEEE, 2013), pp. 1224–1227
9. M. Boujelben, S.B. Rejeb, S. Tabbane, A novel self-organizing scheme for 4G advanced networks and beyond. (IEEE, 2014), pp. 5874–5878
10. I.N.M. Isa, M.D. Baba, Self-organizing network based handover mechanism for LTE networks, in *International Conference on Computer, Communication, and Control Technology*, pp. 11–15 (2015)
11. H.-S. Park, Y.-S. Choi, Taking advantage of multiple handover preparations to improve handover performance in LTE networks. in *International Conference on Future Generation Communication and Networking*, pp. 09–12 (2014)
12. T.-L. Sheu, J.-Y. Sie, A dynamic adjustment scheme in handover thresholds for off-loading LTE small cells, in *IEEE International Conference on Applied System Innovation*, pp. 180–183 (2018)
13. Y.-H. Wang, G.-R. Huang, A handover prediction mechanism based on LTE-A UE history information. (IEEE, 2014), pp. 08–12
14. H. Weiyang, Z. Jihong, Cooperative handover algorithm based on auxiliary carrier in LTE-advanced relay system, in *ICACT*, pp. 1065–1070 (2014)

Two-Level Data Dissemination for Energy-Efficient Routing in IoT-Based Wireless Sensor Network

Roopali Dogra, Shalli Rani and Bhisham Sharma

Abstract The Internet of Things (IoT) is governed by the progressive growth in smart sensors, different communication technologies and various internet protocols. When the sensors are deployed in some remote areas, it becomes cumbersome to provide energy resources for their continual operations in IoT-based WSN (Wireless Sensor Network). Therefore, the main concern in such a network is the network longevity which is predominantly achieved by the energy-efficient routing techniques. Therefore, to resolve the above-said concern, in this paper, the cluster-based routing is considered that selects double Cluster Heads (CHs) in each cluster against the single CH in the conventional routing methods. While doing so, the data dissemination from the cluster members is done by the Primary CH (PCH), whereas the Secondary CH (SCH) helps in forwarding data to the next SCH after collecting it from PCH of its cluster. The multi-hop routing is followed and the nodes are of two-level energy heterogeneous in their configuration. The selection of PCH and SCH is done based on the residual energy, distance and node density. The proposed technique is named as EEDCH (Energy-Efficient Dual CH). It is observed that the proposed scheme outperforms the state-of-the-art protocols comprehensively, in terms of different performance metrics namely, stability period, network lifetime and network remaining energy.

Keywords WSN (Wireless Sensor Network) · Primary Cluster Head (PCH) · Secondary Cluster Head (SCH) · Heterogeneous · Double cluster heads · Routing

R. Dogra · S. Rani (✉)
Chitkara University Institute of Engineering and Technology, Chitkara University, Punjab, India
e-mail: shalli.rani@chitkara.edu.in

R. Dogra
e-mail: Roopali.dogra@chikara.edu.in

B. Sharma
Department of Computer Science & Engineering, Chitkara University School of Engineering and Technology, Chitkara University, Himachal Pradesh, India
e-mail: bhisham.sharma@chitkarauniversity.edu.in

M. Dutta et al. (eds.), *Proceedings of International Conference on IoT Inclusive Life (ICIIL 2019), NITTTR Chandigarh, India*, Lecture Notes in Networks and Systems 116,
https://doi.org/10.1007/978-981-15-3020-3_12

1 Introduction

Over the years, technological advancements have connected every physical object with the Internet so that they can communicate with each other and the terminology behind this is termed as Internet of Things (IoT). There are numerous examples where IoT has shown tremendous growth when it comes to smart city, smart transportation, smart toll plaza, and various other fields as shown in Fig. 1. Furthermore, the rapid growth in MEMS technology and radio technology has led to the production of tiny and low-cost sensors. These sensors are connected to each other and after sensing the data and aggregation the collected data is forwarded to the sink. Such a network is termed as Wireless Sensor Network. The evolution of WSN has been a remarkable achievement to facilitate the numerous applications in which the sensing and communication techniques play an important role. There has not been any limit to the capabilities of WSN as it has revolutionized everything in the modern technical world.

The only constraint on the WSN is the limitation in the power resources so the communication has to be effective and to achieve that routing among the sensor nodes needs to be energy efficient in the WSN. To resolve the concern of battery limitations, cluster-based routing techniques have played a crucial role as it tends to decrease the number of data transmission in the network. The primary concern in clustering is the CH (Cluster Head) selection which collects, aggregates, and forwards the data to the sink or to the other CH in the cluster. Conventionally only one CH is selected in the cluster which puts a huge burden on the CH to collect, aggregate and then forward the data, therefore, an attempt is required to distribute the load of the CH so that energy consumption can be reduced. Therefore, in this paper, the following contributions are reported.

Fig. 1 Applications of IoT-based WSN

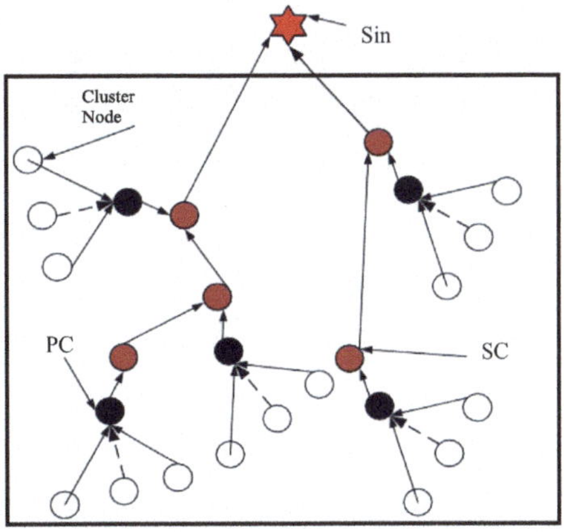

Fig. 2 Data transmission scenario of the proposed work

(a) An energy-efficient clustering is proposed in which two CHs are defined in each cluster; primary CH (PCH) disseminates the data from the cluster and forwards to the secondary CH (SCH). From SCH, it is forwarded to the next SCH or to the sink as shown in Fig. 2.
(b) The selection of SCH and PCH is done based on the significant parameters, namely, residual energy, distance to the sink, and node density and the network is energy heterogeneous considering two-energy level heterogeneity.
(c) The performance evaluation of the proposed scheme is done against state-of-the-art routing protocols.

The rest of the paper is organized as follows: Sect. 2 contains the related work done. Section 3 discusses the proposed protocol and simulation analysis.

Section 4 gives conclusions and future work from the proposed work.

2 Related Work

Since the development of sensor networks its widening applications have given a huge scope for the IoT to expand its horizon of applications.

A number of attempts have been made in achieving the enhanced network lifetime. Numerous researchers have considered cluster-based routing for decreasing the number of data transmissions and distributing the load uniformly across the network. Conventionally, there is only one CH that disseminates the data, but a lot of studies have used double CH in their data collection approach. Table 1 presents some studies that used double CHs and the corresponding research gap is discovered. It is observed that the protocols aimed for network lifetime enhancement. However, the three parameters energy, distance, and node density were not considered collectively for the CH selection. Therefore, energy-efficient CH selection is required

Table 1 Review on techniques employing double CHs in cluster

Ref. study	Name of method	Mode of network	Method used	Research gap
Linping et al. [1]	PDCH	Homo	Chain-based clustering, one CH forwards the data and other collects from cluster members	Hotspot problem exists, Delay is created
Xuegong [2]	DCCCA	Hetero	Assistant CH forwards data to the sink and main CH collects CH rotation is done based on residual energy	Due to uncontrolled multi-hop communication, the hotspot problem is created
Ruihua et al. [3]	PSO-DH	Homo	Vice CH forwards data to the sink and master CH collects CH rotation is done according to PSO with distance and energy parameters	Inefficient multi-hop communication
Fu and Liu [4]	DCHM	Homo	The selection of two CHs is done based on reputation and trust system	Energy efficiency is compromised while handling the security concerns
Li and Liu [5]	DL-LEACH	Homo	Other than selecting two CHs, it also keeps control of the number of CHs in the network	Data transmission from the CH of a cluster is not optimized with respect to the distance from another CH

(continued)

Table 1 (continued)

Ref. study	Name of method	Mode of network	Method used	Research gap
Han et al. [6]	DCE	Hetero	CH selected in two phases, one considers initial energy, residual energy, and distance from the sink. In the next phase, the fused data is forwarded to the sink while involving bit error rate	Complex algorithm, Number of overheads are too high, energy consumption is higher
Akila and Venkatesan [7]	ZCA	Homo	Uses the concept of the geo-cluster head (GCS) that helps in mitigating hotspot issue in the network	The selection of GCS could be improved with the help of more parameters for CH selection
Purkar and Deshpande [8]	EECEP-HWSN	Hetero	Initial energy, hop count, and residual energy are considered for the selection of zone CH. Internal overheads are caused, and therefore boundary locations for the nodes are considered	Distance factor is not considered while selecting CH
Darabkh et al. [9]	BPA-CRP	Homo	Equal sized clustering is done for the network. Load balancing is introduced in the network	Equipping nodes with GPS brings high cost to the network Hot spot problem exists
Darabkh et al. [10]	EA-DB-CRP	Homo	Different components, namely merging cluster algorithm, CH selection, CH replacement, relay node selection algorithm is exploited	Multi-hop communication causes hotspot problem in the network

that considers the above-mentioned parameters. The energy heterogeneity is the least explored in these techniques that exploited the double CH in the cluster.

As the proposed work deals with the energy heterogeneous nodes, therefore a review regarding the evolutionary advancements in the heterogeneous routing protocol is reported.

It started with Stable Election Protocol (SEP) [11] which was the first protocol that incorporated the heterogeneity at two levels. Although the weighted probability is used for the CH selection of different nodes, these nodes are termed as normal and advanced nodes. It is noted that the advanced nodes are more in energy as compared to the normal nodes. These nodes raise the cost of the network, and therefore they are used optimally. It is concluded from the SEP protocol that it doesn't work for the multilevel heterogeneous nodes. Since the advancements of SEP, there have been numerous routing protocols, namely, EEHC which stands for Energy-Efficient Heterogeneous Clustered Scheme was the first protocol of the heterogeneous WSN that used the heterogeneity at the three levels [12]. The CH selection is done based on the energy of the nodes.

With the gradual progress in acquiring the energy-efficient routing, the DEEC protocol which is Distributed Energy-Efficient Clustering was proposed for two-level heterogeneity [13]. Though the protocol considered two-level heterogeneity and considered residual energy for the CH selection. However, the protocol failed for the multilevel heterogeneity. Furthermore, the penalization of the nodes was another concern that was handled in proceeding protocols. To resolve the concern of the aforementioned protocol, DDEEC which stands for Developed Distributed Energy-Efficient Clustering is proposed. Though the protocol is quite similar to the DEEC, it deals with the penalization of advanced nodes [14]. Where in DDEEC, the penalization of nodes is avoided at two levels of energy heterogeneity, but in EDDEEC, i.e., Enhanced Developed Distributed Energy-Efficient Clustering protocol, the same operation was performed at three levels of energy heterogeneity. Subsequently, the heterogeneity level was further enhanced to four levels in BEENISH [15] which was further solved to avoid penalization in iBEENISH [16]. The performance of the iBEENISH protocol was recently improved by the energy-efficient iBEENISH termed as E-BEENISH. Over the years, the main focus of WSN based IoT has been to deliver the optimal network performance in the harsh environment scenarios [17, 18]. The numerous attempts have been registered, but still, there is a lot of scope in enhancing the network performance of IoT-based WSN to bring enhancement in its network longevity. A huge scope exists in the CH selection and for load balancing in the network, dual CH selection is done. MIEEP (Multi-sequenced Energy-Efficient Protocol) enhanced network lifetime but the stability of the network is quite low due to inefficient routing scheme [19]. Dutt et al. in [20] proposed EESCP (Energy-Efficient Sector-based Clustering Protocol), the sectoring-based topology in the WSN.

2.1 Problem Definition

Since the development of WSN, the main task has been the reduction in the energy consumption of the nodes. To perform this task, many clustering techniques have been proposed. However, as discussed in Table 1, numerous attempts have been made to introduce double CHs in the network. Though, many heterogeneous and homogeneous networks have experienced the energy heterogeneity while dealing with double CHs in the cluster. However, the scope of improvement lies in the fact that the selection of these two CHs is yet to be optimized. Therefore, as observed from the Table 1 and the study reported under the section 'related work', the CH selection can be improved if the following parameters are considered, namely residual energy, distance to the sink, and node density or intra-cluster distance.

3 Proposed Protocol: EEDCH

The EEDCH operates like any other cluster-based routing techniques. The network model assumptions for the EEDCH are given below.

3.1 Network Assumptions for EEDCH

Network assumptions for the proposed work are listed as follows.

(a) The network is energy heterogeneous and the two-level energy heterogeneity is considered normal and advanced nodes. Wherein advanced nodes have more energy as compared to normal nodes.
(b) The deployment of the nodes is random but uniform. These nodes are location unaware.
(c) The communication link of the nodes is symmetrical.
(d) The base station is placed outside the network which is taken in a rectangular shaped.
(e) Base station is assumed to be energy unlimited.
(f) There is no recharging source for the sensor nodes once they are dead.

3.2 Operation for the EEDCH

The network model for the proposed work EEDCH is given as follows. In this network model, the heterogeneity among the sensor network is introduced as shown in Fig. 2. There are heterogeneous nodes forming a cluster in different zones that are formed in the network.

3.2.1 Role of Double CHs

In each cluster, despite one, two CHs are taken; among them, one is Primary Cluster Head (PCH) forwarding the data to the Secondary Cluster Head (SCH) of cluster and another one is Secondary Cluster Head (SCH) which leads to the load balancing in the network. After disseminating the data from the cluster member nodes, PCH performs data aggregation which is the removal of redundant data. So that only meaningful information is conveyed. The collected data by the SCH will be forwarded to either sink or another SCH on the way to the sink.

While doing so, multi-hop communication is adopted to avoid long haul transmission. The protocol works in the two phases like any other clustering protocol; set-up phase and steady-state phase. The set-up phase includes the initialization and cluster formation whereas, the steady-state phase includes the data transmission phase in the network.

3.2.2 Set-up Phase

The heterogeneous nodes, normal and advanced nodes, are uniformly but randomly deployed in the network. The network can be attended or unattended depending upon the application. The set-up phase involves the selection of both CHs and sink placement. The process of CH selection is initiated by the *Hello* message from the sink to the network and in return, the unique IDs are shared from each node with the sink. As soon as the sink receives these IDs, it broadcasts all of them in the network. Thereafter through the distributed approach, the CH selection takes place in each cluster through the following process.

The selection parameters for the CH are listed as follows.

(a) **Residual energy**: It is the current energy of the node in each round which is monitored.
(b) **Distance to the sink**: The distance of the node from the sink is checked so that energy consumption is reduced.
(c) **Node density**: The intra-cluster distance which can also be termed as node proximity is also considered.

A. **The selection of PCH and SCH in the cluster among the normal nodes**

It must be noted that a threshold profile is generated for the two most suitable node to be selected as PCH and SCH, respectively. The SCH is selected with rank one with the generated threshold profile and the PCH is actually a node selected with ranked two in its threshold profile. As considered in EESCP, the probability of each node is calculated, which is further used in the calculation of threshold value.

The probability of the normal node is computed by Eq. (1). The notations defined in Eq. (1–4) are referred to in Table 2.

$$\mathrm{P_{NRM}} = \frac{\mathrm{P_{OPT}}}{(1 + \lambda \times \phi)} \tag{1}$$

Table 2 Symbols representations

Symbols	Expansions
P_{NRM}	Normal node's probability to be selected as CH
P_{ADV}	Probability of advanced nodes to be selected as CH
P_{OPT}	Optimum probability for the number of CHs
λ	Advanced fractions for advanced nodes
ϕ	Fraction number of advanced nodes
r	Round
$E_{\mathrm{NRM}}, E_{\mathrm{ADV}}$	Energy of normal and advanced nodes, respectively
$D_{\mathrm{N\_SINK}}$	Distance of a node from sink
$T_{\mathrm{NRM(i)}}$	Threshold for normal nodes
G'	Set of nodes that have not become CH yet
$T_{\mathrm{ADV(i)}}$	Threshold for advanced nodes

$$\mathrm{T}_{\mathrm{NRM(i)}} = \begin{cases} \dfrac{\mathrm{P}_{\mathrm{NRM}} \times \mathrm{E}_{\mathrm{NRM}} \times \mathrm{Node\_\ Den\_\ N}}{(1 - \mathrm{P}_{\mathrm{NRM}}\left(\mathrm{r} \bmod\left(\frac{1}{\mathrm{P}_{\mathrm{NRM}}}\right)\right) \times \mathrm{D}_{\mathrm{N\_SINK}}} & \text{if } \mathrm{S(i)} \in \mathrm{G}' \\ 0 & \text{otherwise} \end{cases} \quad (2)$$

In Eq. (2), $\mathrm{T}_{\mathrm{NRM(i)}}$ is the threshold value for the normal nodes, which is placed in the table. Similarly, the threshold value is computed for all other nodes in the cluster and a table is generated. The node with the highest value of the threshold is ranked one and the second highest is termed as ranked two. These ranked one and ranked two nodes are selected as SCH and PCH, respectively. Furthermore, these threshold values are compared with the random number generated by that node. If the generated random number is less than the 'ranked one and two' threshold, it will be selected as SCH or PCH as explained above. Otherwise, that particular node stays as a normal cluster member node.

It is noted that if these ranks are observed for the complete number of nodes in the cluster until two CHs are selected in the cluster. The whole operation for the CH selection is shown in Fig. 3.

B. **CH selection among advanced nodes**

The advanced nodes have a higher probability of becoming the CH as compared to the normal nodes. Among advanced nodes, the cluster head selection is referred to by using Eqs. (3) and (4).

$$P_{\mathrm{ADV}} = \frac{P_{\mathrm{OPT}}}{(1 + \lambda \times \phi)} \times (1 + \lambda) \quad (3)$$

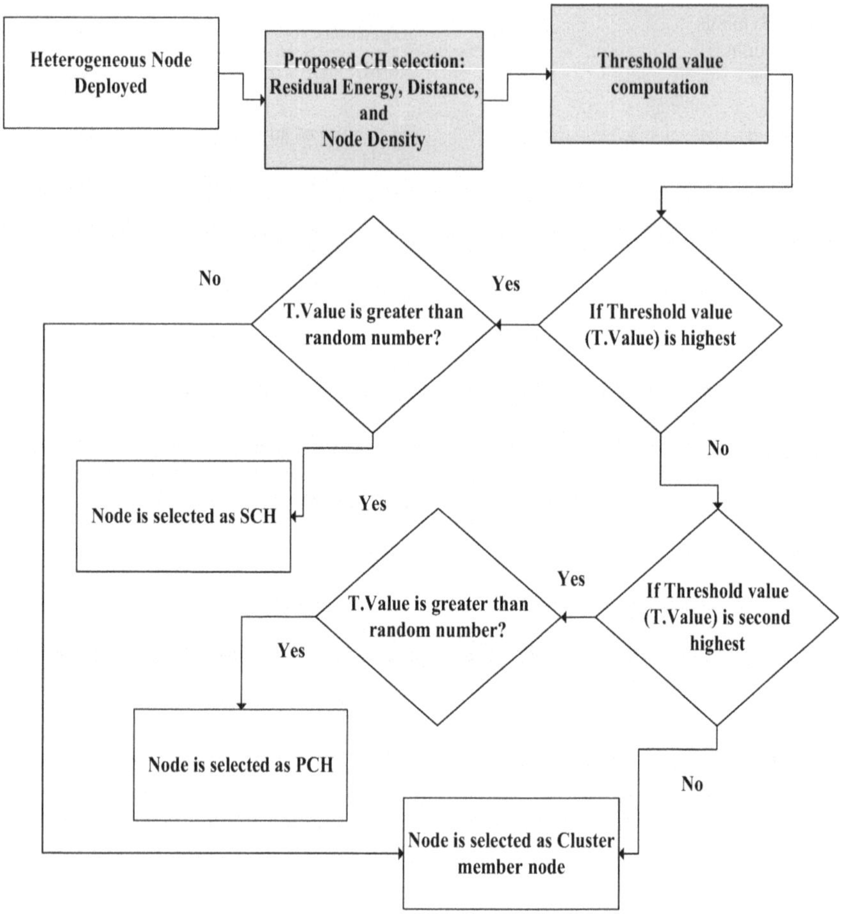

Fig. 3 Flowchart of the working operation of EEDCH

In Eq. (3), the probability of an advanced node is computed. In Eq. (4), $T_{ADV(i)}$ is the threshold value for the advanced nodes, hence the comparison is made with the random number generated by that node.

$$T_{ADV(i)} = \begin{cases} \frac{P_{ADV} \times E_{ADV} \times \text{Node\_ Den\_ N}}{\left(1 - P_{ADV}\left(r \mod \left(\frac{1}{P_{ADV}}\right)\right)\right) \times D_{N\_ SINK}} & \text{if S(i)} \in \text{G}' \\ 0 \text{ otherwise} \end{cases} \quad (4)$$

As explained for the normal nodes, Sect. 3.2.2 (A), a similar ranking system is followed in the case of advanced nodes. If the generated random number is less than $T_{ADV(i)}$, it will be selected as CH else it will be treated as a normal cluster member node.

Table 3 Simulation parameters

Parameter	Value
Network coverage	100 m^2
Data packet size	2000 bits
Initial energy (Quantity)	In Js 0.5
Node number	100
advance fraction (a)	1
Number fraction of advanced nodes (m)	0.1

When the CH selection is done, the nodes will consume their energies according to the equations specified by the radio energy consumption model.

It is noted that the value obtained from the threshold computed in Eq. (4), is compared with randomly generated numbers. If the random number is greater than the threshold value, the node is said to be cluster member node, however, it becomes the cluster member when the random number is smaller than the threshold.

C. **Steady-State Phase**

This phase deals with data transmission at different levels. It includes data transmission from the following.

(a) Cluster member nodes to the PCH
(b) PCH to the SCH
(c) SCH to another SCH if the distance is more than threshold otherwise, SCH to sink directly.

The threshold distance decides about the single-hop or dual-hop communication for the SCH nodes. It is basically equal to the maximum range of CH node which is already defined by the user at the start.

D. **Radio Energy Consumption Model**

The same energy model is used as used by various cluster-based routing [21]. As soon as the steady-state phase is initiated, the nodes consume energy according to this model (Table 3).

3.3 Simulation Analysis

When simulations are performed in MATLAB, the proposed routing scheme, EEDCH is evaluated on different performance metrics. The evaluation is done with respect to a different number of rounds acquired by these metrics. A round is defined as the one iteration in which the data is collected from all the nodes by the sink for one time. The metrics involved are discussed below.

Fig. 4 Performance comparison

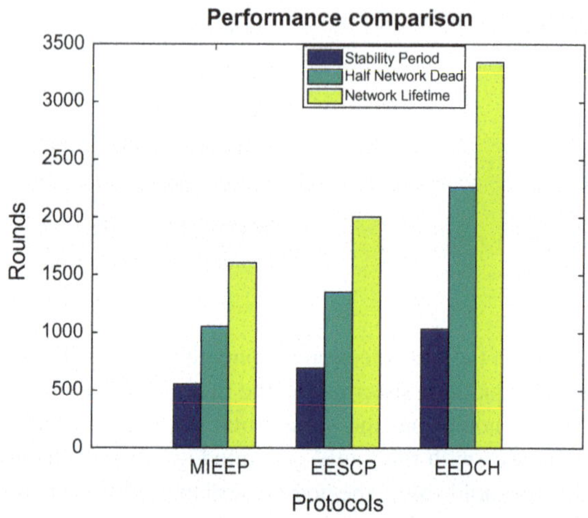

Table 4 Performance comparison of EEDCH as compared to MIEEP and EESCP

Metrics	Protocols		
	MIEEP	EESCP	EEDCH
Stability period	555	693	1033
Half node dead	1053	1350	2259
Network lifetime	1604	2003	3345

(a) **Stability Period**

The number of rounds that is successfully completed before the first node is dead is termed as a stability period. It is one of the essential parameters that ensure the stability and reliability of any routing scheme. As shown in Fig. 4, the stability period of EEDCH is higher than the other protocols. It can be seen from Table 4, the stability period of EEDCH is 1033 rounds whereas it is just 693 and 555 rounds in case of EESCP and MIEEP protocols, respectively. The reason behind such enhancement is the use of network density in the CH selection along with the energy and distance parameters. Furthermore, the enhancement is acquired due to the load balancing approach adopted due to the dual CHs in the cluster. The graph of alive versus round is shown in Fig. 5, which gives the status of alive nodes with the passage of rounds.

(b) **Network Lifetime**

The number of rounds completed before to the moment when all nodes are dead, it is termed as network lifetime. The status of dead nodes versus round in shown in Fig. 6. The first dead node of EEDCH is 1033 rounds, the half-dead nodes are 2259 rounds. It can be observed that the network lifetime of EEDCH is very high as compared to the other protocols. The reason behind such improvement is the distribution of the load of single CH on the second CH which is used in each cluster. Data forwarding

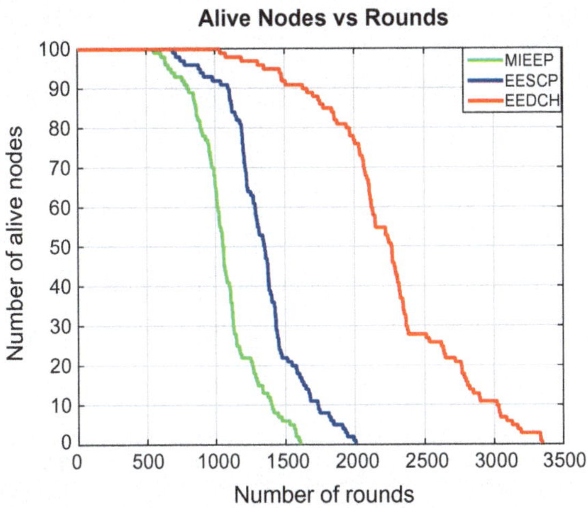

Fig. 5 Alive nodes versus rounds

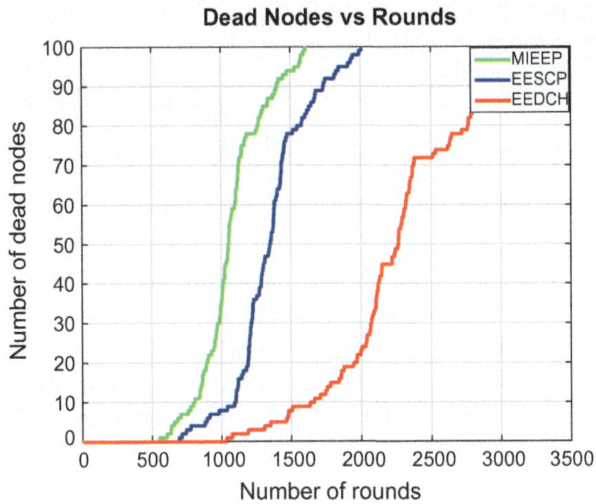

Fig. 6 Dead nodes versus rounds

by the SCH saves a huge amount of energy of PCH in the cluster. That makes them functioning for a longer period of time.

(c) **Networks remaining energy**

The proposed scheme helps in the enhancement of network remaining energy with respect to the number of rounds. The status of the network's remaining energy is acquired during the simulation analysis when the continuous value of the energy of the whole network is updated and the graph given in Fig. 7 is obtained. Due to the energy-efficient CH selection, the network preserves more energy than the other protocols as shown in Fig. 7. The percentage improvement by the EEDCH is shown in Table 4.

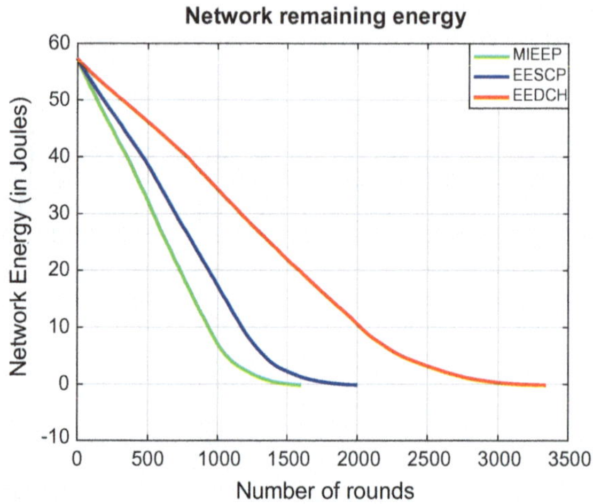

Fig. 7 Network's remaining energy

Table 5 Percentage improvement of EEDCH as compared to others

Metrics	Protocols	
	MIEEP	EESCP
Stability period	86.1	49
Half node dead	114.5	67.33
Network lifetime	108.5	66.99

It is evident from Table 5 that the protocol EEDCH improves the stability period by 49% in comparison to the EESCP and 86.1% in comparison to the MIEEP protocols. Furthermore, the network lifetime is enhanced by 67 and 108.5% in comparison to the EESCP and MIEEP protocols, respectively. The reason for the respective improvement is discussed in the aforementioned metrics.

4 Conclusion and Future Scope

Internet of things (IoT) has made it possible for the things to communicate with each other after when they are connected to the internet. However, the issue of the energy efficiency of the wireless nodes used in IoT-based WSN really hampers the continual functioning of those nodes. In this paper, the extensive literature survey is done that gives a promising glimpse of various methods that have employed the dual CHs scenario. Thereafter, the proposed technique aims to make the routing energy efficient by proposing two CHs in each cluster, i.e., SCH and PCH. The selection of SCH and PCH is done based on the residual energy, distance to the sink and node proximity. It is contemplated that the proposed routing scheme greatly improves network performance in terms of network longevity, etc. It is seen that the proposed scheme improves the network lifetime by 67% and stability period by 49% as compared to EESCP protocol, respectively, in IoT-based WSN. In future work, the hotspot problem will be targeted so that multi-hop communication can be made effective. Furthermore, it will be interesting to note the sink movement in the network through some optimization techniques in IoT-based WSN.

References

1. W. Linping, B. Wu, C. Zhen, W. Zufeng, Improved algorithm of PEGASIS protocol introducing double cluster heads in wireless sensor network, in *International Conference on Computer, Mechatronics, Control and Electronic Engineering*, vol. 1, pp. 148–151 (2010)
2. Q. Xuegong, A control algorithm based on double cluster-head for heterogeneous wireless sensor network, in *2nd International Conference on Industrial and Information Systems*, vol. 1, pp. 541–544 (2010)
3. Z. Ruihua, J. Zhiping, L. Xin, H. Dongxue, Double cluster-heads clustering algorithm for wireless sensor networks using PSO, in *6th IEEE Conference on Industrial Electronics and Applications*, pp. 763–766 (2011)
4. J.-S. Fu, Y. Liu, Double cluster heads model for secure and accurate data fusion in wireless sensor networks. Sensors **15**(1), 2021–2040 (2015)
5. H. Li, J. Liu, Double cluster-based energy efficient routing protocol for wireless sensor network. Int. J. Wirel. Inf. Netw. **23**(1), 40–48 (2016)
6. R. Han, W. Yang, Y. Wang, K. You, DCE: a distributed energy-efficient clustering protocol for wireless sensor network based on double-phase cluster-head election. Sensors **17**(5), 998 (2017)
7. I.S. Akila, R. Venkatesan, An energy balanced geo-cluster head set based multi-hop routing for wireless sensor networks, Clust. Comput. 1–10 (2018)
8. S.V. Purkar, R.S. Deshpande, Energy efficient clustering protocol to enhance performance of heterogeneous wireless sensor network: EECPEP-HWSN, J. Comput. Netw. Commun. **2018** (2018)
9. K.A. Darabkh, M.Z. El-Yabroudi, A.H. El-Mousa, BPA-CRP: a balanced power-aware clustering and routing protocol for wireless sensor networks. Ad Hoc Netw. **82**, 155–171 (2019)
10. K.A. Darabkh, S.M. Odetallah, Z. Al-qudah, K. Ala'F, M.M. Shurman, Energy-aware and density-based clustering and relaying protocol (EA-DB-CRP) for gathering data in wireless sensor networks. Appl. Soft Comput. **80**, 154–166 (2019)
11. G. Smaragdakis, I. Matta, A. Bestavros, SEP: a stable election protocol for clustered heterogeneous wireless sensor networks. Boston University Computer Science Department (2004)
12. D. Kumar, T.C. Aseri, R.B. Patel, EEHC: Energy efficient heterogeneous clustered scheme for wireless sensor networks. Comput. Commun. **32**(4), 662–667 (2009)
13. L. Qing, Q. Zhu, M. Wang, Design of a distributed energy-efficient clustering algorithm for heterogeneous wireless sensor networks. Comput. Commun. **29**(12), 2230–2237 (2006)
14. B. Elbhiri, R. Saadane, D. Aboutajdine, others, Developed distributed energy-efficient clustering (DDEEC) for heterogeneous wireless sensor networks, in *5th International Symposium on I/V Communications and Mobile Network (ISVC), 2010*, pp. 1–4 (2010)
15. T.N. Qureshi, N. Javaid, A.H. Khan, A. Iqbal, E. Akhtar, M. Ishfaq, BEENISH: balanced energy efficient network integrated super heterogeneous protocol for wireless sensor networks. Procedia Comput. Sci. **19**, 920–925 (2013)
16. M. Akbar, N. Javaid, M. Imran, N. Amjad, M.I. Khan, M. Guizani, Sink mobility aware energy-efficient network integrated super heterogeneous protocol for WSNs. EURASIP J. Wirel. Commun. Netw. **2016**(1), 66 (2016)

17. S. Verma, N. Sood, A.K. Sharma, Design of a novel routing architecture for harsh environment monitoring in heterogeneous WSN. IET Wirel. Sens. Syst. (2018)
18. S. Verma, N. Sood, A.K. Sharma, QoS provisioning-based routing protocols using multiple data sink in IoT-based WSN. Mod. Phys. Lett. A, 1950235 (2019)
19. S. Vhatkar, S. Shaikh, M. Atique, Performance analysis of equalized and double cluster head selection method in wireless sensor network, in *Fourteenth International Conference on Wireless and Optical Communications Networks (WOCN)*, pp. 1–5 (2017)
20. S. Dutt, G. Kaur, S. Agrawal, Energy efficient sector-based clustering protocol for heterogeneous WSN, in *Proceedings of 2nd International Conference on Communication, Computing and Networking*, pp. 117–125 (2019)
21. S. Verma, N. Sood, A.K. Sharma, A novelistic approach for energy efficient routing using single and multiple data sinks in heterogeneous wireless sensor network. Peer–Peer Netw. Appl. 1–27 (2019)

A Comparative Study of Cluster-Head Selection Algorithms in VANET

Poonam Thakur and Anita Ganpati

Abstract VANET is a class of ad hoc networks where the ad hoc nodes are the vehicles and the collection of the vehicles and the Road Side Unit (RSU) makes a network. Clustering has been proved to be useful in vehicular ad hoc networks for a number of different issues like reducing the traffic in data propagation, managing the network, load balancing, and target tracking, efficient resource consumption. While designing clusters, there are a number of different concerns like designing stable clusters, deciding cluster members, using double head in one cluster, reducing control overhead but cluster head selection is the most critical concern. Although there are a number of research articles available that discusses the clustering algorithms being used in VANET in this paper, we have discussed for the first time in the literature about the various cluster-head selection schemes being used for VANETs. This paper is going to be a review of various algorithms used for cluster head selection in VANETs. In the end, few open research issues are also given which can be a help to the research community.

Keywords CH · CM · Multi-hop · VANET · Clustering

1 Introduction

VANET is an ad hoc network made up of the moving vehicles on the road and the Road Side Units (RSU). A VANET communication basically consists of V2V or V2I communication, where V2V refers to the communication between the vehicles and V2I refers to the communication between the vehicles and the roadside units. The vehicles in VANETs are provided with an On- Board Unit (OBU), which helps in the communication. The biggest issue of VANET is its dynamic topology and

P. Thakur (✉) · A. Ganpati
Department of Computer Science, Himachal Pradesh University, Shimla, India
e-mail: akku786@gmail.com

M. Dutta et al. (eds.), *Proceedings of International Conference on IoT Inclusive Life (ICIIL 2019), NITTTR Chandigarh, India*, Lecture Notes in Networks and Systems 116,
https://doi.org/10.1007/978-981-15-3020-3_13

continuously changing network due to which a reliable and scalable routing protocol can't be found. The use of clustering can prove to be a help in improving routing scalability and reliability in VANETs [1, 2]. Clustering groups the vehicles on the basis of some parameters like relative velocity or distance or link connectivity or correlated spatial distribution which results in the distributed formation of hierarchical network structures. In addition to these routing benefits, clustering can serve as the foundation for accident or congestion detection, easy dissemination of information and entertainment applications. Clustering has proved to be helpful in target tracking and improving the scalability of the large scale VANET. Clustering also helps in efficient resource consumption/allocation with low overhead and load balancing. In this paper, we have presented a comparison of various strategies for cluster head selection which results in stability in clusters and ultimately proves to be helpful for routing. The rest of the paper is organized as unit 2 discusses the clustering procedure and the types of nodes a cluster can have. Unit 3 gives the comparison of various cluster head selection algorithms. Unit 4 provides discussion and results. In the end, Unit 5 gives the open research issues of VANET.

2 Clustering

A cluster-based network is a kind of distributed network where the underlying nodes are categorized into different types.

2.1 *Types of Nodes*

These are the following types of nodes that can be found in any vehicular cluster model.

1. Cluster Head (CH)—This is the node that is the coordinator or head of the cluster. It takes responsibility for the discovery and maintenance of the routing paths. All the intra-cluster and inter-cluster communication takes place using this node.
2. Secondary Cluster head (CSH)—This is the node selected by the CH on the basis of different parameters. This node acts as a backup or as a subordinate to CH and assists it in the clustering process. CSH may or may not be present in the cluster based on the algorithm being used for clustering and the application of the algorithm.

3. Cluster Member (CM)—The node which is not CH or SCH is a cluster member. These are the nodes which broadcast messages to each other for information exchange.
4. Gateway Node (GW)—This is the node which helps in communication with RSU. It is not necessary that this is present in every cluster.

3 Cluster Model

A cluster architecture based vehicular ad hoc networks can be modeled into two different ways, i.e., single-hop or multi-hop. Both models have their own advantages and disadvantages based on the applications they are used for. A **single-hop model** is one where every node is directly connected to CH. A **multi-hop model** is one where the distance between the CH and the CM can be n-hop. Depending on the distance the multi-hop can be 2-hop, 3-hop, etc. This distance is predefined by a maximum hop count. Multi-hop effectively expands the coverage of a cluster, which can lead to less change in CH and ultimately stable clusters.

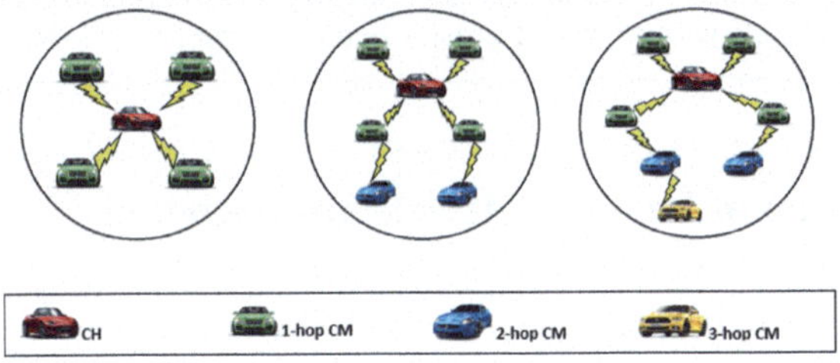

4 Cluster Formation

Different cluster formation procedures are available in the literature for cluster formation. The important phases for cluster formation in VANET includes:

Setup phase—In this phase advertisement messages are sent by nodes for the selection of primary CH, CM and then transmission of regular data packets will take place between them. There can be few strategies included in between the advertisement message transmission and CH selection so as to design a stable cluster.

Cluster maintenance phase—During this phase selection of secondary CH, stable cluster merging, re–clustering, and cluster splitting takes place.

There are a number of issues which need to be addressed to have a global view of clustering in VANETs:

- What are the parameters that decide the role of cluster nodes?
- Which device will be selected as a cluster head?
- Are the selected CH evenly distributed.
- Which model is suitable for highly dense networks-single-hop or multi-hop?
- Is there any requirement of cluster merging?
- Who will initiate cluster merging?
- Cluster splitting is required and when it is required?

4.1 Stability of Clusters

The most important task while designing a cluster-based distributed model is maintaining the stability of the highly dynamic VANET devices [3–6]. The commonly used metrics for measuring the stability of the clusters are cluster head duration, cluster member duration, number of cluster head change [1, 2, 7–9].

5 Comparison of Various CH Selection Schemes

There are a number of research articles available that have given the classification of clustering algorithms used in VANET. But in this paper, the cluster-head selection algorithms for VANET are being discussed. The classification of CH selection algorithms is done. A comparison table for each category of selection is given, which compares the various algorithms in that category on the basis of parameters used for CH selection, stability of the cluster, clustering logic used, network density under consideration, network model (single-hop/multi-hop) used and performance.

5.1 Weight-Based Cluster Head Selection Algorithms

In this category of algorithms, the CH is selected on the basis of weights being assigned to each of the vehicles under consideration on the basis of various parameters like distance, velocity, speed, direction, mobility, etc. The vehicles can be decided as CH by using the highest weight or lowest weight as a decision factor. In [2], a double-head clustering algorithm is proposed where a subordinate node is selected by the CH. The CH is selected on the basis of position, speed, SNR, link expiration time. An enhanced weight-based clustering algorithm is given in [10], where the CH is the node with the highest weight calculated on the basis of speed, distance and connectivity level. In [11], an algorithm is proposed which calculates the total force of a node as based on that decides the CH. The total force is calculated as a sum of force along X-axis and Y-axis. Given below is the table of comparison of various such algorithms available in the literature. Most of the CH selection algorithms fall under this category.

Proposed work	Parameters used for CH selection	Stability	Clustering logic	Traffic Scenario	Network model	Overall performance
Alsuhi et al. [2]	Popularity, relative position, relative speed, average signal-to-noise ratio, average link expiration time	High	Weight based on the parameters given in column 2 here	Highway and urban scenario	Single-hop	Increased cluster stability and efficiency
Tambawal et al. [10]	Node connectivity level, mean speed, mean distance	High	Weight based on the 3 parameters given in column 2 here	Highway scenario	Single-hop	High stability, reduced delay, use of secondary CH
Zhang and El-Sayed [5]	Low aggregate mobility as a function of relative mobility	Low	Packet transmission delay used to calculate relative mobility. Low aggregate mobility	Manhattan mobility scenario	Multi-hop	Efficiency improved as compared to non-clustered networks
Basagni [12]	Mobility, size of the network	Medium	Weight based on different parameters	Dynamic network topology	Single-hop	Efficient partitioning of the nodes of an ad hoc network

(continued)

(continued)

Proposed work	Parameters used for CH selection	Stability	Clustering logic	Traffic Scenario	Network model	Overall performance
Cheng and Huang [13]	Relative mobility	High	Weight based calculation of relative mobility on the basis of speed, relative position, maximum acceleration difference	Highway scenario	Single-hop	High stability and low packet loss rate
Hadded et al. [14]	Direction, ID based on average distance, average speed, number of neighbors	High	ID-based clustering, where ID is calculated using the weight function on three parameters defined	Urban highway scenario	Single-hop	High stability as compared to previous WCA, LID, HD algorithms
Daeinabi et al. [15]	Direction, distrust value of a vehicle based on blacklist/white list, relative position for entropy calculation	Medium	Minimum weighted sum vehicle selected as CH. The priority of the vehicle is calculated on the basis of distrust vale which helps in deciding CH	Highway scenario	Single-hop.	Improved stability, reduced communication overhead, increased network connectivity
Maglaras and Katsaros [11]	Position, relative mobility, relative force between nodes, distance	Medium	Total Force decides the CH. Node having the highest force is elected as CH	Random highway scenario	Multi-hop	Lesser clusters than Lowest-ID and ultimately high stability as compared to Lowest-ID and LPG

(continued)

(continued)

Proposed work	Parameters used for CH selection	Stability	Clustering logic	Traffic Scenario	Network model	Overall performance
Lo et al. [16]	Speed, distance	Medium	Relative Position and Mobility (RPM) calculated for selection of CH using weighted factor. Smallest RPM node is CH	City street map used.	Single-hop	Stable clusters with long life when compared with LID, LCC
Oubabas et al. [17]	Relative speed, relative acceleration, relative distance	High	Weight-based score calculation using degree and mobility similarity for CH calculation	–	Single-hop	Stable and trusted clusters, Better vehicle cooperation, low overhead security when compared to VMCA

5.2 *Priority-Based CH Selection Algorithms*

Under this category comes the algorithms which used the priority of the vehicles as the means for CH selection. In [9], an algorithm is proposed which is a passive clustering approach for clustering during the route discovery process. The priority of the node is determined based on multi-metric election strategy based on various metrics like node degree, transmission count, link lifetime. In [1, 7] other priority-based clustering technique is proposed, where the metrics like link lifetime, velocity, distance is used for prioritizing the node for CH selection.

Proposed work	Parameters used for CH selection	Stability	Clustering logic	Traffic scenario	Network model	Overall Performance
Wang and Lin [9]	Node degree, expected transmission count, link lifetime.	Medium	Passive clustering. Priority based on weighted calculation of metrics	One way multi-lane highway scenario	Single-hop	High packet delivery ratio, high throughput due to stable, reliable clusters
Zhang et al. [1]	Link lifetime, expected transmission counts, node following degree based on position and velocity.	High	Passive clustering. Priority-based node following strategy is used	Two-way road scenario	Multi-hop	Improved reliability, stability robustness
Ji et al. [7]	Link lifetime based on relative velocity and distance	Medium	Conditional-probability-based calculation of Link reliability based on velocity of vehicles	Urban scenario	Single-hop	Longer CH CM duration, low rate of head change, low overhead
Khakpou et al. [18]	Location, velocity, distance, direction	Medium	Tracking failure probability(TFP) is calculated to find the CH. The node with the lowest TFP is selected as CH	–	–	Increased stability, Improved cluster performance for target tracking
Rawashdeh and Mahmud [19]	Mean position, mean distance, standard deviation, mean speed	Medium	Priority of nodes is calculated based on the mobility information of the neighborhood	5 lanes per direction highway	Multi-hop	Increased stability of global network topology, reduced cluster creation rate

5.3 *Fuzzy Logic Based CH Selection*

In this category of CH selection algorithms, fuzzy logic based on multiple attribute decision is used for selection of CH [20]. In [21], a fuzzy logic based algorithm is proposed for improving the performance of the vehicular ad hoc networks. They have optimized the CH selection procedure by using speed, acceleration, distance, and direction. In this algorithm, CH is elected and re-elected in a distributed manner. The algorithm is highly stable due to its adaptability to driver's behavior and its learning process to predict the future speed using a fuzzy logic interference system.

Proposed work	Parameters used for CH selection	Stability	Clustering logic	Traffic Scenario	Network model	Overall Performance
Calhan [21]	Speed, acceleration, distance, direction.	High	Multi attribute based fuzzy logic	Two-way multilane highway	Single-hop	Increased stability, less communication, and coordination overhead
Harrabi et al. [22]	ID based	Low	Optimization agent is selected by the CH using which routing	–	Single-hop	Performs well compared to MAODV routing in terms of throughput, end-to-end delay, packet delivery ratio

5.4 *Nature-inspired CH selection*

In this category of algorithms, the CH selection is done using the strategy of various nature-inspired algorithms like ant colony optimization, particle swarm optimization, bee, etc. in [23] an ant colony optimization algorithm is proposed which selects the CH using the probability of all the nodes available for clustering.

Proposed work	Parameters used for CH selection	Stability	Clustering logic	Traffic Scenario	Network model	Overall Performance
[23]	Transmission range, load balance factor, direction, speed.	Optimal	Ant colony optimization (probability of all available nodes)	Varying road segments and transmission range	–	Best when compared to other swarm optimization algorithm
Hadded et al. [24]	Direction, speed, distance, neighboring vehicles.	Optimal	Election function is decided on the basis of avg distance, avg speed, 1-hop same direction vehicles. Minimum weight is selected as CH	Real-highway scenario	Single-hop	Non dominated sorting genetic algorithm is used for improving the result of MOPSO & MODE in terms of spacing, spread, ratio of non-dominated solutions

5.5 Other CH selection algorithms

The CH selection algorithms which are not covered under the above-defined specific category are defined in this category. In these algorithms, the CH is selected using the ID of the vehicles or some other specific parameter or combination of parameters [20]. VMaSC [4] is the first work to simulate the multi-hop stable clustering algorithm. In this work, a novel mobility management metric is proposed for the highly dynamic scenario. A multi-criteria algorithm is proposed in [25], which corresponds to road traffic better. Here the clusters are designed based on SNR, connectivity level, node reputation and relative node position and its prediction in the future. In [6] a neighborhood follow strategy is used for distributed multi-hop clustering. It is also a kind of passive CH selection algorithm where CH is selected passively based on the following relationship between the two one-hop nodes.

Proposed work	Parameters used for CH selection	Stability	Clustering logic	Traffic scenario	Network model	Overall performance
Kuklinski et al. [25]	Link quality, relative node position, connectivity level	Medium	Multi-criteria clustering algorithm based on clustered and unclustered node concept	Urban scenario	Single-hop	Stability has increased significantly when compared to the previously defined algorithms
Chen et al. [6]	Average relative mobility, number of followers	High	Neighborhood follows the relationship between vehicles	Highly dynamic	Multi-hop	Improved stability
Maslekar et al. [26]	Direction	Medium	Direction based clustering with cluster head switching	Car following with overtaking	One-hop	Better stability and hence better accuracy in density estimation
Zhang et al. [3]	Highest connectivity in terms of signal strength, vehicle mobility and host ID	Medium	Average link expiration time of nodes is compared to find a winning metric. The vehicle with the highest winning metric is CH	Random walk mobility	Multi-hop	Improved network topology scalability, stability, reduced latency and overhead
Ucar, and Ergen [4]	Relative speed	Medium	Least mobility calculation based on the speed of same direction vehicles	Two-lane and two-way road	Multi-hop	High CH CM duration and less change in cluster head
Maglaras and Katsaros [11]	Relative velocity, power loss, position	Medium	Capability of the node to be a CH is decided using RVM and PLM calculated during the CH selection. Stay time prediction is also calculated	Bidirectional multilane city road scenario	–	Compared with MOBIC & LOSC, this scheme gives more stable and load-balanced clusters. Security of clusters in also improved

(continued)

(continued)

Proposed work	Parameters used for CH selection	Stability	Clustering logic	Traffic scenario	Network model	Overall performance
Huo et al. [27]	Velocity, position, speed	High	Selection index of vehicles is calculated using the mean relative distance, relative mobility, relative mean speed which decides CH	Bidirectional multilane highway scenario	Multi-hop	Compared to VMSaC, highest-degree shows a higher stability
Regin and Menakadevi [28]	Reliability, average speed, time to leave the road intersection, stability, vehicle connectivity level	Medium	Vehicle with the least speed and which is in middle of the cluster is selected as CH	Single lane dynamic traffic	Single-hop	Reduced network congestion, stable cluster when compared to VMCA

6 Discussion and Results

Clustering is the technique which has been used in ad hoc networks for a number of different applications like routing, data dissemination, tracking of target, etc., the most important characteristic required for any cluster is stability which is dependent on the CH of the cluster since CH is the coordinator of the cluster which controls all other nodes. Above we have discussed different categories of CH selection algorithms based on which we have found the following points:

1. Most of the CH selection algorithms have used speed, distance, velocity as the major parameters for deciding a CH.
2. CH selection is done using various techniques like priority based selection, weight-based selection, fuzzy logic based selection,
3. A stable CH can lead to a stable cluster and ultimately improves the overall stability and reliability of the network.
4. Techniques of cluster maintenance like cluster merging, cluster splitting, use of subordinate/secondary CH is required for improving the cluster stability.

7 Applications

The applications of clustering include the following:

1. Routing in VANET and selecting a standard routing protocol.
2. Improved efficiency of VANET.
3. Target tracking in vehicular ad hoc networks.
4. Load balancing and reduced communication overhead.
5. Efficient message dissemination in various VANET applications.
6. Improved scalability of the large dynamic vehicular ad hoc networks.

Since CH selection is the primary concern for any clustering process. We have discussed various cluster head selection algorithms in the above section.

8 Open Research Issues

No doubt clustering has been proved to be beneficial for the number of issues in VANET discussed above there are few issues that still need the attention of the research community. These issues are presented below [12]:

1. The use of clustering for handover management in VANET is still very less explored.
2. How stable and efficient clusters can be designed for a specific application requirement is the biggest challenge even today.
3. Multi-level clustering is still an unexplored area. Whether one single CH is sufficient or use of double CH is an advantage.
4. What can be the best technique for CH selection and CM selection? What can be the different attributes which help in selecting a stable and efficient CH.
5. The applicability of clustering for providing QoS and a secure VANET environment requires attention.
6. Multi-homing clustering techniques need to be researched more as these can prove useful for increasing stability.

References

1. D. Zhang, H. Ge, T. Zhang, Y.Y. Cui, X. Liu, G. Mao, New multi-hop clustering algorithm for vehicular ad hoc networks. IEEE Trans. Intell. Transp. Syst. (2018)
2. G.H. Alsuhi, A. Khattab, Y.A. Fahmy, Double-head clustering for resilient VANETs, Hindawi Wirel. Commun. Mob. Comput. 2917238 (2019)
3. Z. Zhang, A. Boukerche, R. Pazzi, A novel multi-hop clustering scheme for vehicular ad-hoc networks, in *Paper presented at the Proceedings of the 9th ACM International Symposium on Mobility Management and Wireless Access*, 2011, pp. 19–26

4. S. Ucar, S.C. Ergen, O. Ozkasap, VMaSC: vehicular Multi-hop algorithm for stable clustering in vehicular ad hoc networks, in *IEEE Wireless Communications and Networking Conference: networks*, pp. 2381–2386 (2013)
5. L. Zhang, H. El-Sayed, A novel cluster-based protocol for topology discovery in vehicular ad-hoc network. Procedia Comput. Sci. 525–534 (2012)
6. Y. Chen, M. Fang, S. Shi, W. Guo, X. Zheng, Distributed multihop clustering algorithm for VANETs based on neighborhood follow. J. Wirel. Commun. Netw. 1–12 (2015)
7. X. Ji, H. Yu, G. Fan, H. Sun, L. Chen, efficient and reliable cluster-based transmission for vehicular ad hoc networks. Hindawi Mob. Inf. Syst. 9826782 (2018)
8. P. Thakur, A. Ganpati, Survey on handover techniques in VANET. Int. J. Comput. Sci. Eng. **7**(6), 236–250 (2019)
9. S.S. Wang, Y.-S. Lin, PassCAR: a passive clustering aided routing protocol for vehicular ad hoc networks. Comput. Commun. 170–179 (2013)
10. A.B. Tambawal et al., Enhanced weight-based clustering algorithm to provide reliable delivery for VANET safety applications. PLOS ONE 1–19 (2019)
11. L.A. Maglaras, D. Katsaros, Distributed clustering in vehicular ad hoc networks. in *International Workshop on Vehicular Communications and Networking*, pp. 593–599 (2012)
12. S. Basagni, Distributed clustering for ad hoc networks, in *Proceedings of International Symposium on Parallel Architectures, Algorithms and Networks*, pp. 310–315 (1999)
13. X. Cheng, B. Huang, A center-based secure and stable clustering algorithm for VANETs on highways. Wirel. Commun. Mob. Comput. (2019)
14. M. Hadded, P. Muhlethaler, R. Zagrouba, A. Laouiti, L.A. Saidane, Using road IDs to enhance clustering in vehicular ad hoc networks, in *IEEE*, pp. 285–290 (2015)
15. A. Daeinabi, A.G.P. Rahbar, A. Khdemzadeh, VWCA: an efficient clustering algorithm in vehicular ad hoc networks. J. Netw. Comput. Appl. 207–222 (2011)
16. S.-C. Lo, Y.-J. Lin, J.-S. Gao, A multi-head clustering algorithm in vehicular ad hoc networks. Int. J. Comput. Theory Eng, pp. 242–247 (2013)
17. S. Oubabas, R. Aoudjit, J.J.P.C. Rodrigues, S. Talbi, Secure and stable vehicular ad hoc network clustering algorithm based on hybrid mobility similarities and trust management scheme. Veh. Commun. (2018)
18. S. Khakpour, R.W. Pazzi, K. El-Khatib, A distributed clustering algorithm for target tracking in vehicular ad-hoc networks, in *DIVANet'13*. (Barcelona, Spain), pp. 145–152 4–8 Nov 2013
19. Z.Y. Rawashdeh, S.M. Mahmud, A novel algorithm to form stable clusters in vehicular ad hoc networks on highways. Wirel. Commun. Netw. (2012)
20. K.A. Hafeez, L. Zhao, Z. Liao, B.N.-W. Ma, A fuzzy-logic-based cluster head selection algorithm in VANETs, in *IEEE ICC*, pp. 203–207 (2012)
21. A. Calhan, A fuzzy logic based clustering strategy for improving vehicular ad-hoc network performance. Sadhana 351–367 (2015)
22. S. Harrabi, I.B. Jaafar, K. Ghedira, A novel clustering algorithm based on agent technology for VANET. Netw Protoc Algorithms (2016)
23. F. Aadil, S. Khan, K.B. Bajwa, M.F. Khan, A. Ali, Intelligent clustering in vehicular ad hoc networks. KSII Trans. Internet Inf. Syst. **10**(8), 3512–3528 (2016)
24. M. Hadded, R. Zagrouba, A. Laouiti, P. Muhlethaler, L.A. Saidane, A multi-objective genetic algorithm-based adaptive weighted clustering protocol in VANET, in *IEEE* pp. 994–1002 (2015)
25. S. Kuklinski et al., Density based clustering algorithm for vehicular ad-hoc networks. Int. J. Internet Protoc. Technol. **4**(3), 149–157 (2009)
26. N. Maslekar, M. Boussedjra, J. Mouzna, A stable clustering algorithm for efficiency applications in VANETs, in *IEEE*, pp. 1188–1893 (2011)
27. Y. Huo, Y. Liu, L. Ma, X. Cheng, T. Jing, An enhanced low overhead and stable clustering scheme for cross roads in VANETs. Wirel. Commun Netw. (2016)
28. R. Regin, T. Menakadevi, Dynamic clustering mechanism to avoid congestion control in vehicular ad hoc networks based on node density. Wirel. Pers. Commun. (2019)

29. R.S. Bali, N. Kumar, J.J Rodrigues, Clustering in vehicular ad hoc networks: taxonomy, challenges and solutions. Veh. Commun. 5 May 2014
30. G. Wolny, Modified DMAC clustering algorithm for VANETs, in *Systems and Networks Communications*, pp. 268–273 (2008)

Health Monitoring Multifunction Band Using IOT

Pooja Nagpal and Sarika Chaudhary

Abstract In the era of IOT, where all devices are controlled remotely, Medical and health area also needs attention. We have created a wearable device that is health monitoring multifunction band. Although we already have many smart bands in market which have multiple functionalities in it. But, at the same time, it cannot be denied that they are too costly and every functions of it are not utilized optimally. Keeping in mind, the medical use of the smart band, we came up with an idea to make a band with minimal features keeping all necessary features like pulse rate monitor, temperature sensor, etc. for medical purpose. This band will be serving the medical institutes and will help the doctors and medical practitioners to monitor their patient's health. We have coined the band as "Health Monitoring Multifunction Band". The band has a "Pulse Sensor" to monitor heartbeat. It also has a "Temperature Sensor" to record the body temperature. The device includes an "OLED" to show the recorded data in real time. Since, we have integrated only the necessary features so the overall price of the band will be very less in comparison to what we already have in the market.

Keywords Wrist band · Health monitoring band · Multifunction band · Medical institutes · Smart device

1 Introduction

In research and health industry, wearable health monitoring devices have attracted a great attention during the last decade [1]. Furthermore, development efforts are adjoining in this field [2, 3]. In today's world, the health-care cost is increasing day by day and rate of growth of population is also increasing, there has been a need to monitor a patient's health status every time even in the absence of a nurse [4]. All of us know that factors like temperature of body, Pulse rate, Heartbeat, and others are primary and important factors to monitor an individual's health. Even our Doctors

P. Nagpal (✉) · S. Chaudhary
Amity University Gurugram, Haryana, Gurugram, India
e-mail: pbnagpal@ggn.amity.edu

M. Dutta et al. (eds.), *Proceedings of International Conference on IoT Inclusive Life (ICIIL 2019), NITTTR Chandigarh, India*, Lecture Notes in Networks and Systems 116,
https://doi.org/10.1007/978-981-15-3020-3_14

and Nurses consider these for the pre-checkup. Truth to be told, nowadays, we are facing a lack in the number of nurses per individual patients. This is the scenario which sometimes leads to the uncared death of unfortunate patients. Nurses keep track of the essential health factors and report to the concerned Doctor for any severe fluctuations. But when a nurse is not present near a patient or is busy dealing others, the readings are not noted or the patient remain unassisted which sometimes lead to the death of the patient. To overcome this, we came up with an idea of developing a Multifunction Health Monitoring Band. This band will be capable of keeping track of the primary health factors. It will be connected to a mobile application. One can see the fluctuation in the readings through the mobile application in real time. The data recorded by the band will be synced to the database. The mobile application will be primarily available on the android platform based smart phones. Apart from this, the multifunction band has other features like an LCD integrated on the top to show the readings and a dedicated button for an emergency alarming system. When in the emergency, the patient presses the button, the concerned nurse will get a ring on her mobile device, and she can rush to the patient immediately. Proposed and implemented model is used to collect data information which will act as data set through health status monitors that would comprise patient's heart rate, blood pressure, and ECG and sends an emergency alert along with current situation and detail medical history to patient's doctor or concern health-care agent. The project can be of great use for the hospitals where the ratio of nurses and patients is unequal, i.e., nurses are less in comparison to that of patients.

2 Related Work

Pantelopoulos and Bourbakis [5] give information about the current existing research and development of wireless biosensors system for effective health-care monitoring. This system consists of wireless sensors using ZigBee wireless technology and ultra-low power technology This system also supports wireless communication for wireless body area networks (WBANs), in which it adapts the individual physiological conditions using artificial neural network. It also uses certain ranges in between 2360 and 2400 MHz band for medical BAN services to avoid interferences from wireless technologies, where these wearable systems must be reliable, multifunctional, and easy to use for the patients monitoring. It must be applicable for real-time usage. Milenkovi et al. [6] spoke about the close monitoring of health-care system, in providing feedback and alert medical person to maintain optimal health-care monitoring. In our system, we have assembled physical sensors, embedded microcontrollers, and radio interfaces on a single chip called as wearable wireless body/personal area network. In addition, it is very cheaper in cost and portable to carry. It also provides an immediate feedback to the user about the health status and updates the medical records in the system. The system supports continuous health monitoring and provides benefits to patient, where there is an improvement needed on quality of service

(QoS) for a wireless communication, reliability of sensors, security, and standardization of interfaces and interoperability. Kumar et al. [7] spoke about the very wide usage of wireless sensor network for remote monitoring of patients, storage of data in cloud environment, and also the patient data are taken, where it is transferred through a wireless network without any interrupt. So that we can monitor the accumulated data from the patient using some smart applications with a comparison to the existing information in the system. Alert Short Message Service is sent to the doctor and to the patient caretaker. To provide the security and privacy to patient data and mobile computing, there is a need for health-care services with high quality and low cost wearable device which can collect data and show readings. Nithin et al. [8] spoke about the sensors that will record not only the current day's data but also the previous days. Data provided by the sensors are longitudinally rich and helpful to the doctor to give precautions. WBAN consists of wearable sensors, which measures various physiological parameters. Sensor transmits the gathered data to a gateway server. Kocabas et al. [9] said that the digital health is the next big revolution when the internet was invented. Here, remote data are widely spread in the system, where system consists of two super layers named the front end and the back end. Here, front end acts as an interface between the patient and the system. Furthermore, back end acts as an interface between the system and the doctor. Here, privacy and the security part have been included, in which it also identifies business opportunity arising from the system during sharing and analytics. Page et al. [10] said that there is no proper method to predict the cardiovascular diseases, chronic heart failure; therefore, there is a need for comprehensive monitoring system, which is required for effective clinical diagnostics. In addition, this system allows continuous monitoring of the patient in which it gives feedback by an automatic alarm for patient's long-term status report. This system is very helpful for patients with high risk of life, like an ECG monitoring, and also in this, a novel virtualization mechanism allows the doctor to monitor the real-time multiple patients. Thus, it requires a need to implement sending the data to a proper database and should have an automatic update from the live data itself.

We can make early prediction and prevention based on data from the system so that it is very helpful for General Hospital Wards. In addition, these predictive systems are being developed for patients for early predictions. So, that these lead to the introduction of bucketing technique which is used for capturing vital signals changes in the system.

3 Proposed Method

The reliance of health care on IoT is expanding step by step to enhance access to care, intensify the nature of consideration, and in particular, to restrict the expense of care. It examines the uses of IoT in customized health care to accomplish astounding healthcare at least expense. IoT based health-care systems play a key role in the development of medical information systems. To upgrade the health-care system tracking, tracing and monitoring of patients are fundamental. However because of

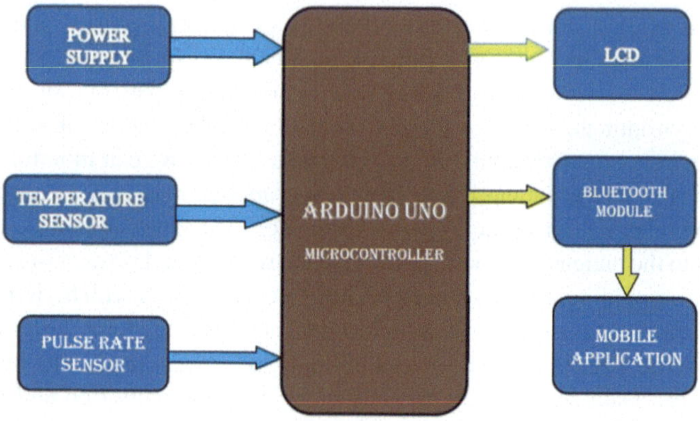

Fig. 1 Flow diagram for multifunction health monitoring band

lacking health-care circumstances, i.e, therapeutic advances, the accessible devices can't meet a similar precision. Figure 1 shows the flow graph of Multifunction Health Monitoring band.

First sensor incorporated in the band is a Heartbeat sensor of 50–60 Hz frequency and the second sensor is temperature sensor for maintaining the record of the patient's overall health. Heartbeat sensor is designed to give a digital output of heartbeat when the band is tied around the wrist. When the heartbeat detector is working, the data will be recorded in the database. All the readings will be synced to the database in real time. The recorded readings can be studied and analyzed through a mobile application. There will also be a backup of data which will help the doctor to study and examine the basis of any pattern recorded. The emergency button will help the patient to call the nurse/doctor by just on a single press of it. It will invoke an alarm and the concerned nurse and doctor will get an emergency notification. Apart from this, we have integrated a dedicated button for home automation system which primarily will be controlling the lights in the ward. Figure 2 shows the circuit diagram of Monitoring Band.

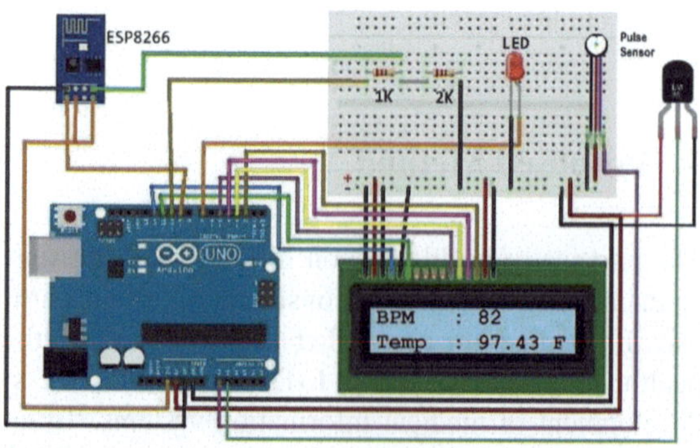

Fig. 2 Circuit diagram of multifunction health monitoring band

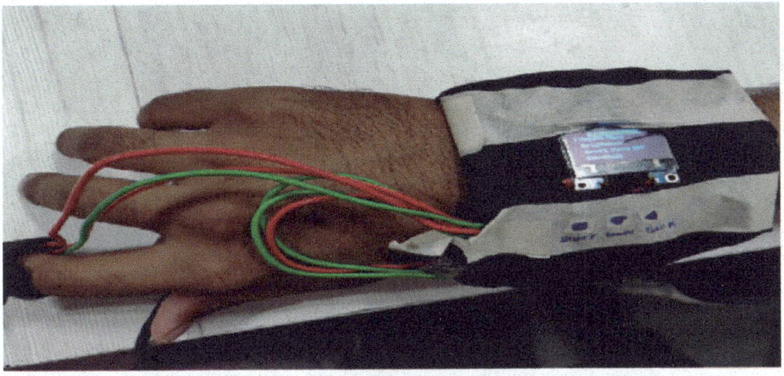

Fig. 3 Patient wearing multifunctioning health band

Fig. 4 Display screen of band

4 Implementation

In this section, we shall show picture of real multifunctional health band. In Fig. 3, patient is wearing health band.

Further in Fig. 4, display is shown separately without band.

In the next picture, Circuit of inner band is shown (Fig. 5).

Figure 6 shows the band's screen with real-time values of temperature and ECG.

5 Result and Analysis

After completion of all connections of circuits and display monitor initially, slightly false reading of temperature were captured but after recalibrating, we received almost precise value. Whereas, the pulse monitor was fairly accurate and was able to receive almost the same reading in comparison to the ECG that we have in the hospitals. We

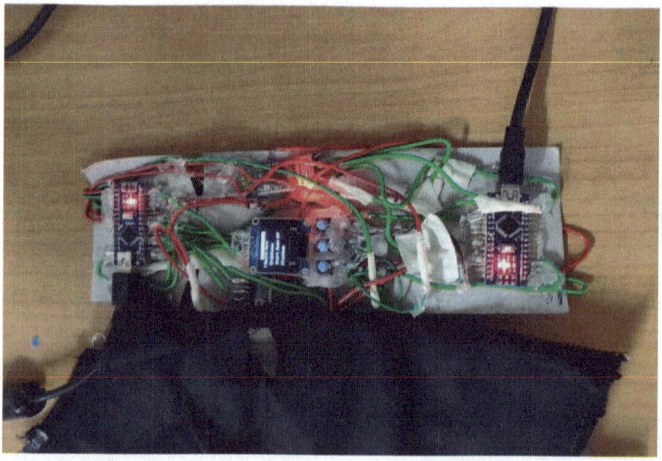

Fig. 5 Inner circuit of band

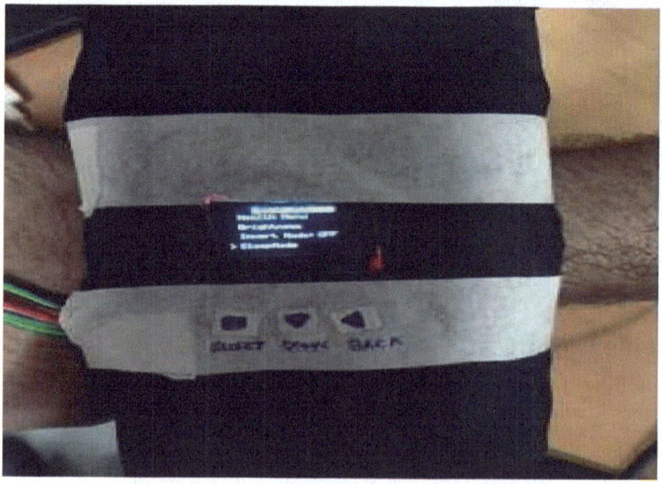

Fig. 6 Band's Screen showing values of temperature and ECG

also added fail-safes to prevent the project from showing false readings. The project was a success and met our goal.

6 Conclusion

From the research and deep understanding of wearable devices for health-care industries, it can be said that a lot more time and lives can be saved by using the technology in the medical sector. The Multifunction Health Monitoring band can be of great use where there is a lack in the proportion of nurses and the patients. In future, the device can be made available for the patients who are not able to move on their own. The device functionalities can be extended with the integration of home automation facilities for more purposes. Secondly, we can shorten the size of band. Because of some resource issues, we could not make it compact.

References

1. L. Gatzoulis, I. Iakovidis, Wearable and portable eHealth systems. IEEE Eng. Med. Biol. Mag. **26**(5), 51–56 (2007)
2. A. Lymperis, A. Dittmar, Advanced wearable health systems and applications, research and development efforts in the European union. IEEE Eng. Med. Biol. Mag. **26**(3), 29–33 (2007)
3. G. Troster, The agenda of wearable healthcare, in *IMIA Yearbook of Medical Informatics*. (Stuttgart, Germany, Schattauer, 2005), pp. 125–138
4. Y. Hao, R. Foster, Wireless body sensor networks for health monitoring applications. Phys. Meas. **29**, R27–R56 (2008)
5. A. Pantelopoulos, N. Bourbakis, A survey on wearable sensor-based systems for health monitoring and prognosis. IEEE Trans. Sys. Man Cybern. Part C Appl. Rev. **40**, 1–2 (2010)
6. A. Milenkovi, C. Otto, E. Jovanov, Wireless sensor networks for personal health monitoring: issues and an implementation. Sci. Direct Comput. Commun. **29**, 2521–33 (2006)
7. P. Kumar, S.V. Prasad, A. Patak, Design and implementation of M-Health. system by using cloud computing. Int. J. Innovat. Res. Sci. Eng. Technol. **5**, 2319–8753 (2016)
8. N.P. Jain, P.N. Jain, T.P. Agarkar, An embedded, GSM based, multi parameter, real time patient monitoring system and control—an implementation for ICU TELEHEALTH—in IoT environment department of ECE, Dr. AIT 2016–17 Page | 66 Patients, in *IEEE World Congress on Information and Communication Technologies*, pp. 987–992 (2012)
9. O. Kocabas, T. Soyata, J.P. Couderc, M. Aktas, J. Xia, M. Huang, Assessment of cloud-based health monitoring using homomorphic encryption, in *Proceedings of the 31st IEEE International Conference on Computer Design (ICCD)*. (Ashville, VA, USA, 2013) pp. 443–6
10. A. Page, T. Soyata, J.P. Couderc, M. Aktas, B. Kantarci, S. Andreescu, Visualization of health monitoring data acquired from distributed sensors for multiple patients, in *2015 IEEE Global Communications Conference (GLOBECOM)*. (San Diego, CA, 2015), pp. 1–7

Particulate Matter Assessment in Association with Temperature and Humidity: An Experimental Study on Residential Environment

Jagriti Saini, Maitreyee Dutta and Gonçalo Marques

Abstract Air pollution refers to the contamination of the breathing environment with the materials that pose a harmful impact on human health and natural operations of the ecosystem as well. Stats reveal that a large part of the population spends 80–90% of their time indoors; hence, are at more risk of serious health consequences such as respiratory illness. To improve the quality of life and overall well-being, it is important to measure the elements associated with indoor air pollution and find ways to reduce their impact on the living environment. This paper provides an experimental study on the status of indoor air pollution at an apartment and describes the relationship between particulate matter (PM_{10}, $PM_{2.5}$) in association with temperature and humidity levels. The experiment was conducted using Arduino Uno, ESP8266 Wi-Fi module, PM sensor, and DHT11 on the hardware side; whereas the data was collected online on ThingSpeak platform. The variations in the concentration levels were observed by varying apartment living conditions and domestic activities. These experiments indicate very poor air quality index, especially during cooking hours and while using some electric appliances. This study motivates the policymakers, pollution control teams, and healthcare professionals to find relevant ventilation mechanisms for traditional and modern homes to avoid poor health consequences.

Keywords Indoor air quality · Indoor air pollution · Internet of things · Particulate matter · Residential environment

J. Saini (✉) · M. Dutta
National Institute of Technical Teacher's Training and Research, Sec-26, Chandigarh 160019, India
e-mail: jagritis1327@gmail.com

M. Dutta
e-mail: d_maitreyee@yahoo.co.in

G. Marques
Instituto de Telecomunicações, Universidade da Beira Interior, Covilhã, Portugal
e-mail: goncalosantosmarques@gmail.com

Instituto Politécnico da Guarda, Guarda, Portugal

M. Dutta et al. (eds.), *Proceedings of International Conference on IoT Inclusive Life (ICIIL 2019), NITTTR Chandigarh, India*, Lecture Notes in Networks and Systems 116,
https://doi.org/10.1007/978-981-15-3020-3_15

1 Introduction

Indoor Air Pollution (IAP) is primarily defined as the existence of some harmful pollutants inside buildings; the list includes chemical, physical, and biological factors along with inorganic compounds, volatile organic compounds, and particulate matter as well [1]. When people breath in such a polluted environment, it leaves a negative impact on their overall health. With such consequences, Indoor Air Quality (IAQ) has become a matter of interest for researchers these days. As per the reports presented by World Health Organization (WHO), residents in the urban areas spend 90% of their routine time indoors; around 70% at the office and 20% at home [2]. Environmental Protection Agency (EPA) reveals that the level of indoor air pollutants nowadays has increased by 2–5 times as compared to outdoor pollutants [3]. IAP is the main cause behind increasing Global Burden of Diseases and leading risks of morbidity and mortality in various developing countries [4].

Particulate Matters (PM) play a major role in IAQ and as a result, is directly related to human health and well-being. Epidemiological studies highlighted a strong relation between IAP and cardiovascular, pulmonary diseases, respiratory health, and it may also lead to symptoms of cancer as well [4]. Some of the most common sources of IAP are poor cooking conditions, tobacco smoking, wood-burning, and kerosene heating. Other than this, some human activities such as dusting, showering, vacuuming, operations of electric motors, and humidifiers also contribute to poor IAQ [5]. The impact of outdoor elements is reported to be 79–83% for $PM_{1.0}$ (particles $\leq 1.0\ \mu m$), 67–76% for $PM_{2.5}$ (particles $\leq 2.5\ \mu m$), and 32–45% for PM_{10} (particles $\leq 10\ \mu m$) [6]. The concentration of PM levels in the indoor environment depends on several factors such as type of human activities, building structure, meteorological factors, and opening and closing of doors. It is also affected by temperature and humidity levels in the premises; hence, while studying the impact of PM levels on human health, it is important to consider temperature and humidity as well [7].

Considering all the harmful impacts of IAP, it is important to find some potential techniques for IAQ management. Several researchers around the world have already proposed indoor air quality monitoring systems (IAQMS) to deal with this issue. Kapwata et al. [6] performed a study on IAQ by monitoring PM levels in the indoor and outdoor air at Rural Limpopo in Sound Africa. Argunhan et al. [1] carried out an experiment at University classrooms in Turkey and collected readings on essential IAQ parameters such as PM, radon, carbon dioxide, temperature, and humidity. Further, the results were examined and analyzed using SPSS 17 statistical program. Sivasankari et al. [8] proposed an Internet of Things (IoT)-based system for monitoring IAQ. They focused on smoke concentration, CO, NO_2, temperature, and humidity to evaluate environmental health. A Raspberry-Pi module was used to generate alarms whenever parameters crossed set thresholds for IAQ. Idrees et al. [9] proposed real-time IAQMS; the prototype for the module was designed using the Arduino microcontroller and IBM Watson IT platform. Authors during this study focused on eight different parameters, such as PM_{10}, $PM_{2.5}$, CO, NO_2, SO_2, O_3, temperature, and humidity. Benammar et al. [10] designed an IAQMS using WSN-based environment that focused on measurement of ambient temperature, relative humidity, chlorine, CO_2, CO, NO_2, SO_2, and O_3. The sensor data was analyzed through a web server, and seamless integration with mobile phones was also established. Several studies have been already proposed; however, few improvements are still needed to design a reliable system for IAQ management. The early systems were mainly based on WSN environment; however, recent developments focus on IoT technology. But most of these researchers either worked on simulated environments or within predefined laboratory settings; results were not applicable to real-time conditions. To address the problems associated with IAQ management, it is important to execute experiments in real-time settings. Moreover, researchers need to focus on the essential needs of rural as well as an urban lifestyle so that quality living environment can be created. The systems are desired to be cost-effective, power-efficient, and must ensure reliable results for real-time conditions.

There is a relevant need to design IAQ monitoring systems with mobile health concept. IoT technology has huge potential to contribute in this direction. It is possible to get instant updates about IAQ conditions via IoT sensor-based network, and the alerts can be directly sent via messages on mobile phones and emails. This paper presents an experimental study carried within a residential building at Chandigarh city in India during the summer season. The proposed monitoring system is based on IoT sensor network that was installed in a residential apartment to measure values of $PM_{2.5}$, PM_{10}, temperature, and humidity levels. The impact of domestic activities on these parameters was observed for one week from June 7, 2019 to June 14, 2019. The main objective of this study is to analyze the impact of human activities on IAQ levels; this information can be further utilized for carrying out valuable epidemiological studies in the future.

2 Materials and Methods

2.1 Data Collection and Sampling Site

Chandigarh is a highly populated Union Territory in the northern part of India with optimal weather conditions in all seasons. In this study, the apartment for data collection and sampling was selected in a residential building with three adult members in the family. It was a two-bedroom apartment with one living room, kitchen, and two bathing areas. The monitoring system was installed in the living room at a height of 2.4 m from the ground. The IAQ for four parameters $PM_{2.5}$, PM_{10}, temperature, and humidity were measured with a sampling period of 15 min, and the results were stored on ThingSpeak platform. The variations in these four parameters were monitored during usual domestic activities such as cooking, use of clothing iron, use of electronic room cooler, cleaning, and dusting activities with door open and closed conditions for 24 h a day. Flowchart for methodology is given in Fig. 1.

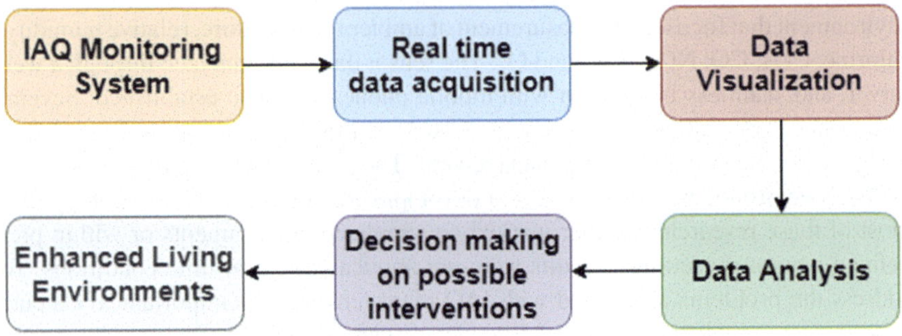

Fig. 1 Flowchart for methodology

2.2 *Monitoring System*

The monitoring system for this study was designed using Arduino Uno microcontroller and the sensors used for measurement were DHT11 and SDS011; where former one provides readings for temperature and humidity levels and later is for measuring $PM_{2.5}$ and PM_{10}. A 16 × 2 LCD module was also attached to the system to display monitored values that were updated every 15 min. ESP8266 Wi-Fi module was used to transfer data from the monitoring system to the ThingSpeak channel that was powered through a home Wi-Fi router. The architecture of the IAQMS is shown in Fig. 2. Four field values were assigned on the ThingSpeak channel, and the sampled data was collected in the form of graphs. The members living in the apartment were allowed to carry on normal life routines so that most accurate values can be obtained from the scenario.

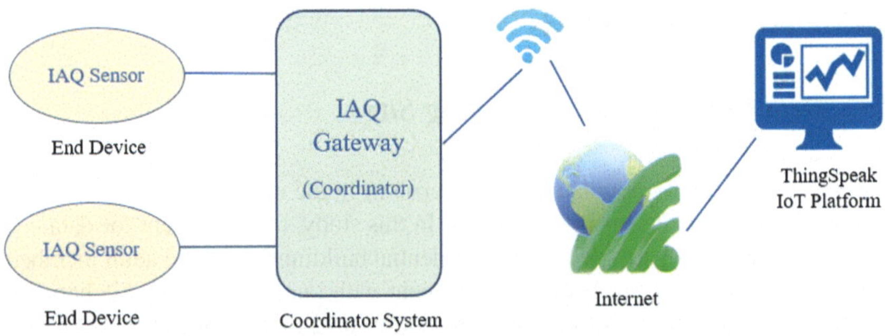

Fig. 2 IAQ monitoring system architecture

3 Results and Discussions

The test on IAQ was performed using the IAQMS as described in Sect. 2. The parameters values were recorded during various domestic activities throughout the day and night.

Chandigarh city is known for optimal weather conditions; however, in the year 2019; when this study was carried, the city achieved peak heat in the summer season. Air conditioners and air coolers are the main necessity of the people living indoors at home or office. This study is carried out in a residential environment where family members relied on air cooler to balance the summer heat. The average outdoor temperature during this study was 44 °C; however, it was reduced to the average value of 34.25 °C in the indoor environment. Average humidity during this 7-day monitoring period was observed to be 35.57 g/m^3.

Environment Pollution Agency (EPA) has defined standard threshold ranges for all parameters/pollutants associated with indoor and outdoor air. The EPA air quality index for $PM_{2.5}$ and PM_{10} is provided in Table 1.

As per the data obtained from proposed IAQMS, the average value of $PM_{2.5}$ for the considered residential environment was 27.02 μg m^{-3}; it belongs to the moderate levels as per EPA standards. Whereas for PM_{10}, the obtained readings show an average value of 71.57 μg m^{-3}, which is unhealthy and can lead to major health complications.

The highest peaks for the PM levels were obtained during cooking hours. However, the residents used cleaner fuels (LPG gas) for cooking needs; the IAQ was affected by a great extent. The sudden rise in PM levels during cooking hours is shown in Fig. 3. Other than this, the PM level rose to considerably higher values during ironing as well as while using a water heater in the premises. The instant rise in PM levels during ironing activity is shown in Fig. 4. Furthermore, the impact of using a water heater is also highlighted in Fig. 5. A moderate impact of gadgets such as mobile phones, tablets, and laptops was observed in the obtained readings.

Figures 2, 3, and 4 show the impact of different household activities on PM levels. When compared to the EPA standards, the $PM_{2.5}$ falls in the unhealthy range whereas PM_{10} raises to the very unhealthy range during cooking activities as well as at the

Table 1 EPA air quality index for $PM_{2.5}$ and PM_{10} [11]

Index values	Category	$PM_{2.5}$ (μg m^{-3})	PM_{10} (μg m^{-3})
0–50	Good	0–15.4	0–54
51–100	Moderate	15.5–40.4	15.5–40.4
101–150	Unhealthy for sensitive groups	40.5–65.4	40.5–65.4
151–200	Unhealthy	65.5–150.5	65.5–150.4
201–300	Very unhealthy	150.5–250.4	150.5–250.4

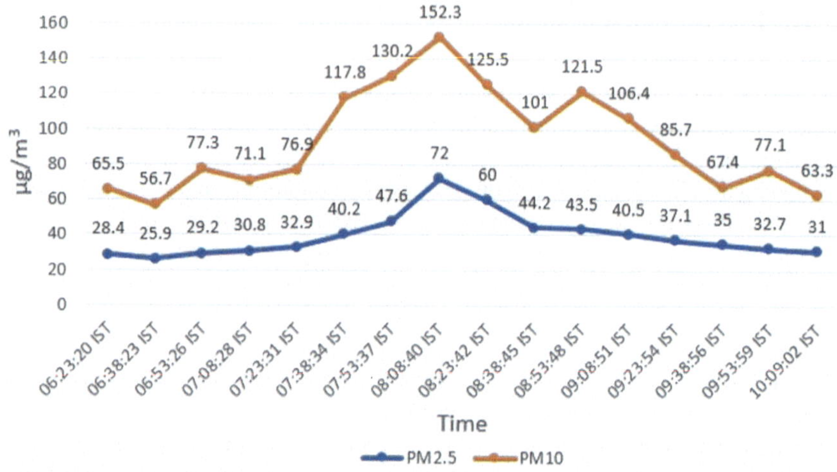

Fig. 3 Impact on PM levels during cooking hours

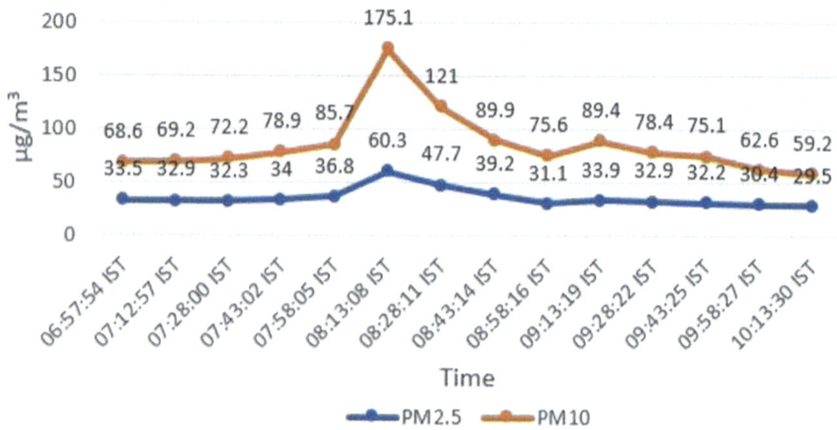

Fig. 4 Impact of ironing on PM levels

time of ironing activities. The status of $PM_{2.5}$ and PM_{10} is unhealthy of sensitive people and unhealthy, respectively, while using water heater in the premises. This pattern shows poor IAQ levels in the apartment and needs some immediate solution to avoid the associated health problems. This study is based on a residential building in the city area where residents make use of cleaner fuels (LPG) for cooking; the cases can be worse in case of rural areas where people are still dependent on biomass fuels. A study proposed in the year 2013 reveals that IAQ index changes as per the usage patterns in different locations [12]. The proposed IoT sensors-based IAQMS can be installed at numbers of locations to observe the changes. The ThingSpeak web portal-based updates can provide instant updates to the residents so that they

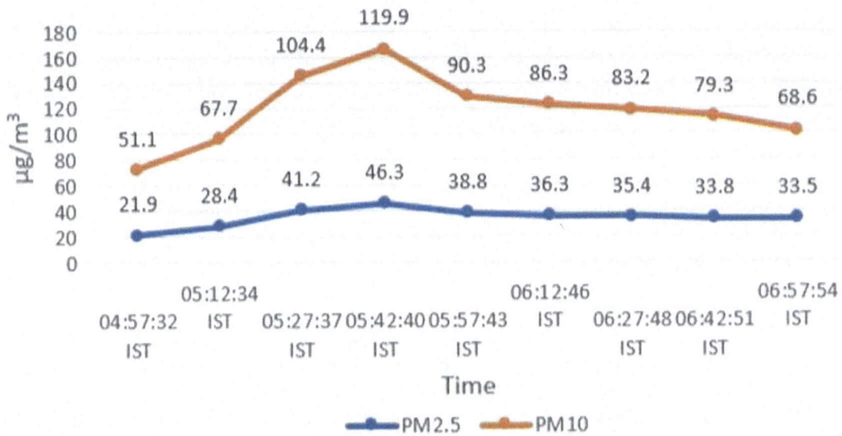

Fig. 5 Impact of using water heater on PM levels

can create ventilation arrangements in the environment accordingly. This IoT-based IAQ monitoring system can be also integrated into smartphone applications to get instant updates about IAQ index in the premises. Experts also advise to make use of advanced ventilation systems to improve IAQ. There are generally three types of natural ventilation systems stack ventilation, pressure-driven flows, and wind-driven ventilation. However, homeowners can also plan to install latest mechanical ventilation systems such as exhaust-only, balanced, and supply-only.

4 Conclusion

This paper presents an IoT monitoring system for IAQ by focusing on four major parameters: temperature, humidity, $PM_{2.5}$, and PM_{10}. The advantages of the proposed system are its calibrated sensor network, Arduino Uno-based controls for reliable operations and cost-effective design. It can be installed at any rural and urban building to monitor IAQ and can provide instant updates about present stats via ThingSpeak-based online web portal. Users can stay up to date about IAQ conditions on their premises even when they are at any other corner of the world. It is a flexible solution for assisted living and smart home applications. The quality of the proposed monitoring system can be further enhanced by using gas sensors and VOC sensors to analyze all essential parameters associated with IAQ index. The results present a relevant contribution to IoT systems for IAQ monitoring. However, the proposed system has some limitations. In the future, the proposed system needs further experimental validation for better calibration and accuracy. Furthermore, hardware and enhancements are planned to adapt the solution to specific use cases such as schoolrooms and hospitals. The collected data can be correlated with the occupant's health problems to support for clinical diagnostics. Real-time IAQ supervision allows to detect unhealthy situations and to plan interventions for enhanced living environments in useful time.

References

1. Z. Argunhan, A.S. Avci, Statistical evaluation of indoor air quality parameters in classrooms of a University. Adv. Meteorol. **2018**, 1–10 (2018). https://doi.org/10.1155/2018/4391579
2. W.H.O.R.O. Europe, For: air quality guidelines for Europe (2000)
3. J.P. Barbara, E.W. Celia, D. Amy, L. Steven, S. Carolyn, S. Donald, Environmental health and safety hazards experienced by home health care providers: A room-by-room analysis. Workplace Health Saf. **63**, 512–522 (2015). https://doi.org/10.1177/2165079915595925
4. T. Li, S. Cao, D. Fan, Y. Zhang, B. Wang, X. Zhao, B.P. Leaderer, G. Shen, Y. Zhang, X, Duan, Household concentrations and personal exposure of $PM_{2.5}$ among urban residents using different cooking fuels. Sci. Total Environ. 548–549, 6–12 (2016), https://doi.org/10.1016/j.scitotenv.2016.01.038
5. R.L. Corsi, J.A. Siegel, C. Chiang, Particle resuspension during the use of vacuum cleaners on residential carpet. J. Occup. Environ. Hyg. **5**, 232–238 (2008). https://doi.org/10.1080/15459620801901165
6. T. Kapwata, B. Language, S. Piketh, C.Y. Wright, Variation of indoor particulate matter concentrations and association with indoor/outdoor temperature: a case study in rural limpopo. S. Afr. Atmos. **9**, 124 (2018). https://doi.org/10.3390/atmos9040124
7. L. Fang, G. Clausen, P.O. Fanger, Impact of temperature and humidity on the perception of indoor air quality. Indoor Air **8**, 80–90 (1998). https://doi.org/10.1111/j.1600-0668.1998.t01-2-00003.x
8. IOT based Indoor Air Pollution Monitoring using Raspberry PI. Int. J. Innov. Eng. Technol. **9** (2017), https://doi.org/10.21172/ijiet.92.03
9. Z. Idrees, Z. Zou, L. Zheng, Edge computing based IoT architecture for low cost air pollution monitoring systems: a comprehensive system analysis. Des. Consid. Dev. Sens. **18**, 3021 (2018). https://doi.org/10.3390/s18093021
10. M. Benammar, A. Abdaoui, S. Ahmad, F. Touati, A. Kadri, A modular IoT platform for real-time indoor air quality monitoring. Sensors **18**, 581 (2018). https://doi.org/10.3390/s18020581
11. Improving national air quality forecasts with satellite aerosol observations. Bull. Am. Meteorol. Soc. **86**(9), https://journals.ametsoc.org/doi/abs/10.1175/BAMS-86-9-1249
12. G. Marques, R. Pitarma, An indoor monitoring system for ambient assisted living based on internet of things architecture. Int. J. Environ. Res. Public Health **13**, 1152 (2016), https://doi.org/10.3390/ijerph13111152

Comparative Study of Ambient Air Quality Prediction System Using Machine Learning to Predict Air Quality in Smart City

Gopal Sakarkar, Sofia Pillai, C. V. Rao, Atharva Peshkar and Shreyas Malewar

Abstract It is a herculean task to predict air quality of a particular area due to indefinite characteristics. As air pollution is a complex mixture of toxic air components that include ozone (O_3), particulate matter 2.5_m (PM2:5), SO_2, RSPM, SPM and nitrogen dioxide (NO_2). These small particles penetrate deep into the alveoli as far as the bronchioles, interfering with a gas exchange within the lungs. Though research is being conducted in environmental science to evaluate the severe impact of particulate matters on public health. The capital city of Maharashtra, Nagpur is used as a case study since nearly ten thousand motor vehicles are being registered in Nagpur on a monthly basis contributing exponentially to air pollution. Various machine Learning-based algorithms are checked to compare and to find out the predictive analysis using available dataset. After comparing seven different machine learning algorithms, Boosted Random Forest algorithm was found out to be the most accurate predictive algorithm, with the maximum coefficient of determination and less mean absolute error.

Keywords Air quality · Forecasting system · Machine learning · Cancer · Forecasting · Ensemble methods · Random forest

---

G. Sakarkar (✉) · S. Pillai · A. Peshkar · S. Malewar
G H Raisoni College of Engineering, Nagpur, India
e-mail: gopal.sakarkar@raisoni.net

S. Pillai
e-mail: sofia.pillai@raisoni.net

A. Peshkar
e-mail: peshkar_atharva.ghrceit@raisoni.net

S. Malewar
e-mail: malewar_shreyas.ghrcecs@raisoni.net

C. V. Rao
Former-Sr. Scientist, NEERI, Nagpur, India

M. Dutta et al. (eds.), *Proceedings of International Conference on IoT Inclusive Life (ICIIL 2019), NITTTR Chandigarh, India*, Lecture Notes in Networks and Systems 116,
https://doi.org/10.1007/978-981-15-3020-3_16

1 Introduction

We humans have been very successful in the domains of technology, commerce and each and every field of human existence, but through a due course of time, we have inflicted an indelible footprint of several sets of polluting factors. Our environment is being deteriorated innumerable factors like Deforestation, Water pollution, Genetic modification, Ozone Layer Depletion, Air pollution, etc. Each and every type of pollution is lethal, but we have considered Air Pollution as the most detrimental type because it gives the least reaction time to a person. Air pollution might cause Pneumonia, Influenza, Bronchitis, etc. Such health complications can be controlled to a certain level by raising the awareness about air quality conditions in urban areas, enabling the citizens to limit their day to day activities in the cases of elevated pollution episodes and planning their routes and schedule to avoid inhaling harmful pollutants, by using machine learning models to forecast air pollution in areas after a certain time duration.

Air pollution is caused for the most part by transportation, fuel ignition in stationary sources, consuming of petroleum derivatives like coal, wood, dry grass and development action. Engine vehicles produce abnormal amounts of Carbon Monoxide (CO) and Hydrocarbons (HC) and Nitrogen Oxides (NO). Development exercises, industrial chimneys, terrible streets and consumption of petroleum products are in charge of dust pollution. Private and commercial exercises additionally add to Air Pollution [1].

2 Literature Review

Following the Rio De Janerio Earth summit in 1992 Environmentalists and Researchers worldwide have been focussing on Air Quality and weather prediction systems. Elia Dragomir proposed a solution to Predict Air Quality using the K Nearest Neighbour Technique. She focussed on SO_2, CO, NO, NO_2 and O_3 pollutants only. She referenced the prediction results to a fuzzy set of Quality Index and concluded that best results are yielded in 10 fold cross-validation. She was unable to reveal the characteristics of the dataset.

Qi Feng in his research paper Improving Neural Network Prediction Accuracy for PM10 Individual Air Quality Index Pollution Levels stressed on pollutants having a diameter less than $<10\ \mu m$ (PM10) in two major cities of China. The reason for the generation of fugitive dust was due to construction activities and was interlinked with Construction Influence Index. His Neural Network Models were based on perceptron, Elman and Support Vector Machine. The dataset was decomposed into wavelet representations and then wavelet representations were predicted.

His predictions were tested between 1 January 2005, and 31 December 2011, at six monitoring stations situated within the urban area of the city of Wuhan, China.

It yielded better results than previous models but he only focussed on pollutants <10 μm (PM10).

Ozone and PM10 were two pollutants which were emphasized Giorgio Corani in research paper Air quality prediction in Milan: feedforward neural networks, pruned neural networks and lazy learning. Feedforward Neural Networks (FFNNs), Pruned Neural Networks (PNNs) and Lazy Learning (LL) were the foundation of entire statistical prediction. Lazy learning provided the best results on the basis of evaluation metrics such as correlation and mean absolute error.

3 Methodology

3.1 Dataset

The data used for comparing the predictive accuracy of the models was accessed from Kaggle uploaded by Shruti Bhargava as 'India Air Quality Data' (https://www.kaggle.com/shrutibhargava94/india-air-quality-data), it is a highly cleaned and compiled version of the 'Historical Daily Ambient Air Quality Data' released by the Ministry of Environment and Forests and Central Pollution Control Board of India under the National Data Sharing and Accessibility Policy (NDSAP).

The columns of the data include Station Code, State, City, Agency, Type of Area, concentrations of Sulphur Dioxide, Nitrogen Dioxide, Respirable Particulate Matter, Suspended Particulate Matter, the Location of the monitoring area, PSI 2.5, Date of recording. The columns in the data are of numeric and string type and contain categorical variables. Thus, it is essential to appropriately encode categorical variables. For that purpose, we have used the LabelEncoder function from Scikit Learn Machine Learning Library.

Since the missing values and outliers can have a great negative impact on the predictive accuracy of the models, the authors have imputed the missing values in the numeric columns (conc. of the pollutants) with the mean values. To identify and remove outliers, the ZScore has been calculated, with −3 and 3 being the threshold to retain the values in the dataset.

3.2 Evaluation Metric

3.2.1 Mean Absolute Error

In statistics, the mean absolute error is the quantity used to measure how close forecasts and predictions are to the actual outcomes. Mean absolute error performs in ways that disregard the directions of over or under prediction [11].

Table 1 Boosted random forest results

Pollutants	MAE
SO_2	1.7091
NO_2	3.8402
RSPM	21.2891
SPM	19.8359
PM2.5	0.0426

A lower value of MAE indicates a small difference between the pollutant concentration predictions by the model and the actual concentrations, averaged over the entire dataset, making the model a good fit for the dataset, whereas a higher value of MAE proves the opposite. MAE is an important metric for this study due to the highly fluctuating nature of the data that we are dealing with.

$$\mathrm{MAE} = \frac{1}{n}\sum_{j=1}^{n} \left| y_j - \hat{y}_j \right|$$

3.3 Fitting Data to Various Regression Models

3.3.1 Boosted Random Forest Regression

An ensemble learning method is implemented for classification and regression by creating a multitude of decision trees during training. Applying a boost algorithm to Forest Regression reduces the bias and variance.

Implementation of Boosted Random Forest regression on the dataset used in the study yielded the following results (Table 1) (Fig. 1).

The spike in the performance can be attributed to the ensemble method, where the final prediction depends on the predictions of individual decision trees while the boosting corrects the wrong predictions. This makes the model much more robust to fluctuations also helping avoid the model from developing bias and variance.

3.3.2 Deep Neural Network

It is a neural network with multiple neural layers which process data by advanced mathematical modelling wherein each mathematical manipulation correlates to a single layer. An object is shown as a layered composition of primitives in a compositional model of DNN architecture.

Implementation of Deep Neural Network on the dataset used in the study yielded the following results (Table 2) (Fig. 2).

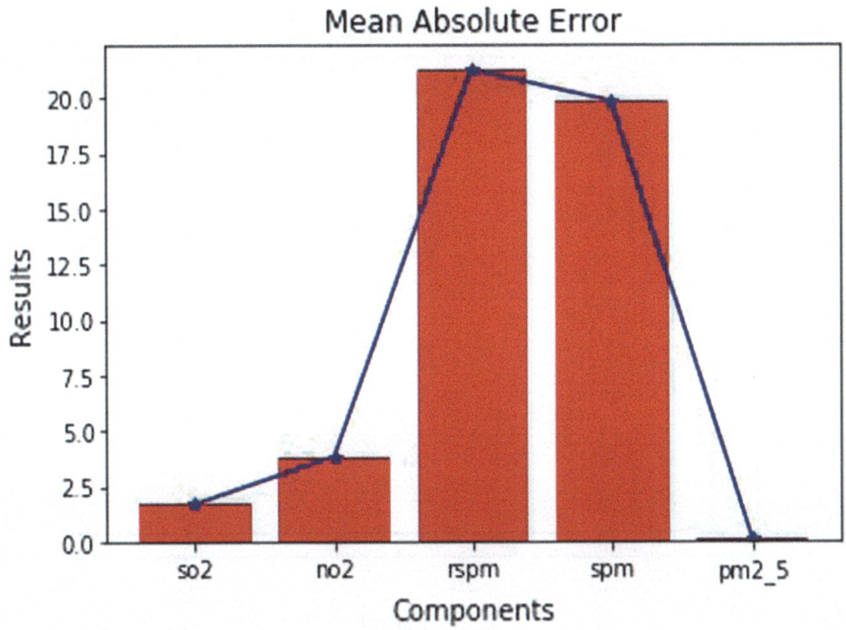

Fig. 1 MAE of BRF

Table 2 Results of deep neural network

	MAE
SO_2	5.9781
NO_2	12.1681
RSPM	94.3660
SPM	162.8340
PM2.5	0.5048

The highly fluctuating nature of the data causes the model to overfit the dataset, thereby giving subpar results on all the pollutants.

3.3.3 Stochastic Gradient Descent (SGD)

Gradient descent is used in Machine Learning algorithms to minimize a cost function to global minima. It is iterated innumerable times to attain the optimal value of desired parameters. But when considering significantly larger datasets the standard gradient descent algorithm ceases to work efficiently, therefore, a batch of values is randomly selected from the entire dataset and then Gradient Descent is applied on the values which prove to be relatively more efficient.

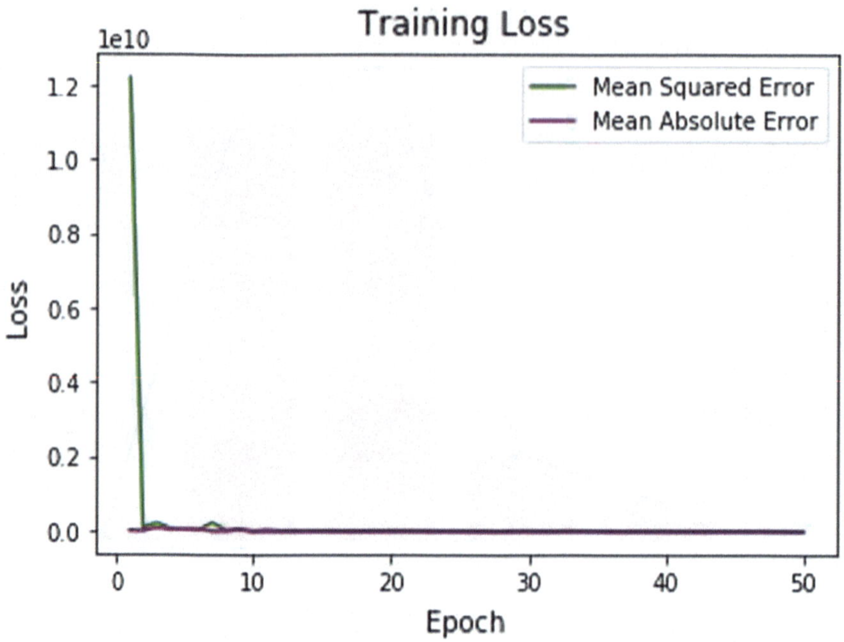

Fig. 2 MAE of DNN

Table 3 Results of stochastic gradient descent

	MAE
SO_2	5.3493
NO_2	1.3719×10^{19}
RSPM	1.24341×10^{19}
SPM	7.7897×10^{19}
PM2.5	7.0088

Implementation of Stochastic Gradient Descent on the dataset used in the study yielded the following results (Table 3) (Fig. 3).

The SGD is unable to learn from the data giving performing the worst at predicting the concentrations of pollutants, which can be inferred by the negative and exponentially large values of COD.

4 Conclusion

The algorithm performing the best at predicting the target pollutant concentration is Boosted Random Forest which can be devised from the high values of coefficient of determination for all the pollutants considered in this study. The performance of

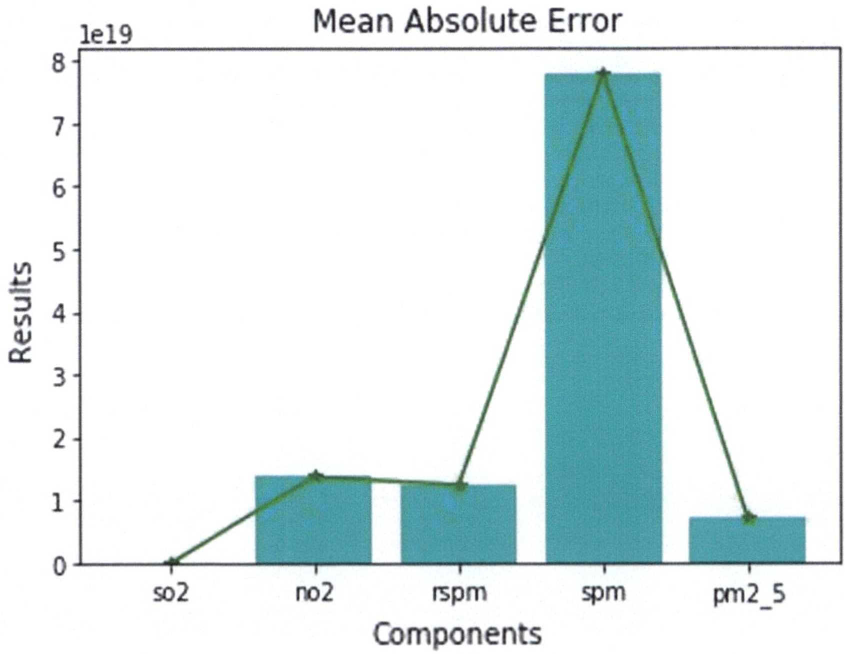

Fig. 3 MAE of SGD

a model can be generalized to all the pollutants. Thus a single well-tuned model performs well on all the pollutants. The model captures the patterns in the gases in a much better way compared to the particulate pollutants like SPM, RSPM and PM2.5. The Boosted Random Forest is a much more robust model for future work.

Acknowledgements This study was supported by Microsoft under the Microsoft AI for Earth Grant.

References

1. T.-C. Bui, V.-D. Le, S.-K. Cha, A deep learning approach for forecasting air pollution in South Korea using LSTM. Environ. Sci., Comput. Sci., Math. (2018)
2. S.B. Kotsiantis, D. Kanellopoulos, P.E. Pintelas, Data preprocessing for supervised learning. Int. J. Comput. Sci. **1**(2), 111–117 (2006)
3. R.G.D. Steel, J.H. Torrie, in *Principles and Procedures of Statistics with Special Reference to the Biological Sciences* (McGraw Hill, 1960)
4. C.J. Willmott, K. Matsuura, Advantages of the mean absolute error (MAE) over the root mean square error (RMSE) in assessing average model performance. Clim. Res. **30**, 79–82, 19 Dec 2005
5. E.L. Lehmann, G. Casella, *Theory of Point Estimation*, 2nd edn. (Springer, New York 1998). ISBN 978-0-387-98502-2. MR 1639875

6. H.L. Seal, The historical development of the Gauss linear model. Biometrika **54**(1/2), 1–24 (1967)
7. R. Quinlan, Learning efficient classification procedures. in *Machine Learning: an Artificial Intelligence Approach*, Michalski, Carbonell, Mitchell eds. by (Morgan Kaufmann, 1983), pp. 463–482, https://doi.org/10.1007/978-3-662-12405-5_15
8. L. Breiman, Bias, variance, and arcing classifiers
9. T.K. Ho, The random subspace method for constructing decision forests (PDF). IEEE Trans. Pattern Anal. Mach. Intell. **20**(8), 832–844 (1998)
10. Y. Bengio, Learning deep architectures for AI, Trends Mach. Learn. **2**(1), 1–127 (2009), CiteSeerX/10.1.1.701.9550
11. S. Mei, A mean field view of the landscape of two-layer neural net-works. Proc. Natl. Acad. Sci. **115**(33), E7665–E7671 (2018)

A Configurable Healthcare Monitoring System

Gurdip Singh, Shravanthi Kallem and Pavani Ayyagari

Abstract Advances in sensing and communication technologies are enabling new applications for providing more effective health care. In particular, systems that focus on prevention by continuously and remotely monitoring patients and their surroundings can improve healthcare quality and ease the task of healthcare professionals. This paper describes a Health IT Testbed that provides a comprehensive framework for monitoring a healthcare facility and provides a visualization interface to peruse this data. The testbed comprises two subsystems, a sensing subsystem and a visualization subsystem. The sensing system is responsible for collecting patient and ambient data, and is organized in a hierarchical fashion. For fine-grained physical entity sensing (e.g., sensing body postures of patients), we use SunSPOT wireless modules and a state machine approach to detect the current state of patients. For system-level sensing (e.g., state of the entire healthcare facility), we have created a Crossbow Motes wireless modules based infrastructure, and used a graph-based approach to detect current state and movement of entities between physical spaces. The monitoring system correlates various sensor data to populate a database with information regarding the state of the patients and the surrounding systems. The visualization system takes its input from the database being populated by the monitoring system and allows users to peruse both real-time and historical data. We also present performance results to evaluate the system.

Keywords Healthcare systems · Sensor networks · Sensor monitoring

G. Singh (✉)
College of Engineering and Computer Science, Syracuse University, Syracuse, USA
e-mail: gsingh06@syr.edu

S. Kallem
S&P Global, Washington DC, USA

P. Ayyagari
RTI International, Raleigh, USA

M. Dutta et al. (eds.), *Proceedings of International Conference on IoT Inclusive Life (ICIIL 2019), NITTTR Chandigarh, India*, Lecture Notes in Networks and Systems 116,
https://doi.org/10.1007/978-981-15-3020-3_17

1 Introduction

Recent advances in networking technologies and wireless communication are enabling new ways of delivering health care, and are being used in conjunction with medical technologies in many healthcare applications [1]. These advances have allowed an increased focus on prevention and early detection rather than operating in a reactive mode. The existing practice of manually monitoring patients involves high usage of resources and is not cost-effective. Hence, there is an increased demand for tools to allow healthcare professionals to monitor live as well as historical information about the patients.

There has been a significant amount of research in the last few years on wearable medical devices, body area networks, and sensor network infrastructure development. A body area network consists of multiple wireless platforms with sensing and communication capabilities [2–5]. Much of the research has focused on developing communication technologies, miniature (low-power), and minimally invasive sensing modules. Communication standards such as those from Zigbee, Bluetooth, and 802.11 g have been studied as candidates for body area networks [6, 7]. A sensor platform with wearable sensors was developed in the Mercury project wearable to analyze patient motion while ensuring long battery lifetime [2]. Other works have focused on protocols [3, 4] and signal processing algorithms [8, 9].

This paper describes the design and implementation of a configurable healthcare monitoring system. The system comprises two main subsystems: a sensing subsystem and a visualization subsystem. The sensing subsystem is responsible for collecting various types of data including patient data, location of equipment/devices and state of the physical locations in the facility. We have designed the system for collecting patient data using off-the-shelf SunSPOT sensing modules. The architecture of the system is hierarchical in nature and allows multiple levels of data aggregation. We propose a *state-machine-based* approach to describe possible transitions between physical states to aid the detection process. For example, at the node aggregating data from sensors on a patient, the state machine may describe possible state transitions between different body postures and guide the process of identifying these postures. At the system level, we model the facility as a *graph* whose nodes correspond to physical locations and the edges describe reachability between adjacent physical locations. This graph is used to capture possible transitions between system states (e.g., a person can move from a room to only one of the adjacent rooms). The state-machine-based approach and graph-modeling are the two technical contributions to enable state monitoring. The second component of the system is a visualization system that allows a user to visualize real time as well as historical data via a flexible interface. This interface displays data in a manner that ensures that the visualization of different physical components is consistent. We have performed extensive testing to evaluate the correctness and performance of the system.

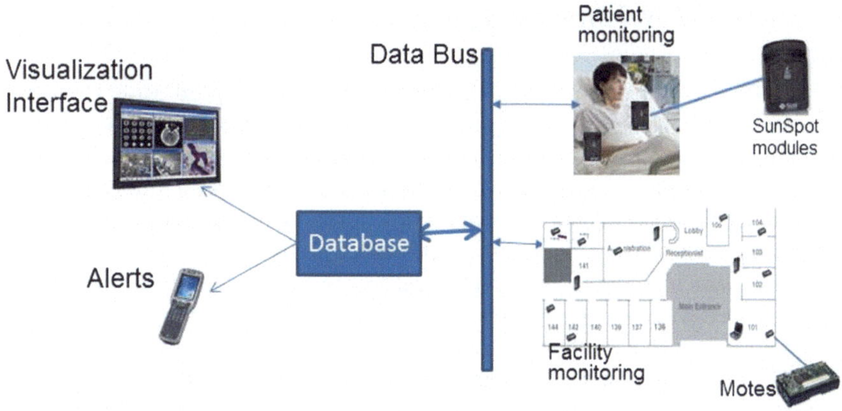

Fig. 1 Architecture of HITT

2 The Health Information Technology Testbed (HITT)

The architecture of HITT is shown in Fig. 1. There are two major components of HITT, the monitoring subsystem and the visualizing/alerting subsystem. The task of the monitoring system is to collect sensory data from the healthcare system and publish the information to a data bus. The data published on this bus is also archived in a database. The visualizing/alerting system is responsible for providing mechanisms for healthcare professionals to effectively use this information for providing patient care. This may be in the form of interfaces to visualize current as well as historical information of patients or providing alerts when certain conditions arise. For instance, the system can be programmed to generate alerts when a patient is in a specific body position. In the following sections, we describe each of these subsystems in more detail.

3 The Monitoring System

The goal of the monitoring system is to collect sensory data and aggregate this data to provide current state information of the entities in the system. The monitoring system operates at two levels: system-level monitoring and physical entity monitoring. System-level monitoring is responsible for collecting state of the physical locations that may include environmental data such as temperature, light intensity, and humidity, and occupancy data such as the set of patients and medical equipment present in each location. The physical entity monitoring component is responsible for monitoring the state of each entity (patient or a medical equipment).

3.1 System-Level Monitoring

System-level monitoring requires that the facility be equipped with a sensing infrastructure to gather various types of data such as current location of each patient/equipment, proximity of patients to healthcare professionals, and attributes such as temperature and light intensity in each location. We have developed such an infrastructure using a collection of Crossbow wireless modules (motes) and Stargate Netbridge modules [10]. The motes use the Zigbee protocol for communication [6]. In this infrastructure, the individual physical spaces have been instrumented with a mote equipped with sensors. To each mote, we have also attached a EasySen WiEye sensor board (which has motion sensors) to detect movement of objects [11]. However, the WiEye motion sensors cannot distinguish between different objects. To simulate detection, each object announces its presence with its id when it moves to a new location. The communication range has been adjusted so that only the mote sensing that particular location is able to receive the object announcement messages. All of the sensed information is communicated using a multi-hop communication protocol to a base station. The base station is attached to a Stargate Netbridge gateway node by a USB cable. The infrastructure has several such clusters of motes, each with its own base station and Stargate Netbridge. The gateway nodes, in turn, use wireless Ethernet communication to publish information to the data bus. The operating system on the motes in TinyOS and are programmed using the nesC language [12]. A previous version of the system is utilized with Cerner's Real-Time Location Service (RTLS). In the RTLS system, real-time location information of patients, hospital staff, and medical equipment is published to iBUS, a middleware based on publish-subscribe paradigm. Since RTLS is based on a proprietary ultrasound tracking technology, we transitioned to our Crossbow-module (employing Zigbee)-based platform described above to have an open-source technology-based solution.

Modeling Cyber-Physical Infrastructure as a Graph: With the infrastructure discussed above, it is possible to collect raw data from each sensing module and publish it on the data bus. However, for this information to be presented in a consistent manner, we utilize the physical topology of the healthcare facility. In particular, we model a facility as a graph $G = (V, E)$, where V represents the set of physical areas in a healthcare facility. These areas are those which are outfitted with sensors and can detect the presence of physical entities. The edges in E represent direct reachability between physical areas. For example, in Fig. 2, we have four rooms equipped with sensors and hence, we have four nodes in the graph shown on the left (each area A_i is represented by node P_i). An edge between P_2 and P_4, for instance, represents the fact that a physical entity can move directly between areas A_2 and A_4. The edges can be annotated with parameters such as the type of edge (door, stairs) and timing information. For example, if two rooms are connected by a corridor which is not monitored, the timing information on the edges connecting these rooms can indicate how long it takes to travel through the corridor. Note that this graph is different from the communication toplogy—even though each room is equipped with a mote, the communication range of a mote may allow it to communicate with motes which are

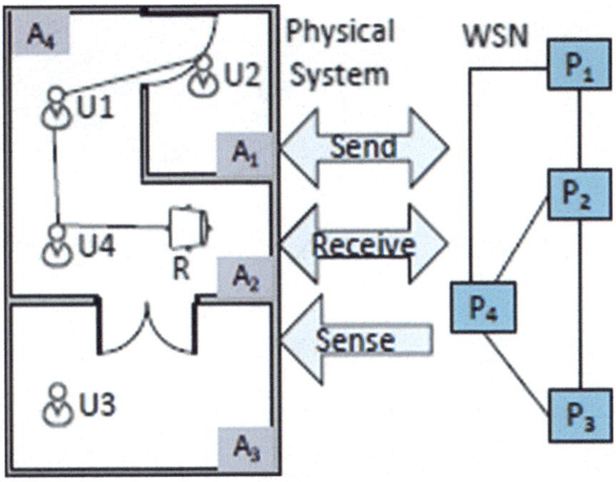

Fig. 2 Physical topology graph

far away (e.g., the mote in A_1 and A_3 may be able to communicate even though they are not physically adjacent). The modeling in terms of a graph provides us a basis to ensure consistency of data when presented in our visualization interface described later. The sensor infrastructure is programmed to report data that is aligned with this graph. Due to lack of space, we have omitted the formal definition of consistency and the corresponding algorithms to ensure consistency.

3.2 Physical Entity Monitoring System

We have designed the physical entity monitoring system to be hierarchical in nature. At the lowest level, we use data from sensors to detect postures of various parts of the body (e.g., whether the left leg is horizontal, vertical, or bent). At the next level, we use information from these components to determine the state of the patients (e.g., whether the patient is standing, sitting, or in the process of getting up). At each level, we use a state machine specification that describes possible transitions to guide the detection process. An Aggregation component is used at each level that takes this state machine as input and is responsible for combining the sensor data to detect the required conditions. The goal of this hierarchical arrangement is to provide flexibility and to reduce the amount of communication by local data aggregation. For instance, rather than each of the sensors on the legs sending all of the messages to the base station, a local device can aggregate this information and send summarized data back to the base station.

We use a collection of SunSPOT wireless modules for patient monitoring (Fig. 3). This module is a wireless platform manufactured by Sun Microsystems. The sensors on this module include a 3D accelerometer, a light sensor, and a temperature sensor. It also measures the orientation of the SPOT with respect to gravity.

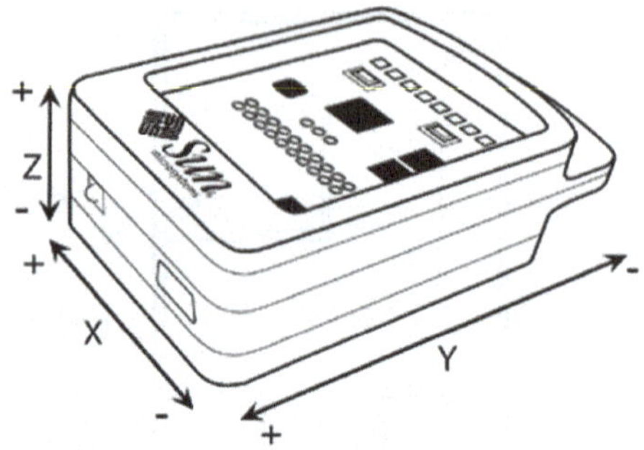

Fig. 3 SunSPOT module

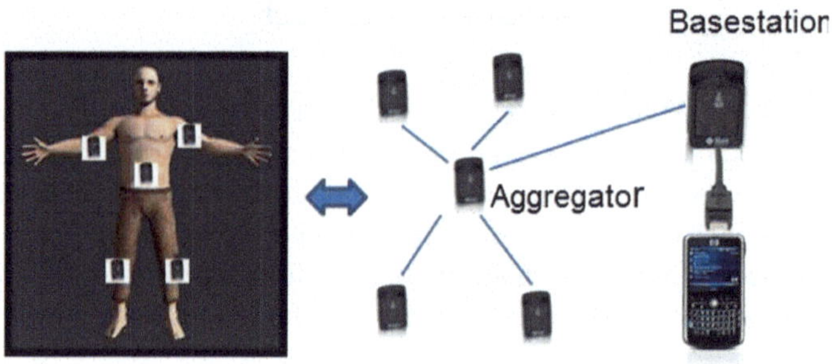

Fig. 4 Patient monitoring system

In our prototype system, we use a set of SUNSPOT sensors arranged in a star network (one-level hierarchy) to monitor a patient's state (see Fig. 4). One of these modules is distinguished as the Aggregator that is connected to the base module, which in turn is connected to a laptop or a handheld device. We connect the base station to a handheld device (smartphone) that can wirelessly communicate the data to the database to allow more patient mobility.

State–Machine-based Monitoring: A posture is calculated based on the sensor readings of the thigh, leg, and chest sensors. Each sensor gets the following values about its position in terms of tilt and acceleration in each of the three dimensions. The average of these values is taken over readings from a previous time interval. The Aggregator uses *state machine* as input which represents reachability between postures. Figure 5 shows a portion of such a state machine. The nodes in the state machine are possible postures that can be detected by the system and an edge from one posture to another represents reachability between these postures. The edges of the state machine are also labeled with conditions which specify when transitions can be safely inferred. The Aggregator uses this state machine diagram as a base to determine valid reachable postures and to detect abnormality in physical activity. The Aggregator maintains an ActionPath String to record the sequence of postures detected. Every time when a new posture is detected, it validates if the posture is a

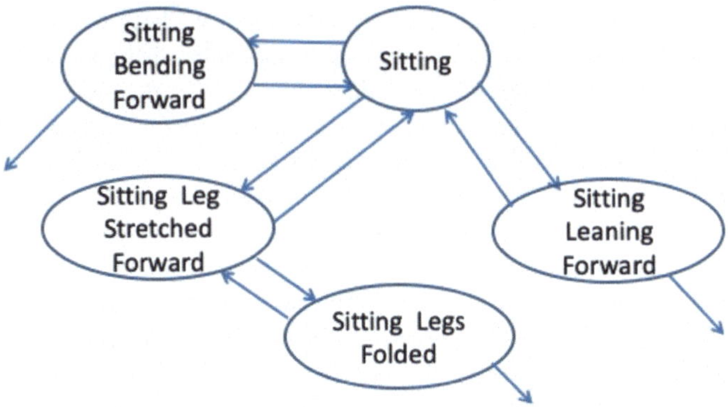

Fig. 5 Posture state machine

valid reachable posture from the previous posture. To determine whether a transition is valid, it must also check whether the readings reported by the sensing modules satisfy the condition specify for that transition in the state machine. Thus, the state machine provides a flexible mechanism to aggregate raw data values from the sensors to meaningful transition between postures. We have extensively tested this system to detect possible transitions specified in the state machines by having a person outfitted with sensors that go through different physical posture sequences. We are currently developing machine learning-based techniques to learn from past data to better detect transitions.

4 The Visualization System

The visualization system uses the database populated by the monitoring system to provide information to the end users. In addition, it also uses a database of images to reflect the current states of the physical entities. For instance, the image database has a floor map or a blueprint model of the healthcare facility. While it is possible to generate a representative floor map from the physical topology graph, we have chosen to use a predefined image instead. The image database also has predefined images for possible postures of the patient. Again, although it might be possible to dynamically generate images for such postures from sensor readings, we rely on preexisting images.

An important component of the visualization system is the Interpolation component. Due to factors such as unsynchronized clocks, it is possible that attributes such as patient-location, room temperature, and light intensity are recorded in an unsynchronized manner. These values, for instance, might not be recorded at the same instance or interval the patient postures are being recorded. However, when the overall state is being presented to the end user, they have to be presented with a consistent view. Attributes such as temperature and humidity that vary in a continuous manner can be interpolated so that the values are synchronized. We have written an interpolation module that does these computations as visualization is being performed. As the pointer on the slider progresses with time, the values of the entities change and have to be updated. Hence, the action event associated with the slider is called every time its position changes and the entities being displayed such as the images, location coordinates, and status of lights are updated from their respective arrays. The action event is also associated with a button along with a window listener for each patient (top left window). On clicking the button, the respective patient's postures can be viewed. All these values are always in synchronization with the slider. Figure 6 shows a snapshot with separate windows for each patient whose posture is being monitored. This figure shows the application with six patients with the postures of patients 2, 3, 4, 5 being displayed in individual windows. The postures in these windows change dynamically based on the current state of the patients.

We have conducted extensive performance testing of the visualization interface. In Fig. 7, we show the time taken to retrieve values from the database as the number of patients in the database is varied. Each patient has a separate table in which the patient's activity is recorded. Hence, increasing the number of patients will result in an increase in the number of tables from which the values are being retrieved. Figure 8 shows impact on the time taken to display the values on the GUI as number of patients is increased. Every patient in the system is associated with a separate window with window listeners to view the patient's activity. Hence, every patient's window is to be updated periodically during the visualization. Hence, more number

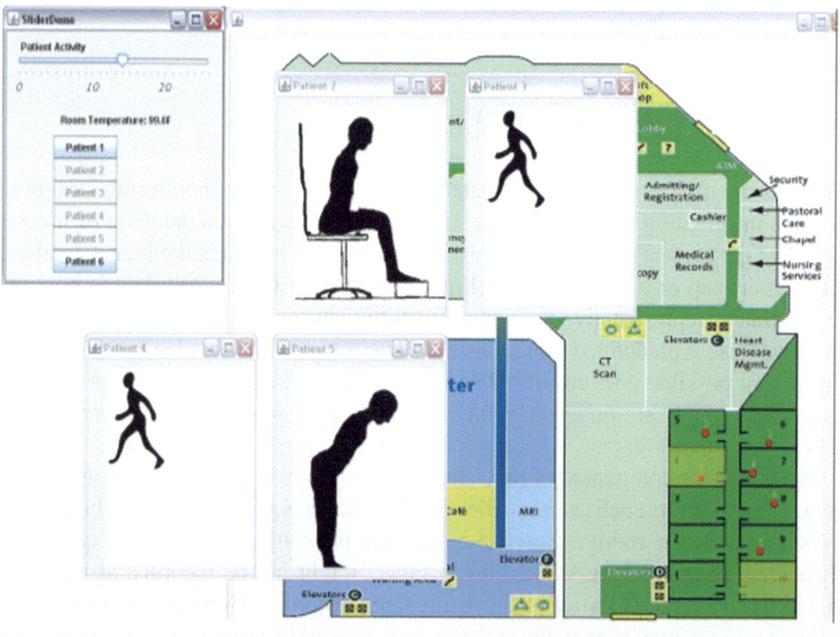

Fig. 6 Screen shot with patient postures

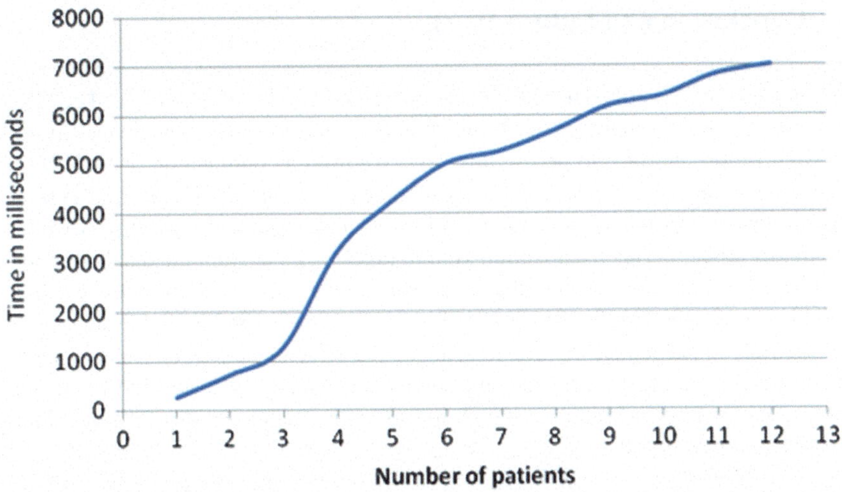

Fig. 7 Impact on database access time

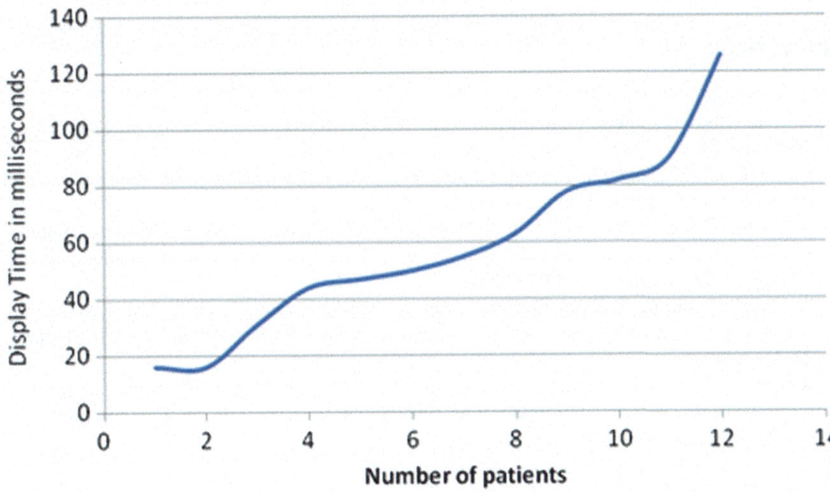

Fig. 8 Impact on display time

of patients in the system will require more time to update the frame for every patient. As can be in Fig. 8, the time taken increases linearly with the number of patients.

5 Conclusion and Future Work

In this paper, we described the design and implementation of Health IT Testbed. This testbed provides a comprehensive framework for monitoring a healthcare facility and a visualization interface to peruse this data. We used a combination of SUNSPOT wireless modules and Crossbow Motes modules to design the monitoring system. The monitoring system uses a state machine and a physical topology graph to guide the detect process. The visualization system takes its input from the database being populated by the monitoring subsystem and allows users to view both real-time and historical data. Our future work will involve extensive end-to-end evaluation of our testbed. In particular, we plan to evaluate the effectiveness of the state machine approach in detecting posture transitions in real time. We will also develop algorithms leveraging the physical topology graph for problems such as state recording, object tracking, and event ordering.

Acknowledgements This work was supported by NSF grants 1135564 and 17220579.

References

1. L. Gatzoulis, I. Iakovidis, Wearable and portable e-health systems. IEEE Eng. Med. Biol. Mag. **26**(5) (2007)
2. C. Doukas, I. Maglogiannis, Advanced patient or elder fall detection based on movement and sound data, in *Pervasive-Health* (2008)
3. D. Curtis, E. Pino, J. Bailey, E. Shih, J. Waterman, S. Vinterbo, T. Stair, R. Greenes, J. Guttag, L. Ohno-Machado, Smart: an integrated wireless system for monitoring unattended patients. J. Am. Med. Inform. Assoc. **15**(1) (2008)
4. A. Wood, G. Virone, T. Doan, Q. Cao, L. Selavo, Y. Wu, L. Fang, Z. He, S. Lin, J. Stankovic, Alarm-net: wireless sensor networks for assisted-living and residential monitoring, IEEE Netw. **22**(4) (2008)
5. K. Lorincz, B. Chen, G. Challen, A. Chowdhury, S. Patel, P. Bonato, M. Welsh, Mercury: a wearable sensor network platform for high-fidelity motion analysis, in *7th ACM SenSys* (2009)
6. Zigbee Alliance (2009), www.zigbee.org
7. Blueooth (2009), www.bluetooth.com
8. T.R. Burchfield, S. Venkatesan, Accelerometer-based human abnormal movement detection in wireless sensor networks (HealthNet, New York, 2007)
9. J. He, H. Li, J. Tan, Real-time daily activity classification with wireless sensor networks using Hidden Markov model. EMBS (2007)
10. The MEMSIC website, http://www.memsic.com/
11. WiEye Website, http://www.easysen.com/WiEye.htm
12. D. Gay, P. Levis, R. von Behren, M. Welsh, E. Brewer, D. Culler, The nesC language: a holistic approach to networked embedded systems, in *ACM SIGPLAN 2003 PLDI* (2003)

Evaluation of Hand Movement Using IoT-Based Goniometric Data Acquisition Glove

Prashant Jindal, Rashi Aditi Ranjan, Poojita Garg, Pranav Raj, Parneet Kaur, Varun Karan, Ishan Madhav and Mamta Juneja

Abstract Assessment of movements across various finger joints is essential for assisting physiotherapists in the detection of the harm & impairment caused due to injuries on the human hand and to determine its recovery. A soft hand glove with flex sensors has been developed to measure angular finger MetaCarpoPhalangeal (MCP) and ProximalInterPhalangeal (PIP) joint movements of the human hands. These gloves are designed for both hands and are used to record real-time information of the joint angular movements of all four fingers and the thumb of both sides, through an electromechanical interface. The data is stored on cloud which is accessible by both the doctor and the patient through an app. The data is further used to quantify the impairment and recovery rate of patients after physiotherapy sessions. The data can further be standardized for the comparison between healthy and unhealthy individual suffering from joint related disease.

Keywords Arduino · Flex sensors · Flexion · Bluetooth · IoT

P. Jindal · R. A. Ranjan · P. Garg · P. Raj · P. Kaur · V. Karan · I. Madhav · M. Juneja (✉)
UIET, Panjab University, Chandigarh, India
e-mail: mamtajuneja@pu.ac.in

P. Jindal
e-mail: jindalp@pu.ac.in

R. A. Ranjan
e-mail: rasee.aditi@gmail.com

P. Garg
e-mail: poojita8garg@gmail.com

P. Raj
e-mail: rajpranav2806@gmail.com

P. Kaur
e-mail: parneetrana373@gmail.com

V. Karan
e-mail: varunkaran19@gmail.com

I. Madhav
e-mail: imadhav77@gmail.com

M. Dutta et al. (eds.), *Proceedings of International Conference on IoT Inclusive Life (ICIIL 2019), NITTTR Chandigarh, India*, Lecture Notes in Networks and Systems 116,
https://doi.org/10.1007/978-981-15-3020-3_18

1 Introduction

The hand motion of a person may be affected due to various injuries, and the movement of fingers is an ideal way to determine the extent of damage or its rehabilitation during and after therapy. Analysis of the movement of human fingers and the relationship between different finger joints has been a vital area of scientific research for many years [1]. Apart from therapy, studies have been carried out to understand various patterns of finger movements for applications in different fields like virtual reality, therapeutic training, sign language, gestures, entertainment, and so on [2–5]. Hand gesture has emerging applications in robotic control system like robotic arms which can be further used in various fields like defense, health and medicine, assembly lines, etc.

Our finger consists of three joints, namely, MetaCarpoPhalangeal (MCP) joint, Proximal InterPhalangeal (PIP) joint, and Distal InterPhalangeal (DIP) joint. Different types of hand-related manual tasks like gripping, holding, and pressing require analysis of human finger movements and establishing a relationship between the movements of the three joints [6].

The analysis of finger movements about different joints can be very useful in choosing a particular therapy. Stiffness of the PIP and MCP joint may cause disability in movement, which may be due to both traumatic and atraumatic conditions. These joints are hinged joint with unique anatomy, which makes it stiff and suitable for flexion and extension [7]. Häger-Ross et al. [8] aimed to determine the motion of other digits when one digit was moved and measured the degree at which the movement in each number occurred. They concluded that motion of the thumb, middle, and index fingers was more individuated compared to the other fingers.

Several attempts have been made to transfer the hand movements directly to the computer systems to avoid human efforts and any uncoordinated hand movements that may arise due to the use of intermediate devices such as keyboard, mouse, or joysticks [9]. The development of glove-based devices dates back to around the middle of twentieth century and has continued to grow since then [10]. A glove-based system is defined as a system comprising of multiple sensors placed atop on surface or within the glove that can be worn on the hand and electronic equipment for data acquisition connected to a power supply. The first glove-based systems to be developed include the Sayre glove, the MIT LED glove, and the digital data entry glove that used flexible tubes with a light source, LED-based system and proximity, knuckle-bend-measuring, inertial sensors, respectively, in their design [9, 11]. The first commercial data glove was launched in the US market in 1987, which was made of thin plastic tubes consisting of LED lights and detectors to detect motion along joints [12].

Flexion and abduction of fingers were checked by studying the coordination between three joints of the fingers. Despite various experiments conducted on hand movements using multiple techniques; there are limitations and difficulties in applying and interpreting these techniques [13]. Our work in this paper was primarily

Fig. 1 Current design of glove

focused on the finger movements about the PIP and MCP joints, as it gives the maximum deflection under flexion, thus, providing us with a wide range of angles to work and analyze. This study was carried out by designing a hand glove as shown in Fig. 1 to detect the finger movements about the PIP joint without any constraints. It was used to conduct different studies related to maximum flexion movements of the fingers and the thumb of both hands. The extent of this joint movement with respect to other fingers can provide useful data for standardization of angular movements among population spread across different age groups and genders. This data can also assist physiotherapists in early detection of diseases, planning the rehabilitation course of action and checking the rehabilitation status for patients suffering from various joint-related diseases. The device makes the evaluation session easy and accessible for both doctor and patients as it takes up the data value and displays directly through app along with comparison from previous data in graphical form. It is made to ease the work of physiotherapist by making a glove that takes up the angular movement and displays it through an app. Physiotherapists use the conventional goniometer for assessment of various finger joint movement, measuring each joint of each finger manually which is time-consuming and cumbersome. Further, the device gives an advantage of automatic data entry over the manual entries by the physiotherapists.

2 Materials and Methods

2.1 Glove

A soft cotton material based glove of standard adult size was selected for both hands. Flex sensors were adhered on each finger of these gloves, which could capture data based on strain developed in each finger during finger movements, i.e., flexion or bending. The electrical resistance of these sensors was ~10 KΩ.

2.2 Electronic Circuit

In order, to capture the bending strains (flexion) of the sensors, an electronics-based circuit was developed. Arduino Nano was used to program the flex sensors to obtain proper readings. Proper adjustments and calibrations of the sensors were done, at room temperature, by comparing the digital values generated by the device, with the angular deflections at 0°, 15°, 30°, 45°, 60°, 75°, and 90°. The bending strain induced in the sensor, upon flexion, generates a change in resistance of the sensors which are reflected as digital values by the arduino microcontroller. The actual angular deflections were measured in degrees using goniometer.

2.3 Method

The sensors on bending give resistance value which is further mapped to angular movement value through various calibration curve for each joints of each finger. The data is then sent to arduino and shown on the app interface which provides a user-friendly experience. The app contains interfaces for doctor and patients wherein the patient will give all of his details like name, age, gender, previous medical history, problem, etc. This data is stored as that patient's profile and will be accessible by doctor. The individual's data will be recorded first to measure the amount of impairment with respect to standard data depending upon that individual's profile. The patient's recovery is then quantified after regular intervals. As the readings are to be taken on routine basis thus, the hassle of recording and maintaining the database of each patient's readings is circumvented by providing a provision of direct storage on Google sheets. Also, a graphical representation of subsequent recordings is generated for providing a convenient way of judging the progress. The overall process is shown in Fig. 2.

Fig. 2 Flow diagram representing data collection and storage

2.4 Subject

This study was carried out on 30 different healthy subjects including 23 males and 7 females with age ranging from 18 to 45 years guided by the expert physiotherapist. A consent form was duly obtained from each subject declaring their volunteer participation and data measurement used for research purposes. Each subject was made to sit comfortably on a chair with their back and shoulders straight.

2.5 Procedure

Each subject was asked to wear the glove for each hand. It was ensured that each subject was sitting properly with his arm and hand placed in the correct position. Then the hand was placed on the table in front of the subject, to maintain the stationary position of the fingers without any bend that is at an angle of 0°. Based on the requirement of the study, different fingers were moved about PIP joints only. Tests were also carried out on all the subjects for repeatability and analysis of the glove sensors. The data was also collected to measure the divergence of values for each individual. Calibration of the flex sensor was done by comparing the angular movements with goniometer. A difference of ±5° was recorded which was within acceptable limits. The data was calibrated, and the best fit was used to reduce the error in values.

2.6 Repeatability

To check the repeatability of the glove, the same procedure was performed on one subject repeatedly. The subject was made to repeat the experimental procedure 5 times with a gap of 1 h in between. Each time the flexion and extension of each finger were checked about the PIP joint at different angles ranging from 0° to 90°. Data in the form of electrical resistance (Ω) was recorded for different finger positions at 0°, 15°, 30°, 45°, 60°, 75°, and 90°.

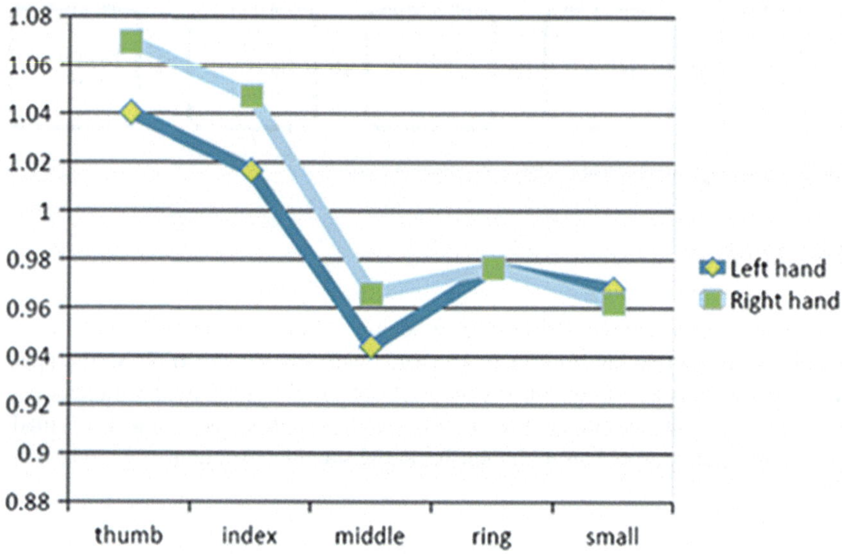

Fig. 3 Variation of flexion ratio

3 Results

The reading of maximum flexion of 30 healthy individuals was taken for both hands PIP joint. The readings of left and right hand were compared for males and females. The variation of the ratio of maximum flexion of male and female of each finger PIP joint is shown in Fig. 3. This shows that the flexion of males' thumb and index finger is greater than that of females while it is less for middle, ring, and little fingers. Also, the male to female ratio of thumb, index, and middle of right hand is greater than left hand. For ring finger, both hands have the same ratio and in case of little finger, left hand shows more flexion ratio than the right hand.

4 Discussion

A hand glove was developed to measure flexion of finger joints. For testing, PIP joint movement was evaluated specifically due to its significance for assessing future courses of action related to rehabilitation for patients suffering from joint-related problems. A total of 30 subjects (23 males and 7 females) in an age group of 18–45 years were considered for this study and gloves were developed for both left and right hands. Sensors were calibrated effectively by comparing the finger angular movements with goniometer. Angular finger movements were recorded by the sensor through App interface wherein the complete angular movement of the PIP joints was obtained in real time for all the fingers and was also stored in Google sheets. All the angular movements were translated into electrical resistances. Individuation and Stationarity indexes were obtained for all the subjects.

5 Conclusion

The rise of Internet of things (IoT) has engendered a whole new revolution of medical devices. This paper aims to highlight the resulted benefit of integrating the IoT with goniometric data acquisition glove for providing a less cumbersome and user-friendly experience. The gloves employ flex sensors to measure angular finger MCP, PIP joint movements of both the hands and after that, recording the real-time measurements into Google sheets. The users are provided with the convenience of viewing the readings on an app interface. These gloves can also be used for remote data recordings wherein a physiotherapist's physical presence may not be required to monitor the records. These gloves were used on healthy subjects to assess PIP joint movements for studying individuation of the subjects. Since the data recorded for both left and right hand was repeatable, therefore, this glove could be used effectively for this study.

Acknowledgements The authors are grateful to the Ministry of Human Resource Development (MHRD), Govt. of India for funding this project under Design Innovation Centre (DIC) sub-theme Medical Devices and Restorative Technologies.

Conflict of Interest The authors have no conflict of interest.

References

1. J.R. Napier, The prehensile movements of the human hand. J. Bone Joint Surg. Br. Vol. **38**(4), 902–913 (1956)
2. S. Ueki, H. Kawasaki, S. Ito, Y. Nishimoto, M. Abe, T. Aoki, Y. Ishigure, T. Ojika, T. Mouri, Development of a hand-assist robot with multi-degrees-of-freedom for rehabilitation therapy. IEEE/ASME Trans. Mechatron. **17**(1), 136–146, 23 Dec 2010
3. I. Dimbwadyo-Terrer, F. Trincado-Alonso, A. de los Reyes-Guzmán, M.A. Aznar, C. Alcubilla, S. Pérez-Nombela, A. del Ama-Espinosa, B. Polonio-López, A. Gil-Agudo, Upper limb rehabilitation after spinal cord injury: a treatment based on a data glove and an immersive virtual reality environment. Disabil. Rehabil.: Assist. Technol. **11**(6), 462–467, 17 Aug 2016
4. J. Gałka, M. Mąsior, M. Zaborski, K. Barczewska, Inertial motion sensing glove for sign language gesture acquisition and recognition. IEEE Sens. J. **16**(16), 6310–6316, 22 June 2016
5. J.K. Tang, G.Y., Ng, B.K. J.H. Leung, Hui, A. Kong, W.M. Pang, VR-MMA: a virtual reality motion and muscle sensing action game for personal sport. in *Proceedings of the 13th International Conference on Advances in Computer Entertainment Technology* (ACM) p. 40, 9 Nov 2016
6. M. Nakamura, C. Miyawaki, N. Matsushita, R. Yagi, Y. Handa, Analysis of voluntary finger movements during hand tasks by a motion analyzer. J. Electromyogr. Kinesiol. **8**(5), 295–303, 1 Oct 1998

7. G.C. Comer, S.J. Clark, J. Yao, Hand therapy modalities for proximal interphalangeal joint stiffness. J. Hand Surg. **40**(11), 2293–2296, 1 Nov 2015
8. C. Häger-Ross, M.H. Schieber, Quantifying the independence of human finger movements: comparisons of digits, hands, and movement frequencies. J. Neurosci. **20**(22), 8542–8550, 15 Nov 2000
9. D.J. Sturman, D. Zeltzer, A survey of glove-based input. IEEE Comput. Graph. Appl. **14**(1), 30–39 (1994)
10. L. Dipietro, A.M. Sabatini, P. Dario, A survey of glove-based systems and their applications. IEEE Trans. Syst. Man, Cybern. Part C (Appl. Rev.) **38**(4), 461–482, 20 June 2008
11. G.J. Grimes, inventor; Nokia Bell Labs, assignee. Digital data entry glove interface device. U.S. Patent No. 4,414,537, 8 Nov 1983
12. TG. Zimmerman, inventor; VPL Res Inc, assignee. Optical flex sensor. U.S. Patent No. 4,542,291, 17 Sep 1985
13. R. Richard, I.S. Parry, A. Santos, W.S. Dewey, Burn hand or finger goniometric measurements: sum of the isolated parts and the composite whole. J. Burn Care Res. **38**(6), e960–e965, 1 Nov 2017

CoTusk: IoT-Based Tooth Shade Detecting Device

Mamta Juneja, Jannat Chawla, Sumindar Kaur Saini, Divya Bajaj and Prashant Jindal

Abstract IoT technology has gained a lot of importance due to its efficient and accurate data collection, processing and analysis feature. Tooth shade determination has become an important part of dental aesthetics as it is required while making dental crowns. This paper describes the cost-effective tooth shade detecting IoT device named CoTusk and discusses the shade matching procedure used for assessing the proper tooth color. According to dentists and dental technicians, the tooth has a shade not just a color. A shade has three dimensions—hue, chroma and value. Various visual instruments such as Vitapan Classical Shade Guide, are available in the market, which is commonly used by dentists for shade matching. But that procedure is not accurate because it solely depends upon the person's ability to perceive the colors. So, a camera-enabled IoT device is developed for quantitative measurement of tooth color. The shades of the VITA Classical shade guide are taken as reference shades. The image of each shade is clicked and stored in the database on Azure. The patient's tooth image is clicked which is compared with the database images already stored on cloud. The MDNS is used to connect MQTT server which eventually uploads the feed from pi to cloud. The proposed methodology has increased the accuracy in determining the color of the tooth to 89.9% as compared to the previous approaches. Thus, this paper aims to use IoT device for detection of tooth shade efficiently, effectively and accurately on a large scale.

M. Juneja (✉) · J. Chawla · S. K. Saini · D. Bajaj · P. Jindal
UIET, Panjab University, Chandigarh, India
e-mail: mamtajuneja@pu.ac.in

J. Chawla
e-mail: chawla.jannat98@gmail.com

S. K. Saini
e-mail: sumindarkaursaini@gmail.com

D. Bajaj
e-mail: divyabajaj1997@gmail.com

P. Jindal
e-mail: jindalp@pu.ac.in

M. Dutta et al. (eds.), *Proceedings of International Conference on IoT Inclusive Life (ICIIL 2019), NITTTR Chandigarh, India*, Lecture Notes in Networks and Systems 116,
https://doi.org/10.1007/978-981-15-3020-3_19

Keywords IoT (Internet of Things) · Tooth shade detector · Vita shade guide · Aesthetic dentistry

1 Introduction

Tooth shade detector in the field of aesthetic dentistry is a topic that is not only concerned with the professionals who are concerned with detecting the color of the tooth for dental restorations but also with people who want to improve their teeth and brighten the world with their smile. The challenges in the aesthetic dentistry are color assessment and reproduction. The face is the most observable part of our body. The tooth shade detector aims to serve dentists in determining the proper match or shade of the tooth for restoration. This has helped to reduce the interhuman differences based on color. Basically, shade matching is an important attribute that involves many variations [1]. It is a task that involves art and science. Matching the accurate shade needed for restoration is based on various factors which include light, surroundings and most importantly, the receiver's eye [1, 2]. To overcome this shade match, tooth shade detector is introduced to produce the best shade matching results. The most common method to determine tooth shade is making comparison using shade guides [3, 4] with patient's teeth. A report of the National Dental Practice-Based Research Network [5] demonstrated that 98% of 365 dentists indicated that they frequently use commercial shade tabs for their patients' shade selection. However, even when shade guides are made from the same material as those used in restorations, they do not necessarily replicate the same color [6, 7]. Tooth shade detector has been more effective as compared to the visual methods.

To determine the dental shade of the tooth according to the Munsell system, three dimensions of color are defined in this order: Hue, Chroma and Value [8, 9]. It is important to make a selection in this order while using a shade guide. Color is the most important part of the light. Objects reflect a wavelength of that color that enters our eyes. Hue is described as a diverse form of color. It can be yellow, red, etc. The saturation of the hue is known as chroma and the value is the relative darkness or lightness of that particular color. The value-based shade guide is $B1$, $A1$, $B2$, $D2$, $A2$, $C1$, $C2$, $D4$, $A3$, $D3$, $B3$, $A3.5$, $B4$, $C3$, $A4$, $C4$. This paper aims at using a Vita classical $A1$–$D4$ shade guide to find the accurate match of the tooth. The dental shade guide is traditionally based on an increasing hue, which states that A is reddish-brown, B is reddish yellow, C is gray and D is reddish gray. This shade guide produces the most accurate tooth shade measurement under its environment, light conditions and gives the natural look with the restoration. The efficiency of this guide is high and different shades can be matched in order to obtain high results with accuracy. Digital image technology makes use of dental tooth shade scanner to produce the best shade match available. This digitization helps to solve the key challenges in the field of dentistry such as matching the tooth shade to be restored with the adjacent one. Hugo et al. in 2005 evaluated the performance of shade matching devices [10]. The SpectroShade device, the ShadeVision device and the Digital Shade Guide DSG4 were assessed with respect to their agreement with the color perception of three examiners looking at 57 test persons with a dataset of six teeth each for a total of 342. The shades were reported in Vita Classical shades. In many cases, computer-aided color shade determination of natural teeth seems not to reflect human perception. In 1970, Culpepper [11] already demonstrated visual shade selection is often unreliable and imprecise. The best agreement of the evaluated devices was obtained—generally as well as among the human testers—by the X-Rite ShadeVision system, followed at a statistically significant distance by the MHT SpectroShade and the Rieth DSG4.

2 IoT Methodology Used

The determination of tooth shade is an important step in dental aesthetics and it is practically difficult to accurately determine the region of interest along with the tooth color, hence the new technique determines the required area and color for making the accurate dental crowns. The components used in the methodology are shown in Fig. 1.

The following section represents the proposed methodology for tooth shade detection.

Step 1: Collection of images

The VITA Classical shade guide is taken as reference for different shades. The reference images of the tooth are collected using the Raspberry Pi camera of size 40×60 and stored in the database on google cloud. These images are in jpeg/jpg format. The raspberry pi camera has a sensor with a native resolution of 5 megapixels and has a fixed focus lens onboard. The camera is able to generate 2592×1944 pixel static images along with supporting the videos for implementation of the detector. The image of four teeth is taken that is fed into a segmentation network and the resultant is the region of interest for one tooth that is highlighted in Fig. 2 and used for further steps. A power bank is used for power supply.

Step 2: K-means Clustering technique

K-means Clustering technique is used to find the shade of the tooth of the patient. The RGB value is much nearer for every tooth color code; therefore, YIQ (Luminance(Y),

(a) Raspberry pi (b) Raspberry pi cam Module (c) VITA Classical shade guide

Fig. 1 Components used in methodology

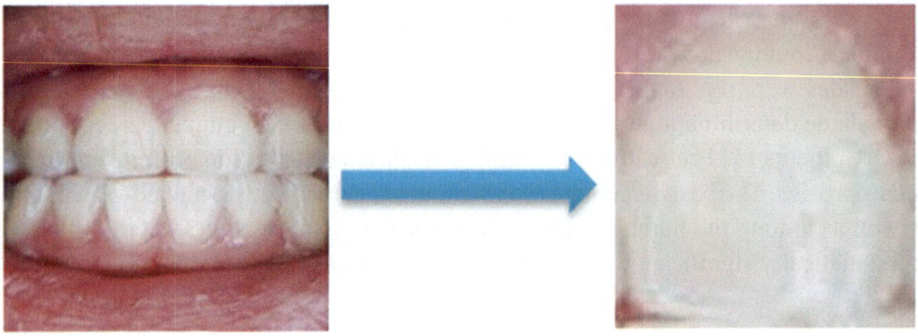

Fig. 2 Segmentation of the region of interest

In-phase, Quadrature) color space is used. The advantage of this model is that more bandwidth can be assigned to the Y-component (luminance) to which the human eye is more sensible than to color information. In order to find out the shade of every tooth, 16 clusters were used. The collected data was analyzed using various techniques out of which K-means clustering technique gave the best results having an accuracy of 89.9%.

Step 3: Application of conditions

The segmented tooth acts as the input to the code. The YIQ coordinates are then calculated by the code and matched with the YIQ coordinates of the reference images. The best match is then given as the output shade.

Step 4: Technologies used

The technology stack is used to combine the overall technologies that have been implemented in the proposed methodology. It combined all the coding frameworks and then the clustering algorithm gave the actual required results. This algorithm is implemented using scikit-learn. The csv file of the tooth information is read using the Pandas library. The Numpy library is used to make an array of all the color values and that is followed by performing the calculations on the required region of interest of the tooth Color sys is used to change the RGB to YIQ color code. Cv2 captures the image from the device camera and PIL crops the image into the size of 40×60. The various technologies used in the proposed methodology are shown in Fig. 3.

Step 5: Cloud computing

Data is sent from raspberry pi to cloud using Wi-Fi. Authentication is done via JWT instead of user/password as it's more secure. To handle that, the library is pyjwt, which depends on a Python library called cryptography. The Wi-Fi module is connected to it and stores the data on the cloud. Any IoT platform can be used to store data. Azure being easy to use, more reliable and secure was used to store data. MDNS is used to connect MQTT server that is eventually used to send the feed from pi to the cloud. Thus, the comparison of clicked image is made with the images already stored in database on cloud. The best possible shade is displayed as the output on the LCD (Fig. 4).

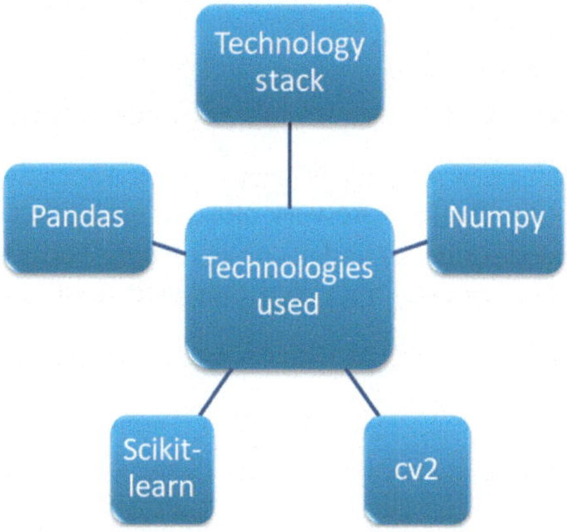

Fig. 3 Various technologies used in methodology

Fig. 4 Steps involved in the methodology

3 Results and Discussions

The practical implementation was performed in Python 3.7.2 using scikit and OpenCV libraries. To overcome the problem of surrounding light a tunnel was used which was 3D printed. The YIQ color scheme was used. The Y-channel contains luminance information while the I and Q channels (in-phase and in-quadrature) carry the color information. The reference shades are taken from the Vita Classical shade guide. The image of each reference shade is taken and stored in the dataset on google cloud. The YIQ coordinates of each reference shade are internally calculated and stored on Azure dataset. Thus, the patient's tooth is captured and its YIQ coordinates are matched with the YIQ coordinates of the reference images that are stored already in the database on Azure. The MDNS is used to connect to the MQTT server that uploads the image from pi to cloud. The best possible match is given as the output shade. The output shade is displayed on the LCD.

3.1 Dataset Description

VITA classical *A*1-*D*4 shade guide is used as a reference. The arrangement of the shades in the VITA classical family of shades is as follows:

$A1$—$A4$ (reddish-brownish)
$B1$—$B4$ (reddish-yellowish)
$C1$—$C4$ (grayish shades)
$D2$—$D4$ (reddish-gray)

The RGB values of tooth shade are very close to each other. As an alternate, the YIQ color scheme is used for better and accurate results. The YIQ coordinates of the input image are matched with the YIQ coordinates of the reference images to give the optimum shade match.

3.2 *Tunnel Description*

Color matching is a complicated process because of the subjectivity of color perception [12, 13] and environmental factors such as lighting conditions [14].

To overcome the problem of surrounding light a tunnel was constructed. 3D plastic PLA (Polylactic Acid) filament was used to print the tunnel having infill ranging from 30 to 50%. FDM-based 3D printer was used for printing. The tunnel is illuminated with white LEDs. This helps to eliminate the interference of surrounding light.

3.3 *Performance Metrics*

The CoTusk device has been tested on 30 patients successfully. The results in most of the cases were in accordance with the dentist's visual results. The accuracy obtained in the exact determination of the tooth color is 89.9%. This resulted accuracy is better than the previous approaches and along with the tunnel and cloud computing, various other additions have been installed for the tooth color determination that makes it more feasible and practical approach for usage.

4 Conclusion

Cotusk aids the dentist to determine the exact shade of the tooth so that no human or environmental factors may intervene. It helps people who want to improve their teeth and brighten the world with their smile. This study was done to evaluate the performance of Cotusk, a tooth shade detecting device which is a simple, portable, lightweight and user-friendly device for tooth color detection. Cotusk produces the most accurate matching for tooth as it involves complete standard for environment, light conditions and thus, gives a natural look with restoration. It can also be extended to detect the thickness of the enamel layer on tooth. The efficiency of this device is high and it reduces human error. The future scope of this methodology is to increase the accuracy in determining the color and to make the steps more productive and feasible.

Acknowledgements The authors are grateful to thank MHRD for funding this project of Design Innovation Center under the subtheme "Medical Devices & Restorative Technologies" for their consistent support in completing this work.

References

1. G. Wyszeki, W.S. Stiles, *Color Science*, 2nd edn. (Wiley, New York, 1982)
2. M.N. Alkhatib, R. Holt, R. Bedi, Age and perception of dental appearance and tooth colour. Gerodontology **22**(1), 32–36 (2005)
3. A.A. Barrett, N.J. Grimaudo, K.J. Anusavice, M.C.K. Yang, Influence of tab and disk design on shade matching of dental porcelain. J. Prosthet. Dent. **88**, 591–597 (2002)
4. T.P. van der Burgt, J.J. ten Bosch, P.C.F. Borsboom, W.J.P.M. Kortsmit, A comparison of new and conventional methods for quantification of tooth color. J. Prosthet. Dent. **63**, 155–162 (1990)
5. National Dental Practice-Based Research Network T, Quick Poll Results—Getting the color just right. 2016, 7 Oct 2016
6. R.D. Paravina, J.M. Powers, R.-M. Fay, Color comparison of two shade guides. Int. J. Prosthodont. **15**, 73–78 (2002)
7. C.L. Groh, W.J. O'brien, K.M. Boenke, Differences in color between fired porcelain and shade guides. Int. J. Prosthodont. **5**, 510–514 (1992)
8. A. Joiner, I. Hopkinson, Y. Deng, S. Westland, A review of tooth colour and whiteness. J. Dent. **1**(36), 2–7 (2008)
9. L.L. Odioso, R.D. Gibb, R.W. Gerlach, Impact of demographic, behavioral, and dental care utilization parameters on tooth color and personal satisfaction. Compend. Contin. Educ. Dent. (Jamesburg, NJ, 1995). Supplement 2000(29), S35–S41
10. B. Hugo, T. Witzel, B. Klaiber, Comparison of in vivo visual and computer-aided tooth shade determination. Clin. Oral Invest. **9**(4), 244–250 (2005)
11. W.D. Culpepper, A comparative study of shade-matching procedures. J. Prosthet. Dent. **24**, 166–1735 (1970)
12. A. Joiner, Tooth colour: a review of the literature. J. Dent. **32**(Suppl), 3–12 (2004)
13. J.J. Ten Bosch, J.C. Coops, Tooth color and reflectance as related to light scattering and enamel hardness. J. Dent. Res.
14. N. Corcodel, P. Rammelsberg, O. Moldovan, J. Dreyhaupt, A.J. Hassel, Effect of external light conditions during matching of tooth color: an intraindividual comparison. Int. J Prosthodont. **22**, 75–77 (2009)

Data Science and Computational Intelligence

Multi-agent Based Recommender System for Netflix

Harjot Kaur, Harsharandeep Kaur and Amitpal Singh

Abstract The main aim of this work is to build a multi-agent based recommender system that comprises heterogeneous software agents, where various member agents interact among themselves to accomplish various tasks and achieve objectives of the system. The main objectives of this work are: first, it presents some basics of multi-agent based recommender system. Second, it reviews the main research developments and works previously performed in the field of recommender systems with machine learning. Third, it introduces a multi-agent based recommender system framework based on a collaborative approach by considering various agents to recommend movies to users, who have similar interests. Furthermore, the proposed framework has been experimentally assessed by implementing a cosine similarity algorithm to measure user–user similarity based on movie ratings. The multi-agent programming environment NetLogo is used to simulate the results.

Keywords Recommender system · Machine learning · Netflix · Cosine similarity algorithm

1 Introduction

As information expands, it becomes possible to search and select information from tons of available options. Users may not have enough time or even knowledge to access this information. Recommender System (RS) (shown in Fig. 1) is a possible solution to resolve this problem, which uses information filtering techniques to

H. Kaur (✉) · H. Kaur · A. Singh
Department of Computer Science and Engineering, GNDU Regional Campus, Gurdaspur 143521, Punjab, India
e-mail: harjotkaursohal@rediffmail.com

H. Kaur
e-mail: harshdeep9469@gmail.com

A. Singh
e-mail: apsohal@yahoo.com

M. Dutta et al. (eds.), *Proceedings of International Conference on IoT Inclusive Life (ICIIL 2019), NITTTR Chandigarh, India*, Lecture Notes in Networks and Systems 116,
https://doi.org/10.1007/978-981-15-3020-3_20

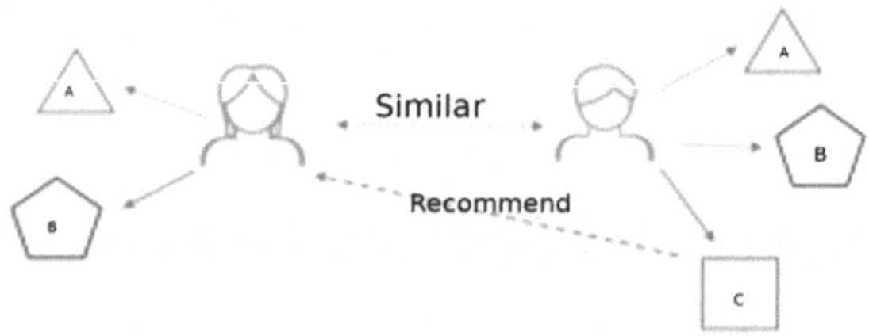

Fig. 1 Recommender system

assist users and gives them a personalized recommendation [5]. There are many recommender systems proposed for a different domain. Many of these applications are based on Multi-agent Systems (MASs). Therefore, Multi-agent based Recommender Systems is a system that comprises a set of agents, which interact (usually cooperate, coordinate, and negotiate) among themselves to fulfill the desired objectives of a recommender system [15].

1.1 Types of Recommender Systems

Different recommendation techniques, which may be applied in recommender systems are as follows:

1. **Content-based Filtering**: These techniques are based on how an item is described and what are various user preferences. It is not only based on users' preferences. In this system, items are described by using keywords, and taking into consideration various items liked by a particular user, his/her preference profile is built.
2. **Collaborative Filtering**: This technique builds a user item preference matrix for choices by users. Then, various users with relevant interests and preferences are matched based upon the computation of their profile similarity. This matrix is further used for recommendations. Such users build a group called a neighborhood [5, 8].
3. **Hybrid filtering**: This technique utilizes the blend of content-based and collaborative filtering for creating a hybrid recommendation system.

1.2 Netflix

Recommender systems are usually used in websites, which have millions of visitors, where users can receive a recommendation for movies, books, music, and videos. These days, the recommender system is getting popular in online streaming services like Netflix [2]. It is a Video on Demand (VOD) service. In Netflix, one may watch the Movie/TV show whenever one likes to watch (unlike TV channels). 60% of the movies on Netflix are selected based on personalized recommendation.

1.3 Machine Learning

According to Mitchell [11], machine learning can be defined as, "*A computer program is said to learn from experience E with respect to some classes of tasks T and performance measure P if its performance at tasks in T, as measured by P, improves with experience E.*" Broadly, machine learning can be categorized into two types, i.e.,

- **Supervised Learning**: It is the type of learning, which is guided by an instructor. The dataset acts as an instructor to train the model or the machine. Once the training of the model is complete, it begins to make predictions, whenever it is fed with a test dataset [10].
- **Unsupervised Learning**: This type of learning in models is performed through observation and later on structures are found from/within the data. Once a dataset is fed to the model, patterns and relationships are automatically diagnosed within the dataset by the model by constructing clusters in the same [10].

This paper is organized as follows. A concise overview of background work, which helped us to understand the basics of the MAS and how to implement them to recommend movies in a recommender system, is mentioned in Sect. 2. Section 3 defines the proposed Multi-agent based recommendation system framework and its algorithm and presents a simulation of the recommender system in NetLogo. The results for experimental analysis are present in Sects. 4 and 5, respectively. Our conclusion is drawn in Sect. 6.

2 Background Work

Derakhshan et al. [5] introduced a general design for an agent-based mobile recommender system. The proposed system employed a content-based filtering formula for a recommender system, which implemented a mobile application for Tabriz tourism. Lorenzi et al. [8] discussed multi-agent based recommender systems, which learnt from prior recommendations for updating profiles of users and generate implicit evaluation so that recommendations can be improved as per user feedback.

Morais et al. [12] demonstrated a recommender system with exploitation agents using two algorithms (i.e., associative rules and cooperative filtering). The authors proposed and evaluated a hybrid algorithm based on these two for alleviating user's contentment. Birukov et al. [3] introduced a tendency to propose multi-agent based recommender system for aiding groups of users in looking out the web by suggesting the most popular search with a search engine.

Lommatzsch et al. [7] introduced an agent-based recommender approach that integrated various heterogeneous recommender agents. The results presented in this work supported a personalized coefficient scheme for the movie domain. Moon et al. [14] presented the planned system, in which a market-based learning mechanism has been enforced to watch the predilections of various users for recommending them the applicable products.

Rosaci and Sarne [16] presented a brand new multi-agent system, referred to as ARSEC (Adaptive Recommender System for e-Commerce), where the e-Commerce website is related to a merchant agent and supported the similarity existing among the worldwide profiles. In this system, the sets of customers are divided into clusters, each being controlled by a counselor agent. The recommendations generated in ARSEC were as a consequence of the coordination between the vendor agent and a few counselor agents related to the client. Andronico et al. [1] studied an integrated multi-agent based recommendation system that implied academic resources to students for mobile learning.

Kurapati et al. [6] investigated an advanced multi-agent TV recommender system that encapsulated three user data streams, i.e., implicit read history, explicit preferences, and assessment information on particular shows to render TV shows recommendations. Xu and Qiu et al. [21] introduced the analysis of client segmentation by proposing a hybrid recommendation system with a multi-agent system to serve the customers with a high profit.

Takeuchi et al. [18] demonstrated a unique real-world recommendation system that made endorsements of retailers supported by users' past location knowledge history. Skocir et al. [17] proposed a recommendation algorithmic program, which supported the knowledge gathered from users' interaction within a game.

Chan et al. [4] represented a multi-agent design for mobile health observation, including a group of intelligent agents that collected patient information and prescribed actions to patients and medical employees in mobile surroundings. Macho et al. [9] addressed the matter of conference arrangement for many participants taking into consideration constraints for private agendas and transportation schedules.

Walter et al. [19] presented a trust-based recommender system framework in a social network. Further, the researchers investigated how the dynamics of trust among agents affected the efficacy of the system, by comparing it to a frequency primarily based recommender system.

3 Research Gaps

This paper works in the direction of proposing a framework for a multi-agent based recommender system for Netflix with the machine learning algorithm because of few reasons, first, Netflix has over 137 million streaming subscribers and one of the problems is preserving their userbase and keeping them entertained. The Netflix recommendation algorithm solves the problem by advertising shows and movies that pertain to the user interest and also for subscribed user retention. The other is to give the advantage to the business value because more members equal more data and that is going to make the company at a better peak.

4 Proposed Model

In this work, a general model for a multi-agent based recommender system for Netflix (as shown in Fig. 2) has been proposed considering three main constituent agents, i.e.,

User agent: This agent runs on the client-side and gets input from users and presents appropriate results to users [5].
Recommender agent: It uses users' preferences to generate a recommendation from the server-side [5].
Activity agent: It is used for storing and removing information from the repository. It updates the information for a recommendation with a change of interest of the user [5].

In this model, we use collaborative filtering for evaluation of better recommendation. The mechanism of the same follows the steps as given below:

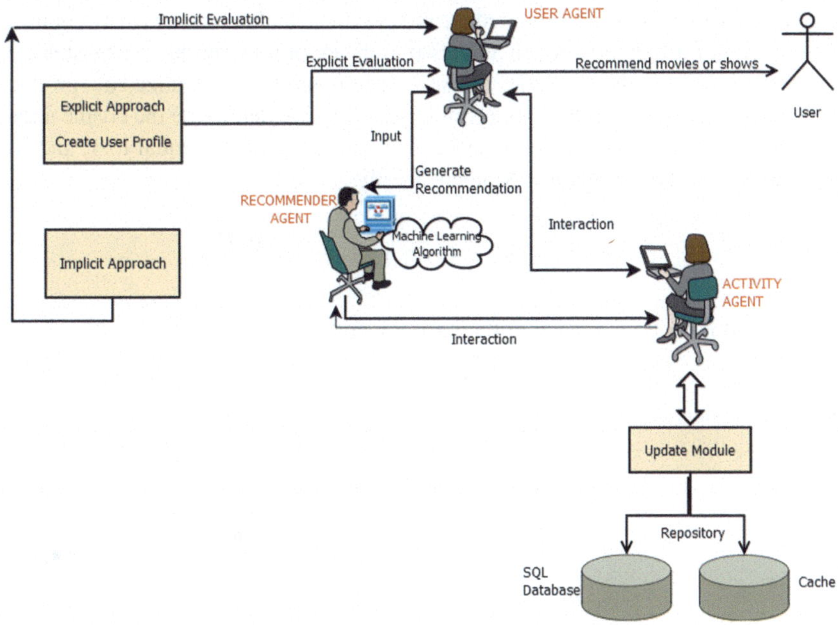

Fig. 2 Schema of multi-agent based recommender system for Netflix

- *Information gathering*: There are two main approaches for gathering information, i.e., explicit and implicit approaches. The explicit approach uses data or information that is provided in an intentional manner, i.e., input explicitly taken from users such as movie ratings. On the other hand, in an implicit approach, user data is gathered in an implicit data manner from available data streams like search history, clicks, access time and similar user profile, etc.
- *Storing data*: The activity agent is used to store and update the data on the basis of collaborative filtering which is based on the individual user.
- *Filtering the data*: After the data is collected and stored, it is filtered so that the required and relevant information can be extracted for making the final recommendations. The recommender agent uses the user profile vector to recommend various attractions or movies or TV shows to the user with their preferences.

4.1 Algorithm Used for Recommender System

The algorithm used in this work for the recommendation in Netflix is supervised machine learning KNN (K-nearest neighbor) with cosine similarity. Cosine similarity (Fig. 3) is the similarity metric used to calculate movie recommendations on the basis of movie ratings given by the users. Then, it calculates the cosine angle similarity between each user. If the user unrated the movies, then it takes an average rating from that user [20]. Further, how much alike two users are or how similar their tastes are, is measured by using the *similarity measure*. It computes the normalized dot product of the likes of two users. The cosine similarity is used to determine the cosine angle between two objects. The trigonometric cosine measure of 0 is equal to 1, and for any other value, it is less than 1. Mathematically,

$$\text{similarity}\ (A, B) = \text{Cos}\ (A, B) = \frac{A \cdot B}{||A|| \cdot ||B||}$$

where A and B are components of the vector, respectively.

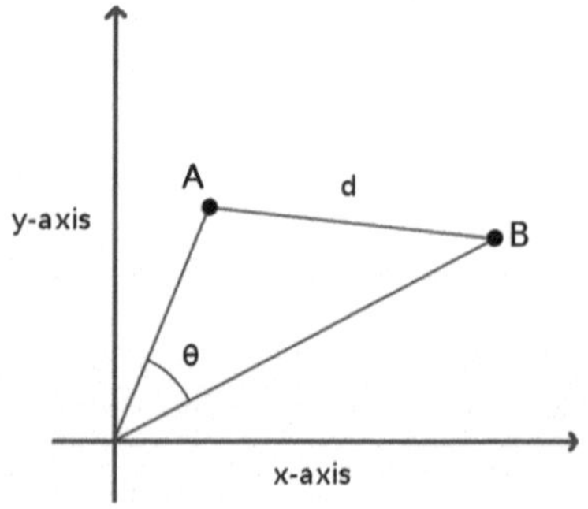

Fig. 3 Cosine similarity

Table 1 Cosine similarity data

User	M1	M2	M3	M4	M5	M6
A	3	2	4	5	4	1
B	1	4	3	5	4	2

$A = (3, 2, 4, 5, 4, 1)$
$B = (1, 4, 3, 5, 4, 2)$

For better understanding, let us assume user *A* and user *B* who give ratings to five movies (Table 1). Based on their ratings, we can measure the cosine similarity [20]. Thus,

$$\text{Similarity } (A, B) = \frac{3.1+2.4+4.3+5.5+4.4+1.2}{\sqrt{(3^2+2^2+4^2+5^2+4^2+1^2)} \times \sqrt{(1^2+4^2+3^2+5^2+4^2+2^2)}} = 0.91$$

The main advantage of cosine similarity is that even if the ratings of two similar users are far apart by the Euclidean distance, chances of they may still be oriented closer together. Also, the smaller is the angle, the higher is the cosine similarity.

5 Simulation of Recommender System in NetLogo

In this paper, a multi-agent based recommender system has been presented which is designed in NetLogo (Fig. 4), in which the movies are recommended to various users by measuring their similarity using cosine similarity machine learning algorithm.

In the presented experimental analysis, the data set formed is collected for various users and is present in the form of CSV file, which contains attributes related to user, movie, movie title, ratings, year, genre, and timestamp. Each row in the dataset stands for one user rating. The column *user-id* consists of the identity of the user that has given the ratings. The column *movie id* contains the identity number assigned to the movie; the column *rating* comprises various users assigned ratings. The values of ratings range from 1 to 5. Each user gives ratings for every movie and a *timestamp* is assigned to the rating as well. The time at which the user left the ratings is referred to as timestamp.

The four different modules named *setup, sign up, search for movies, and get a recommendation* function as a part of a recommender system. The *setup* module creates three agents named as a user agent, activity agent, and recommender agent, respectively, linked with each other. The *signup* module allows the user to enter the username and password to authenticate the former. The data, which user enters at the time of the *signup* is written into CSV file automatically The third module, i.e., *search for movies* allows the user to select the movies from the drop-down list with the genre according to their interest. Eventually, for a top recommendation to a user, the *recommendation* module is used and the movie interest of the user is matched

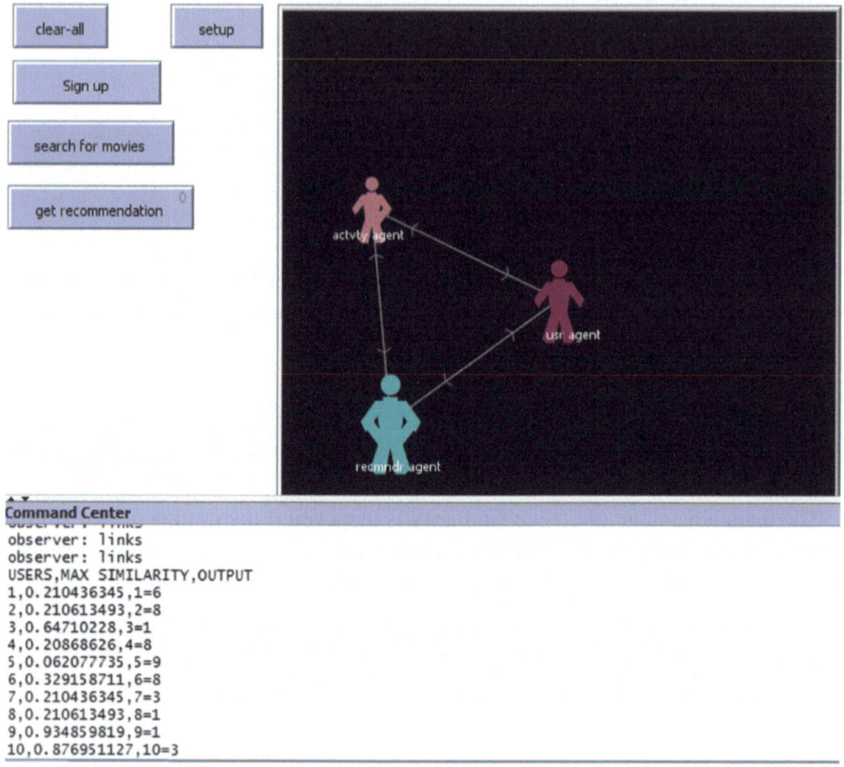

Fig. 4 NetLogo world for multi-agent based recommender system

with the rest of the users. Measuring the cosine similarity will make the closest match and the prediction of the movie to a given user. It shows the user–user similarity.

6 Results

After simulating the recommender system modeled in NetLogo on real-time data, which have been collected from the Internet for various users who watched movies regularly. The results have been recorded and evaluated for movie recommendation and they can be graphically presented in the form of Figs. 5 and 6, respectively.

In Fig. 5, the graph represents the horizontal and vertical axes indicating movies and their ratings, respectively. Therefore, a user assigns integer values in the range (1–5) as ratings. Furthermore, it has been concluded that good movies usually have a high average rating because they are watched by a large number of people and thus are assigned a higher average rating. This graph shows that not all movies have the

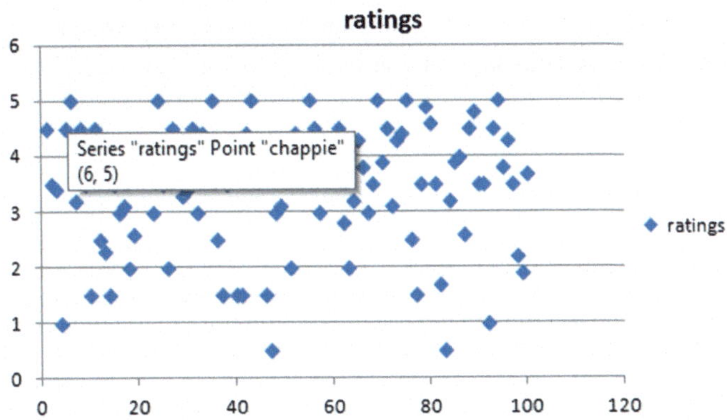

Fig. 5 Movies with high average rating

Fig. 6 User–user similarity

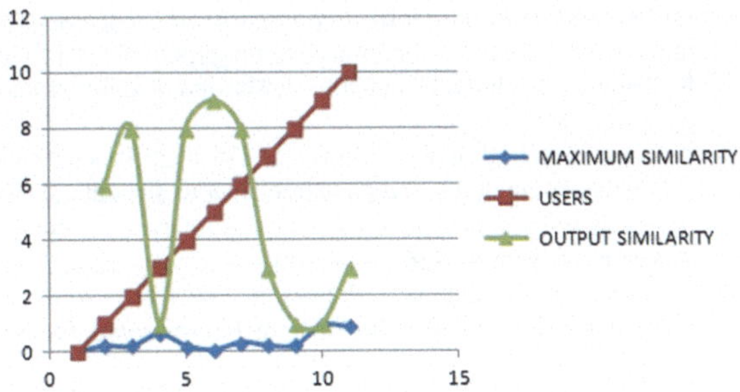

same ratings [13]. Therefore, more users as compared to movies with lower average ratings have watched movies with higher ratings.

In Fig. 6, the graph shows the maximum similarity among various users. The similarity of one user with the other users is measured by using the cosine similarity algorithm. It also made a top recommendation by matching the similar interest of users for the movie.

7 Conclusions

In this paper, we proposed a multi-agent based recommender system for Netflix with a supervised machine learning algorithm (cosine similarity algorithm) which considered three agents, i.e., activity, user and recommender agents. These agents coordinate among themselves to make the top endorsement for movies to various users. The cosine similarity algorithm used finds the cosine distance between various movie vectors on the basis of ratings and then matches the movie interest of one user with the rest of the users to recommend the movies for Netflix. The presented research can be extended in future by applying the proposed approach to predict the top recommendation of a movie for a user on the basis of a genre.

References

1. A. Andronico, A. Carbonaro, G. Casadei, L. Colazzo, A. Molinari, M. Ronchetti, Integrating a multi-agent recommendation system into a mobile learning management system (2003)
2. A. Bandyopadhyay, How Netflix deploys open source AI to reveal your favorites. https://itsfoss.com/netflix-open-source-ai/ (2018)
3. A. Birukov, E. Blanzieri, P. Giorgini, Implicit: an agent-based recommendation system for web search, in *International Joint Conference on Autonomous Agents and Multi-agent Systems*, pp. 618–624 (2005)
4. V. Chan, P. Ray, N. Parameswaran, Mobile e-Health monitoring: an agent-based approach. in *IEEE Conference IET Communication*. vol. 2, pp. 223–230 (2008)
5. F. Derakhshan, M. Parandeh, A. Moradnejadm, An agent-based mobile recommender system for tourisms, in *Research World International Conference* (Berlin, Germany, 2016), pp. 30–34. http://www.worldresearchlibrary.org/up_proc/pdf/177145508804 83034.pdf
6. K. Kurapati, S. Gutta, D. Schaffer, J. Martino, J. Zimmerman, A multi-agent TV recommender, pp. 1–8 (2001)
7. A. Lommatzsch, B. Kille, S. Albayrak, An agent-based movie recommender system combining the results computed based on heterogeneous semantic datasets, in *Proceedings of the 13th GI International Conference on Innovative Internet Community Systems and the Workshop on Autonomous Systems I2CS'13* (2013)
8. F. Lorenzi, B. Fontanella E. Prestes, A. Peres, How to improve multi-agent recommendations using data from social networks?, in *International Florida Artificial Intelligence Research Society Conference*, pp. 63–68(2014)
9. S. Macho, M. Torrens, B. Faltings, A multi-agent recommender system for planning meetings, pp. 1–9 (2002)
10. Machine Learning. https://www.edureka.co/blog/what-is-machine-learning/#SupervisedLearning
11. T. Mitchell, in *Machine Learning* (McGraw Hill), p. 2. ISBN 978-0-07-042807
12. A.J. Morais, E. Oliveira, A.M. Jorge, A multi-agent recommender system. in *Distributed Computing and Artificial Intelligence. Advances in Intelligent and Soft Computing*, eds. by S. Omatu J. De Paz Santana, S. González, J. Molina, A. Bernardos, J. Rodríguez, vol. 151, pp. 281–288 (2012)
13. Movie Recommendation. http://www.webpages.uncc.edu
14. S.K. Moon, T.W. Simpson, S.R.T. Kumara, An agent-based recommender system for developing customized families of products. **20**, 649–659 (2008) https://doi.org/10.1007/s10845-008-0154-9
15. U. Pakdeetrakulwong, P. Wongthongtham, State of the art of a multi-agent based recommender system for active software engineering ontology. Int. J. Digit. Inf. Wireless Commun. **3**(4), 29–42 (2013)
16. D. Rosaci, G.M.L. Sarné, A multi-agent recommender system for supporting device and captivity in e-Commerce. J. Intell. Inf. Syst. **38**, 393–418 (2012) https://doi.org/10.1007/s10844-011-0160-9
17. P. Skocir, L. Marusic, M. Marusic, A. Petric, The MARS—A multi-agent recommendation system for games on mobile phones, in *Agent and Multi-Agent Systems. Technologies and Applications. KES-AMSTA 2012*, eds. by G. Jezic, M. Kusek, N.T. Nguyen, R.J. Howlett, L.C. Jain, Lecture Notes in Computer Science, vol. 7327 (2012)
18. Y. Takeuchi, M. Sugimoto, City Voyager: an outdoor recommendation system based on user location history. in *Ubiquitous Intelligence and Computing. UIC 2006*, J. Ma, H. Jin, L.T. Yang, J.J.P. Tsai, eds. by, Lecture Notes in Computer Science, vol. 4159 (2006)

19. F.E. Walter, S. Battiston, F. Schweitzer, A model of a trust-based recommendation system on a social network. J. Auton. Agents Multi-Agent Syst. **16**, 57–74 (2008). https://doi.org/10.1007/s10458-007-9021-x
20. N. White, Movie recommendations with k-Nearest neighbors and cosine similarity. https://neo4j.com/graphgist/movierecommendations-with-k-nearest-neighbors-and-cosine-similarity
21. M. Xu, J. Qiu, Y. Qiu, Mining the profitability of customers and making right recommendations, in *International Conference on Machine Learning and Cybernetics*, pp. 1990–1994 (2003)

Review of Various Sentiment Analysis Approaches

Ishana Attri and Maitreyee Dutta

Abstract In this new world of social media, the trend of sharing views or opinions on several web portals such as blogs, Instagram pages, Facebook pages, Twitter, YouTube is becoming so popular and common. The data on these sites is humongous which attract the researchers to focus on sentiment analysis. Sentiment analysis is a method of determining positive, negative, neutral, sarcastic, ironical comments. Researchers tend to develop a system or a valid method for accurate prediction of opinions of users. There are various techniques that are used to detect the emotions from the online data such as supervised learning approaches, unsupervised, and many more. In this paper, we present a study of the methods used in the assessment of sentiments.

Keywords Supervised learning · Opinion Mining (OM) · Opinion · Sentiment Analysis (SA)

1 Introduction

Sentimental Analysis (SA) is a process that involves NLP and computer science techniques to keep in track with the mood or sentiments of the people about any latest news or any trending topic or any important news.SA keeps track of the writer's point of view from a text. The main or important function in SA or OM (opinion mining) shall inspect the polarity of the document at different stages, i.e., sentence level, entity and aspect level, or in sentence level.OM is a process that involves collection of data from various sites and then further examines its polarity. SA is a very useful process and this process is used in marketing, e.g., Nykaa which is a famous online shopping site for beauty product launches a new lipstick and they want to check whether or not it is a successful product then they check it by the reviews they got

I. Attri (✉) · M. Dutta
Computer Science and Engineering, National Institute of Technical Teachers Training and Research, Chandigarh 160019, India
e-mail: ishanaattri@gmail.com

M. Dutta et al. (eds.), *Proceedings of International Conference on IoT Inclusive Life (ICIIL 2019), NITTTR Chandigarh, India*, Lecture Notes in Networks and Systems 116,
https://doi.org/10.1007/978-981-15-3020-3_21

from the lipstick users on their sites further they can enhance or add new shades as per the customer requirement.

In today's era, online media and social networking sites (SNS) is taking the region of offline media unexpectedly, which inspire more people to take part in various latest discussions such as political discussions, movies reviews, etc. and ensure the people that their opinion is worth and their opinion counts. Social media provides the open platform for extensive passing of their thoughts and inspiring other human beings for organizational debates with open perspectives. Social networking sites offer higher means to get short reaction and opinion on various worldwide issues and entities with the help of text, tweets, comments, pictures/images, movies. That's why, it could be helpful to analyze peoples' evaluations for gaining knowledge of the customer behavior, market patterns, and fashion of society. It has been found that Twitter has almost 325 million monthly energetic users and it deals with almost a half a million tweets every day. That's why, it used as a great source to analyze and extract heterogeneous critiques published by means of different human beings from different societies for unique functions like the development of nice of services and products, prediction of customer's request and experiences, etc.

SNS are very helpful to show or express opinions. These opinions are extracted from blogs, Twitter, Facebook, discussion forums. The extracted data comprise highly unstructured information in the form of text with image, lingos, photos, animations, and films which are beneficial in making public aware about various issues.

SA is also known as OM which is a subfield of text mining. This is a process used to analyze the point of view of a writer. This is a very fascinating process because it is tough to analyze the mood of a person with the help of text that's why it attracts the attention of researchers and hence it becomes so popular these days. Sentiment analysis process consists of various steps; they are data preprocessing, review analysis, sentiment classification, and analyzing the result. Figure 1 shows a basic sentiment analysis process.

- **Data acquisition**: First of all, we have to collect the data from different online sources.
- **Data preprocessing/data preparation**: It is a process of cleaning and preparing text for further use. In this step, raw data is converted into information.
- **Review analysis**: Review analysis is one of the main processes. It conducted in 6 broad dimensions. They are aspect extraction, review usefulness measurement, sentiment classification, lexicon creation, subjectivity classification, opinion spam detection.
- **Sentiment classification**: In this step, sentiment are classified into binary (positive, negative) or multiclass (positive, negative, sarcastic, neutral, ironic, angry). This can be done by using various classifiers, i.e., SVM, naïve Bayes, etc.
- **Result**: Finally the predicted result is generated and stored for further analysis.

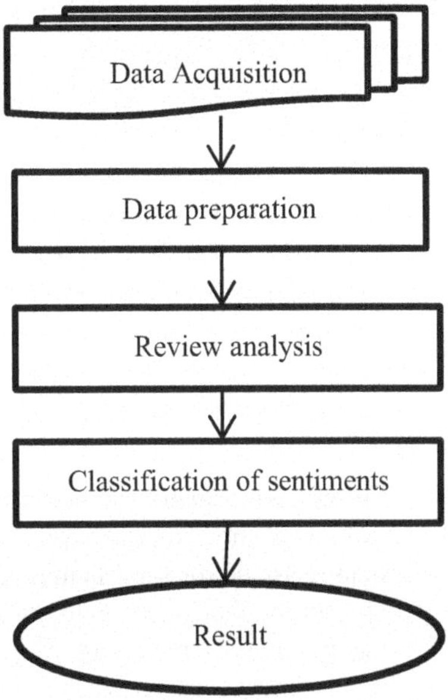

Fig. 1 Sentiment analysis process

2 Data Sources

There are various sources from where we collect; they are shopping sites, social networking sites, datasets, forums, microblogs.

2.1 **Shopping sites**: There are various online websites where users can buy any product by checking various reviews from other buyers. The reviews on the shopping sites such as Amazon, Flipkart, Myntra, etc., are the opinions of the buyers about a specific product. They mainly are in the unstructured format [1].

2.2 **Social networking sites**: Facebook, Instagram, Qzone, Tumblr, YouTube are various sites from where we collect our data to analyze the reviews of the users.

2.3 **Dataset**: There are various datasets available online; they are related to movies, political reviews, multi-domain sentiment datasets, Amazon datasets, etc.

2.4 **Forum**: To check the data in single domain, we use forums or message boards. In forums, the conversation about a specific topic is posted and hence this data can be used for sentiment analysis.

2.5 **Micro-blogs**: Twitter is a famous micro-blogging site. Here each Tweet is of maximum 40 words and there are more than 250 million users and hence for sentiment analysis, we can use twitter data.

3 Approaches for Sentiment Analysis

There are 3 main important basic approaches used for sentiment analysis; they are as follows:

The first one is the supervised system learning method where various algorithms examine the facts from formerly annotated data, permitting them to arrange new, unlabeled records [3]. The next one is unsupervised learning approach, in which the process depends on a set of regulations and heuristics achieved from linguistic expertise [2]. The last one is semi supervised learning based approach; in this, both supervised and unsupervised algorithms are used for sentiment analysis. There is additionally growing variety of studies reporting the successful mixture of each method [4–6]

3.1 **Supervised learning based approach**: It is a process of training a computer using well-labeled information and is also known as machine learning or statistical methods for sentiment analysis and classification, composed of different algorithms that investigate simple patterns from example records [7], which means records whose magnificence or label or a tag is thought for every instance, that we must use later when we arrange new untagged facts. Normally, the steps involve system learning and technology of features to represent the element whose beauty is to be expected, and then used in its illustration as feedback for the rule set. Few features regularly utilized in sentiment analysis are time period frequency, POS tags, terms of sentiments, laws of view, transition in sentiment and syntactic reliance [4, 8].

In [9], the researchers have been the first one to use supervised sentiment approach. They compared NB, Maximum Entropy classification, and SVM tactics, and found that the usage of unigrams as capabilities yielded precise effects.

Pak and Paroubek [10] focus their research work on Twitter satisfied and unsatisfied emoticons to construct a classified training corpus. They later use 3 classifier methodologies: NB Classifier, CRF, and SVM, and find that the NB yielded the satisfactory outcomes. In [11], as with the emoticon labels, Davidov, Tsur, and Rappaport used hash tags to train an algorithm close to k-Nearest Neighbors, the magnificence of untagged tweets is predicted (Table 1).

3.2 **Unsupervised learning based approach**: Often known as semantic predominantly based techniques, the use of set guidelines and heuristics derived from linguistic knowledge helps to assess the polarity of textual content. The first step is to analyze the sentiment shifters and their scope, to mark each sentence with its sentiment polarity, i.e., positive, negative with the use of a lexicon, the next step. Sense shifters are also referred to as valence shifters. These are the words or phrases that affect a sentence or phrase's polarity. In the next step, we check how the final sentiment rating was affected by the clauses [8]. Subsequent steps should include a summary of opinion and visualization. The first examination to address Opinion Mining in an unmonitored manner changed to [22], in which the author created a set of rules for the extraction of features using bigrams and

Table 1 Supervised learning approach

Ref. no.	Technique used	Objectives	Outcome
[12]	Naïve Bayes	Extract opinions or sentiments from a unstructured text	Authors develop a 2 stage Markov blanket classifier by using NB classifier. With the help of this new classifier, the accuracy of the system is enhanced up to 92%
[13]	SVM	Twitter generic feature set is domain transferable and to reduce the feature set	Reduce the feature set into 7 meta features. TSLS domain transferable and the model gives higher accuracy when it used with DAN2
[14]	Multiclass SVM	Compare the multiclass SVM with J&L's, CRF with interdependencies and CRF without interdependencies	Multiclass SVM have accuracy of 61%, precision of 61.9%, and the recall value is 93.4%
[15]	SVM	Text identification by using Fisher Discriminant Ratio	The experiment shows that the FD ratio based on word frequency is 83.3% accurate by using 1006 car review documents
[16]	Naïve Bayes, SVM, CRF	A classifier to check whether a tweet is negative, neutral, or positive	The proposed Classifier based on multinomial Naïve Bayes classifier that uses N-gram and POS-tags as features. The accuracy of the classifier is good
[17]	Naïve Bayes, Maximum entropy classifier	Compare Naïve Bayes and maximum entropy classifier	Max entropy classifier with Google translator provide with better accuracy, i.e., 74.04%
[18]	Naïve Bayes, Maximum entropy, SVM	Comparative study between various classifiers	The accuracy of SVM classifier is 86.4%, Maximum entropy classifier is 85.4%, and accuracy of Naive Bayes is 85.8%
[19]	Naïve Bayes, SVM	Sentimental analysis in Cantonese dialect by using Naïve Bayes and SVM	Naïve Bayes gives comparable and sometimes better result than SVM. Here concept based bigrams give better result than trigrams

(continued)

Table 1 (continued)

Ref. no.	Technique used	Objectives	Outcome
[20]	SVM, Naïve Bayes	Extraction of emotions from multi-lingual unstructured text	Proposed a improved framework for analysis of emoticons of web users in multi-lingual data. The result shows that the proposed method is more efficient than corpus driven approach
[21]	SVM, Naïve Bayes, N-gram	Sentiment classification of online reviews to travel destinations by supervised machine learning approaches	SVM model and character based N-gram model provided the best result. But in general, all three algorithms provide the accuracy up to 80%

grammatical rules, then checking the polarity by PMI and finally calculating the common polarity of each extracted bigram to estimate the general polarity of an overview. Hu and Liu [23] created a list of sentiment terms using WordNet [24] that later predicts sentence orientation, i.e., positive, negative, with the help of determining the usual orientation of the expression. Later, in [25], researchers used intensification words (very, a little, good, pretty) and denial words (no longer) to test the polarity of the affected words' feelings. Vilares et al. in [26]. Likewise, the analysis of syntactic dependencies was integrated with terms of negation and intensification to help result in coping with adverse clauses (Table 2).

3.3 **Semi Supervised Learning Approach**: Semi supervised learning algorithm is a category of system learning algorithm of obligations and techniques that still employ unlabeled facts for learning—usually a small amount of classified facts with a big amount of unlabeled data. Semi supervised falls in between unsupervised and supervised learning (Table 3).

4 Conclusion

In this paper, we provided a very brief introduction to sentiment analysis and three approaches of sentiment analysis viz. supervised, semi supervised, and unsupervised learning approaches. There are various papers reviewed on the basis of which we can conclude that there are various methods to achieve better accuracy in sentiment analysis but the more effective in case of supervised learning approach SVM, multiclass SVM gives better results and accuracy. In case of unsupervised learning approach, SentiWordNet and clustering algorithm performed better and in case of semi supervised approach, SVM with clustering or lexicon based model provides the

Table 2 Unsupervised learning based approach

Ref. no.	Technique used	Objective	Outcome
[27]	Unsupervised Lexicon based algorithm	SA in various social networking sites i.e., MySpace, Twitter, and Forum	It does not require any training and hence can be used for a wide range of data. It provides robust and reliable solution
[28]	Noble unsupervised SA framework	Detection of human emotions or opinions from social media images	More effective to solve the problem of lack of proper labels and helpful to find the sentiment gaps
[29]	CN, CN2, CN3, Google EN, Yahoo EN, DictEN	Sentiment Analysis in Chinese language	Improve the overall accuracy
[30]	CDA model	To develop a very easy, simple and flexible, method domain and language independent	The proposed algorithm method satisfies all the objective and gives accuracy the same as semi supervised method
[31]	Spectral Clustering, K-means Clustering	To find polarity in any tweet	Divide the data into two clusters, i.e., positive and negative. Behavioral analysis help to enhance the accuracy
[32]	SentiWordNet	Sentiment analysis and classification of data in positive and negative sentences	Very simple and domain independent method for sentiment analysis, accuracy is also improved

(continued)

Table 2 (continued)

Ref. no.	Technique used	Objective	Outcome
[33]	Opinion digger	To proposed a unsupervised and effective method for sentiment analysis	Improve the effectiveness by extracting the product aspects and estimating the aspect ratio
[34]	PMI-TFID	Product feature extraction by unsupervised learning method	Feature oriented opinions lexicons are superior to general opinion lexicons for feature oriented opinion determination and hence PMI-TFID shows better distinction ability
[35]	SentiWordNet 4.0 and SpanishSentimentWordNet	Unsupervised polarity detection techniques based on the lexicon of opinion words	More Accurate as compare to other models
[36]	Improved SO-PMI	To expand the reference words to sets of words, to introduce balancing factor, and to detect neutral sentiments	More balanced results and accuracy exceeds up to 62%

Table 3 Semi supervised learning approaches for SA

Ref. no.	Technique used	Objectives	Outcomes
[37]	Filter algorithm, Bayesian n/w structure	To solve the real-world multidimensional sentiment analysis problem	Accuracy is better than unidirectional approaches. In large datasets, semi supervised approaches give better result
[38]	SentiWordNet with SVM	Focus on problem data scarcity, inadequate data, performance and domain dependencies	Give solution of every problem and focus on the importance of nouns and used them as semantic features
[39]	LCCT (lexicon based and corpus based co-training)	To propose a sentiment analyzer for both Lexicon as well as corpus level information	Capable of handling both domain specific and domain independent knowledge and gives better accuracy and performance in English and Chinese
[40]	Novel semi supervised sentiment prediction algorithm	To propose a model for incorporating lexical information as well as unlabeled data	Outperformed various supervised and semi supervised techniques
[41]	K-medoids and Naïve Bayes	To propose an effective semi supervised approach to gain deeper insights of movie review	Manual labeling gives more effectiveness over cluster based approaches
[42]	Semi supervised approach with Dynamic Subspace Generation	To propose a semi supervised learning model for imbalance sentiment classification	Proposed model outperformed the traditional static subspace generation model
[43]	SB DAE, SB DAE^{+}	To check the usage of auto encoders in modeling textual data	Proposed model gives better result than all the existing methods
[44]	LSAR is applied using SVM	Domain specific lexicon model	Achieve better accuracy as compared with AFINN i.e., 0.94
[45]	Graph based semi supervised model	Semi supervised model deals with social relation and text similarity	Proposed model perform better than state of art model
[46]	Vector space model and SVM	Model for sentiment analysis in unstructured Big data	Significant improvement in emotion recognition and polarity detection

users with better accuracy. Sentiment analysis becomes the latest trend in research area and hence in future various new hybrid technologies, methods, or tools may be built which improve the accuracy and effectiveness of opinion mining or sentiment analysis.

References

1. M. Malik, S. Habib, P. Agarwal, A novel approach to web based review analysis using opinion analysis, in *International Conference on Computational Intelligence and Data Science (ICCIDN)* (2018), pp. 1202–1209
2. D. Vilares, M.A. Alonso, C. Gomez-Rodrguez, A syntactic approach for opinion mining on Spanish reviews. Nat. Lang. Eng. **21**, 139–163 (2015)
3. B. Pang, L. Lee, S. Vaithyanathan, Thumbs up sentiment classification using machine learning techniques, in *Empirical Methods in Natural Language Processing*, Association for Computational Linguistics, Philadelphia, vol. 10 (2002), pp. 79–86
4. M. Joshi, C. Penstein Rose, Generalizing dependency features for opinion mining, in *4th International Joint Conference on Natural Language Processing*, Association for Computational Linguistics (Suntec, Singapore, 2009), pp. 313–316
5. Y. Wu, Q. Zhang, X. Huang, L. Wu, Phrase dependency parsing for opinion mining, in *Conference on Empirical Methods in Natural Language Processing*, Association for Computational Linguistics, vol. 3 (Singapore, 2009), pp. 1533–1541
6. T. Nakagawa, K. Inui, S. Kurohashi, Dependency tree-based sentiment classification using CRFs with hidden variables, in *The Annual Conference of the North American Chapter of the Association for Computational Linguistics: Human Language Technologies*, Association for Computational Linguistics, Los Angeles, CA, USA (2010), pp. 786–794
7. V.L. Rebolledo, G. L'Huillier, J.D. Velásquez, Web pattern extraction and storage, in *Advanced Techniques in Web Intelligence-I*, (Springer, Berlin, 2010), pp. 49–77
8. B. Liu, Sentiment analysis and opinion mining. Synth. Lect. Hum. Lang. Technol. **5**, 1–167 (2012)
9. B. Pang, L. Lee, S. Vaithyanathan, Sentiment classification using machine learning techniques, in *Conference on Empirical Methods in Natural Language Processing*, vol. 10 (2002), pp. 79–86
10. A. Pak, P. Paroubek, Twitter as a corpus for sentiment analysis and opinion mining, in *7th International Conference on Language Resources and Evaluation*, European Language Resources Association, Valletta, Malta, vol. 10 (2010), pp. 1320–1326
11. D. Davidov, O. Tsur, A. Rappoport, Enhanced sentiment learning using Twitter hashtags and smileys, in *23rd International Conference on Computational Linguistics*, Association for Computational Linguistics, Beijing, China (2010), pp. 241–249
12. X. Bai, R. Padman, Markov blankets and meta-heuristic search: sentiment extraction from unstructured text, Lecture Notes in Computer Science, vol. 3932 (2006), pp. 167–187
13. M. Ghiassi, S. Lee, A domain transferable lexicon set for twitter sentiment analysis using a supervised machine learning approach. Exp. Syst. Appl. (2018) (manuscript). Elsevier
14. K. Xu, S.S. Liao, J. Li, Y. Song, Mining comparative opinions from customer reviews for competitive intelligence. Decis. Support Syst. **50**, 743–754 (2011)
15. S. Wang, D. Li, X. Song, Y. Wei, H. Li, A feature selection method based on improved Fisher's discriminant ratio for text sentiment classification. Expert Syst. Appl. **38**, 8696–8702 (2011)
16. A. Pak, P. Paroubek, Twitter as a corpus for sentiment analysis and opinion mining, in *7th International Conference on Language Resources and Evaluation (LREC 2010)*, vol. 10, European Language Resources Association, Valletta, Malta (2010), pp. 1320–1326
17. A.K. Soni, Multi-lingual sentiment analysis of twitter data by using classification algorithms, in *Second International Conference on Electrical, Computer and Communication Technologies (ICECCT)* (2017), pp. 1–5

18. R. Xia, C. Zong, S. Li, Ensemble of feature sets and classification algorithms for sentiment classification. Inf. Sci. **181**, 1138–1152 (2011)
19. Z. Zhang, Q. Ye, Z. Zhang, Y. Li, Sentiment classification of Internet restaurant reviews written in Cantonese. Expert Syst. Appl. **38**, 7674–7684 (2011)
20. V.K. Jain, S. Kumar, S.L. Fernandes, Extraction of emotions from multilingual text using intelligent text processing and computational linguistics. J. Comput. Sci. **21**, 316–326 (2017)
21. Q. Ye, Z. Zhang, R. Law, Sentiment classification of online reviews to travel destinations by supervised machine learning approaches. Expert Syst. Appl. **36**, 6527–6535 (2009)
22. P.D. Turney, Thumbs up or thumbs down semantic orientation applied to unsupervised classification of reviews, in *40th Annual Meeting on Association for Computational Linguistics*, Association for Computational Linguistics, Philadelphia, PA, USA (2002), pp. 417–424
23. M. Hu, B. Liu, Mining and summarizing customer reviews, in *International Conference on Knowledge Discovery and Data Mining* (ACM, 2004), pp. 168–177
24. G.A. Miller, WordNet: a lexical database for English. Commun. ACM **38**(11), 39–41 (1995)
25. M. Taboada, J. Brooke, M. Tofiloski, K. Voll, M. Stede, Lexicon based methods for sentiment analysis. Comput. Linguist. **37**, 267–307 (2011)
26. L. Zhou, P. Chaovalit, Ontology-supported polarity mining. J. Am. Soc. Inform. Sci. Technol. **59**, 98–110 (2008)
27. G. Paltoglou, M. Thelwall, Twitter, MySpace, Digg: unsupervised sentiment analysis in social media. ACM Trans. Intell. Syst. Technol. **3**, 1–19 (2012)
28. Y. Wang, S. Wang, J. Tang, H. Liu, B. Li, Unsupervised sentiment analysis for social media images, in *Twenty-Fourth International Joint Conference on Artificial Intelligence Unsupervised Sentiment Analysis for Social Media Images* (2015), pp. 2378–2379
29. X. Wan, Using bilingual knowledge and ensemble techniques for unsupervised Chinese sentiment analysis, in *Empirical Methods in Natural Language Processing* (2008), pp. 553–561
30. S. Brody, N. Elhadad, An unsupervised aspect-sentiment model for online reviews, in *Human Language Technologies, Annual Conference of the North American Chapter of the ACL* (2010), pp. 804–812
31. M. Unnisa, A. Ameen, S. Raziuddin, Opinion mining on Twitter data using unsupervised learning technique. Int. J. Comput. Appl. **148**, 12–19 (2016)
32. V. Soni, M.R. Patel, Unsupervised opinion mining from text reviews using SentiWordNet. Int. J. Comput. Trends Technol. **11**, 234–238 (2014)
33. S. Moghaddam, M. Ester, Opinion digger, in *19th ACM International Conference on Information and Knowledge Management* (2010), pp. 1825–1828
34. C. Quan, F. Ren, Unsupervised product feature extraction for feature-oriented opinion determination. Inf. Sci. **272**, 16–28 (2014)
35. M. Amores, L. Arco, C. Borroto, Unsupervised opinion polarity detection based on new lexical resources. Computación y Sistemas **20**, 263–277 (2016)
36. G. Wang, K. Araki, An unsupervised opinion mining approach for Japanese Web-log reputation information using an improved SO-PMI algorithm. IEICE Trans. Inf. Syst. **91**, 1032–1041 (2008)
37. J. Ortigosa-Hernández, J.D. Rodríguez, L. Alzate, M. Lucania, I. Inza, J.A. Lozano, Approaching sentiment analysis by using semi-supervised learning of multidimensional classifiers. Neurocomputing **92**, 98–115 (2012)
38. F.H. Khan, U. Qamar, S. Bashir, A semi-supervised approach to sentiment analysis using revised sentiment strength based on SentiWordNet. Knowl. Inf. Syst. **51**, 851–887 (2016)
39. M. Yang, W. Tu, Z. Lu, W. Yin, K.-P. Chow, LCCT: a semi-supervised model for sentiment classification, in *The 2015 Annual Conference of the North American Chapter of the ACL*, Human Language Technologies (2015), pp. 546–555
40. V. Sindhwani, P. Melville, Document-word co-regularization for semi-supervised sentiment analysis, in *Eighth IEEE International Conference on Data Mining* (2008), pp. 1025–1030
41. D. Anand, D. Naorem, Semi-supervised aspect based sentiment analysis for movies using review filtering. Proc. Comput. Sci. **84**, 86–93 (2016)

42. S. Li, Z. Wang, G. Zhou†, S.Y. Lee, Semi-supervised learning for imbalanced sentiment classification, in *Proceedings of the Twenty-Second International Joint Conference on Artificial Intelligence* (2011), pp. 1826–1831
43. S. Zhai, Z.M. Zhang, Semisupervised autoencoder for sentiment analysis (2015), pp. 1394–1400
44. S.O. Alhumoud, Semi supervised sentiment analysis of consumer reviews. Imam J. Appl. Sci. **3**, 41–47 (2018)
45. T.J. Lu, Semi-supervised microblog sentiment analysis using social relation and text similarity, in *International Conference on Big Data and Smart Computing (BIGCOMP)* (2015), pp. 194–201
46. A. Hussain, E. Cambria, Semi-supervised learning for big social data analysis. Neurocomputing **275**, 1662–1673 (2018)

Hyperparameter Tuning and Optimization in Machine Learning for Species Identification System

Sofia K. Pillai, M. M. Raghuwanshi and M. Gaikwad

Abstract Hyper parameters are regulated parameters that are selected for training a model that controls the training process itself. Such a set of differently configured models can be achieved through the normal process of developing the network and tuning its hyper parameters. During this process, each model can be saved and a subset of the better models is selected for the set. A random forest creates many decision trees called forests and combines them together to obtain more accurate and stable forecasts. The proposed work presents hyper parameter tuning of random forest and its parameter when it achieves highest accuracy. The proposed work presents the tuning and selection of the best parameters specially for the bird species identification system. Birds play an important role in maintaining the ecological balance of the earth.

Keywords Ensemble methods · Random forest · Hyper parameter tuning

1 Introduction (Hyper Parameter Tuning)

A hyper parameter of the model is a procedure that is outside the model and whose worth cannot be evaluated from the information. For a given issue, you cannot decide the best incentive for the hyper parameter of the model. You can utilize the standard, duplicate qualities that were utilized for different issues or quest for the best incentive by experimentation. At the point when the AI calculation is tuned to a particular issue,

S. K. Pillai (✉)
G. H. Raisoni College of Engineering, Nagpur, India
e-mail: pillaisofia@gmail.com; sofia.pillai@raisoni.net

M. M. Raghuwanshi
Department of Information Technology, Y.C.C.E Nagpur, Nagpur, India
e-mail: m_raghuwanshi@rediffmail.com

M. Gaikwad
Department of Information Technology, G. H. Raisoni College of Engineering, Nagpur, India
e-mail: mahendra.gaikwad@raisoni.net

M. Dutta et al. (eds.), *Proceedings of International Conference on IoT Inclusive Life (ICIIL 2019), NITTTR Chandigarh, India*, Lecture Notes in Networks and Systems 116,
https://doi.org/10.1007/978-981-15-3020-3_22

you basically alter the model's hyper parameters to decide the model parameters that lead to the most complete expectations.

They are likewise part of the model that reaches out from recorded preparing information. Spare the parameters of capacities utilized during programming. For this situation, the parameter is a contention to a capacity that can have a scope of values. In AI, the utilized model is a capacity and expects parameters to foresee new information.

It is a piece of the model that is pulled from the authentic preparing information. In old style AI, we can think about a model as a theory and parameters as an adjustment of a speculation to a particular arrangement of information. Parameters of the model are assessed utilizing the improvement calculation, which gives a proficient inquiry to conceivable parameter esteems.

In AI, the issue of streamlining is to pick a lot of ideal hyper parameters for the learning calculation. A hyper parameter is a parameter whose worth is utilized to control the learning procedure. Be that as it may, the estimations of different parameters (for the most part loads of hubs) are found out. A similar kind of AI model may require distinctive weight points of confinement or learning rate to sum up various examples of information.

For instance, to prepare a profound neural system, before preparing the model, set the quantity of shrouded layers in the system and the quantity of hubs in each layer. The reason for investigation is to scan for various arrangements and to discover the design that works best. Principle speaking, the procedure of hyper parameters investigation is carefully manual on the grounds that the hunt space is huge, and the assessment of every design can be costly.

2 Ensemble Learning

The ensemble learning systems can be gathered by an alternate component, e.g., for instance preparing information, model and forecast techniques are joined. The presumption made here that all forecasts are totally free is to some degree extraordinary on the grounds that they are relied upon to relate The ensemble is the specialty of interfacing various students (singular models) so as to extemporize the security and prescient intensity of the model. Evaluation of the ensemble's expectation generally requires a greater number of computations than estimations of the forecast of a solitary model, so ensembles can be viewed as an approach to make up for frail learning calculations, doing numerous extra figurings. Quick calculations, for example, choice trees, are frequently utilized in ensemble strategies, (for example, irregular trees), albeit more slow calculations may utilize ensemble procedures.

You can use the standard, copy characteristics that were used for various issues, or journey for the best motivator by experimentation. Exactly when the AI figuring is tuned to a specific issue, you fundamentally change the model's hyper parameters to choose the model parameters that lead to the most complete desires. It is a strategy that makes numerous models, and amasses it to make a framework which is preferable in execution over the past. It gives out the best answer for the proposed methodology.

3 Random Forest

Presently, when we have a chance to assess our model, we have to discover how to pick the parameters that best sum up the information. A random forest picks a random subset of capacities and makes numerous choice trees. The random forest has a few parameters that can be changed to improve the speculation of the expectation.

In the wake of embeddings capacities and marks, a few standards are produced for the choice tree, which enable you to anticipate whether the picture is which species. For compari-child, the random forest calculation randomly chooses perceptions and capacities to make different choice trees, and after that gathers the results. In most cases, random forests counteract this by making random subsets of articles and making littler trees with those subsets.

Random forest has nearly the equivalent hyper parameters as the choice tree or sink classifier. A thick forest adds an extra fortuitous event to the model when the trees develop. Rather than searching for the most significant element when parting a hub, a random subset of the capacity is scanned for the best highlight. Along these lines, in a random forest just a random subset of capacities is considered by the calculation of dividing a hub.

It is said that the more trees there are, the more grounded the forest. Random forests make choice trees for randomly chosen information tests, download forecasts for each tree, and pick the best arrangement by democratic. Random forests have numerous utilizations, for example, such as proposal motors, picture arrangement, and determination of capacities.

Each tree in a random forest gains from a random example of preparing perceptions. The models are made utilizing the underlying burden. This implies that a few models are utilized on various occasions in a solitary tree. The fact of the matter is that via preparing each tree on various examples, albeit each tree can have a high fluctuation regarding a particular arrangement of preparing information, the general forest commonly has less difference, however, not to the detriment of expanding the contortion.

4 Experiments and Observations

Random forest classifier was chosen with four parameters for tuning namely

- N_ESTIMATORS-Total no of trees in the forest
- MAX_FEATURES = max no of features required to split a node
- MAX_DEPTH = Max levels in each decision tree

- MIN_SPLIT-Min no of data points before split RFC-Random Forest Classifier.

RFCGSE- Random Forest Classifier with Grid Search Enabled (Table 1).

The system was checked with RFC and compared with boosted (RFCGSE). Different values of n_estimators were recorded and observations were carefully noted (Fig. 1).

As seen Boosted RFCGSE gives better results as compared to RFC (Tables 2, 3, and 4).

The max_features were tested on two options auto and sqrt (Fig. 2).

The max_features were tested on two options auto and sqrt (Figs. 3 and 4).

Table 1 Observations for n_estimators

n_estimators	RFC	Time (s)	Boosted RFCGSE	Time (s)
	Mean validation score		Mean validation score	
20	0.937 (std: 0.017)	6.96	0.968 (std: 0.012)	1.52
40	0.968 (std: 0.012)	33.39	0.946 (std: 0.023)	3.62
100	0.972 (std: 0.008)	66.75	0.975 (std: 0.004)	3.61
200	0.948 (std: 0.016)	135.73	0.974 (std: 0.010)	3.63

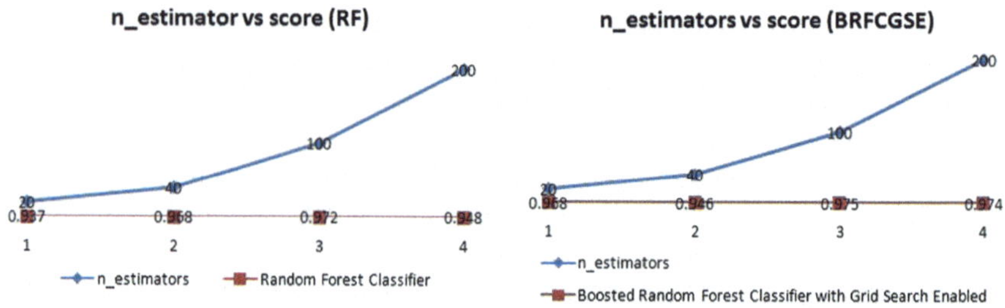

Fig. 1 Comparison of RFC with boosted RFCGSE foe n_estimators

Table 2 Observations for max_features

max_features, n_estimators = 20	RFC	Time (s)	Boosted RFCGSE	Time (s)
	Mean validation score		Mean validation score	
auto	0.931 (std: 0.026)	7.07	0.967 (std: 0.008)	1.57
sqrt	0.934 (std: 0.019)	16.39	0.976 (std: 0.005)	3.74

Table 3 Observations for max_depth

max_depth, n_estimators = 20	RFC	Time (s)	Boosted RFCGSE	Time (s)
	Mean validation score		Mean validation score	
None	0.935 (std: 0.024)	6.14	0.974 (std: 0.008)	1.60
30	0.938 (std: 0.024)	6.14	0.974 (std: 0.008)	1.60
40	0.939 (std: 0.017)	6.14	0.974 (std: 0.008)	1.60
60	0.940 (std: 0.022)	6.14	0.974 (std: 0.008)	1.60

Table 4 Observations for max_depth

min_samples_split, n_estimators = 20	RFC	Time (s)	Boosted RFCGSE	Time (s)
	Mean validation score		Mean validation score	
2	0.935 (std: 0.022)	5.42	0.978 (std: 0.006)	3.67
3	0.930 (std: 0.036)	5.42	0.978 (std: 0.006)	3.67
3	0.933 (std: 0.018)	5.42	0.978 (std: 0.006)	3.67

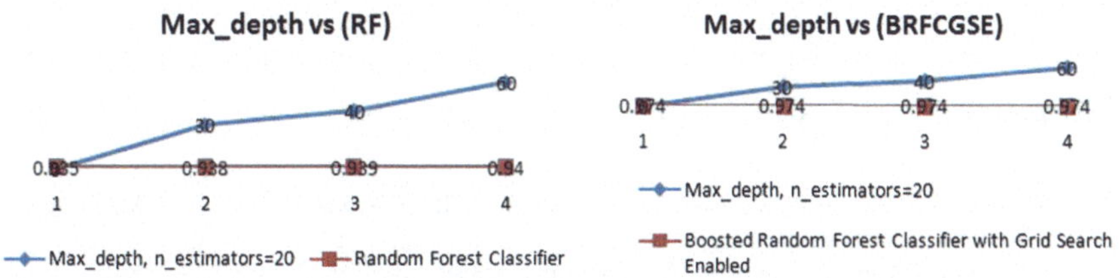

Fig. 2 Comparison of RFC with Boosted RFCGSE for max_depth

Fig. 3 **a** Correctly classified ımage and **b** Misclassified ımage

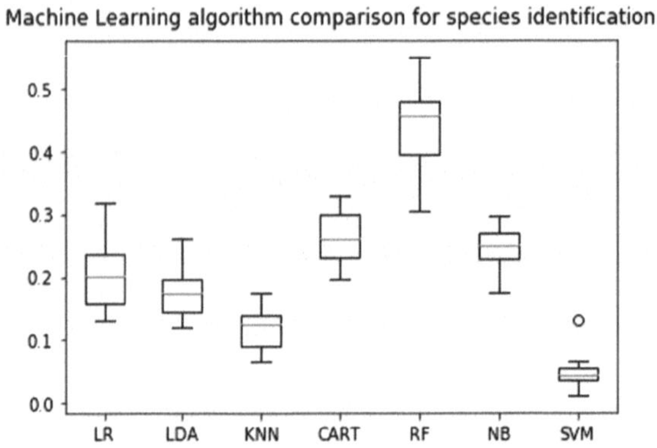

Fig. 4 Comparison of machine learning algorithm with RFC

5 Conclusion

It is concluded that for n_estimators, value is more, it gives better result, but the computational cost becomes high and system takes more time for execution. Auto and sqrt for max_features gives moreover same result. For max_depth, assuming none, at that point then nodes are extended so that all leaves are pure or all leaves contain less than the values. It is also concluded that Boosted RFCGSE always gives the best result as compared to the random forest in all parameters.

References

1. L.K. Hansen, P. Salamon, Neural network ensembles. IEEE Trans. Pattern Anal. Mach. Intell. **12**(10), 993–1001 (1990)
2. R.E. Schapire, The strength of weak learnability. Mach. Learn. **5**(2), 197–227 (1990)
3. A. Krogh, J. Vedelsby, Neural network ensembles, cross validation, and active learning, in *Advances in Neural Information Processing Systems 7*, ed. by G. Tesauro, D.S. Touretzky, T.K. Leen (MIT Press, Cambridge, MA, 1995), pp. 231–238
4. L.I. Kuncheva, C.J. Whitaker, Measures of diversity in classifier ensembles and their relationship with the ensemble accuracy. Mach. Learn. **51**(2), 181–207 (2003)
5. Y. Freund, R.E. Schapire, A decision-theoretic generalization of on-line learning and an application to Boosting. J. Comput. Syst. Sci. **55**(1), 119–139 (1997)
6. L. Breiman, Bagging predictors. Mach. Learn. **24**(2), 123–140 (1996)
7. D.H. Wolpert, Stacked generalization. Neural Netw. **5**(2), 241–260 (1992)
8. L. Breiman, Random forests. Mach. Learn. **45**(1), 5–32 (2001) Ensemble Learning
9. E. Bauer, R. Kohavi, An empirical comparison of voting classification algorithms: Bagging, Boosting, and variants. Mach. Learn. **36**(1–2), 105–139 (1999)
10. K.M. Ting, I.H. Witten, Issues in stacked generalization. J. Artif. Intell. Res. **10**, 271–289 (1999)
11. D. Opitz, R. Maclin, Popular ensemble methods: An empirical study. J. Artif. Intell. Res. **11**, 169–198 (1999)
12. Z.H. Zhou, J. Wu, W. Tang, Ensembling neural networks: many could be better than all. Artif. Intell. **137**(1–2), 239–263 (2002)

13. A. Strehl, J. Ghosh, Cluster ensembles—a knowledge reuse framework for combining multiple partitionings. J. Mach. Learn. Res. **3**, 583–617 (2002)
14. T.G. Dietterich, Machine learning research: Four current directions. AI Mag. **18**(4), 97–136 (1997)
15. S.K. Pillai, M.M. Raghuwanshi, U. Shrawankar, Book computing and network sustainability, lecture notes in networks and systems 75, in *Proceedings of IRSCNS 2018* (Springer, Singapore), (2019) pp 291–298. https://doi.org/10.1007/978-981-13-7150-9_31

Major Convolutional Neural Networks in Image Classification: A Survey

Navdeep Kumar, Nirmal Kaur and Deepti Gupta

Abstract Deep learning is a vital technique to implement artificial intelligence and a significant part of machine learning. In the last decades, deep learning gained enormous popularity due to the remarkable enhancement in computational ability and machine learning experimentation. The high computational time taken by the processes in deep learning because of large data sets can be compensated by increased computational ability. In deep learning, Convolutional Neural Network (CNN) or ConvNet is among the eminent approaches used for image classification. In recent image recognition competitions, CNN is outperforming other techniques of image classification. In this review paper, we have been discussed the basics of CNN, and significant developments in the history of CNN concerning image classification.

Keywords Image classification · Deep learning · Computer vision · State-of-the-art CNN architectures · ImageNet dataset

1 Introduction

Computer vision is a technique that focuses on the extraction of useful information or data for digital images or videos. This technique can help us to make a machine understand that how to perform an intelligent task if used with another method like deep learning.

Artificial intelligence (AI) [1] has been blossoming in recent years though it is not equal to the quality of a human mind. AI and deep learning have a vast list of applications. However, these have not surpassed the capacity of human intelligence. Human intelligence has the capability to quickly use all the small or large information

N. Kumar (✉) · N. Kaur · D. Gupta
University Institute of Engineering and Technology, Panjab University, Chandigarh, India
e-mail: navdeepsandhu8054@gmail.com

N. Kaur
e-mail: nirmaljul19@gmail.com

D. Gupta
e-mail: deeptigupta@pu.ac.in

M. Dutta et al. (eds.), *Proceedings of International Conference on IoT Inclusive Life (ICIIL 2019), NITTTR Chandigarh, India*, Lecture Notes in Networks and Systems 116,
https://doi.org/10.1007/978-981-15-3020-3_23

available in its surrounding, to make a decision. However, AI and deep learning need huge data to clear its aim completely. Since the last decade, we can see deep learning is an emerging area of research. It has the capability to solve problems like pattern recognition, natural language processing, and a lot more. Image classification is a problem in which machine classifies or recognizes images according to the learning by model, in other words, it tells us which object is in the image. For the task of identification of a specific object, image classification with deep learning is a proper technique and is efficiently used now. Image classification works as a base for other problems such as object detection, image segmentation, and image localization [2]. It is the property of machine learning that it can perform the tasks according to the data provided to the machine. Data provided to the model or machine strongly affects the working, efficiency, and accuracy of the machine learning model.

As we are discussing the problem of image classification in deep learning, we will briefly explain the methods of image classification used previously. Before CNN was in trend, traditional methods such as Histogram of Oriented Gradients (HOG) [3] and Scale Invariant Feature Transform (SIFT) [4] were used for the task of image recognition or classification. And as the availability of extensive data to CNN will not be a factor in the future, then the traditional methods might become outdated. There was another method, Speeded Up Robust Features (SURF) [5], which can be considered as speed up version of SIFT. However, these methods were not so useful in generalization as compared to CNN techniques. Before CNN, SIFT was used to deal with the challenge of ImageNet Large-Scale Visual Recognition Competition (ILSVRC) [6].

Traditionally, in a technique like SIFT, there was a two-phase approach followed in which, firstly, you extract features from the image using feature descriptor, and then feed it to a trainable classifier. In these methods, feature extractor stage majorly affects the classification accuracy. However, now CNN is dominating in all the areas of image classification. Even in ILSVRC competition, it is the most used technique among the winning techniques. Initially, Lenet [7] inspired deep learning innovation which led to the emergence of one of the first published CNN model AlexNet [8]. Later on, the "network in network" paper [9] worked well in the journey of deep learning. Also, the development of blocks like inception block and ResNet block in the network gives an excellent boost to the accuracy and generalization ability of network and achieve another milestone in the journey of deep learning.

Neural network [10] is a prominent part of deep learning and AI. The neural network tries to copy the human brain like operations. It calculates the weights and bias to reach a specific target or output. It consists of several layers as well as neurons. Moreover, in the CNN [11], generally, there are three types of layers: convolutional layer, pooling layer, and fully connected layer. Convolutional layer takes the input from the previous layer, which is a rectangular-shaped grid of many neurons. In pooling layers, it takes the data from the previous convolutional layer and performs sub-sampling. There may be a different way of pooling like average-pooling, max-pooling, etc. In fully connected layers, as the name implies, it takes the data from the previous layer and connects the input to all the neurons it has.

Moreover, the fully connected layer is generally connected to the output layer. Some other factors like dimensionality of CNN, optimizer and activation function used in the model, plays a vital role in the output of a model. Dimensionality reduction principles [12] can enhance productivity and reduce the computational cost of the model.

Deep learning [13] gained great achievements since it was used on powerful computers, algorithms, and big data. Deep learning methods work on the high dimensional input–output model. It generates descriptors for this input–output model through these CNN architectures. In recent developments, [14] gave a model which produces astounding features and created a foundation for various computer vision applications which are, image classification, scene recognition, attribute recognition, fine-grained recognition and image retrieval, etc.

In previous surveys, the architectures of CNN were not discussed comprehensively. They [13] did not focus on the interpretability issue which was mainly dealt by ZFNet [15]. And in some papers, they did not cover all the CNN winners in the history of ImageNet challenge. In [16] they discussed the introduction of building blocks like residual block and inception block. Usually, the survey is available in the fragmented way [17]. In our paper, we try to cover all the networks which acted as milestones in the history of CNN in image classification focusing on ILSVRC. We discuss deep convolutional neural network architecture, made up of hierarchical layers and building blocks, focusing on image recognition and ILSVR challenge, interpretability issue and importance of fine-tuning in CNN as in ZFNet [15].

2 Related Work

In recent years, it has been observed that CNN is the most efficient technique for image classification and other computer vision tasks like localization, segmentation, detection. ILSVRC (2012) winner [8] made a remarkable change in the history of the CNN for computer vision. We will discuss some prime state-of-the-art CNN architectures that perform remarkably well in image classification competitions like ILSVRC.

2.1 LeNet

LeNet given by Lecun et al. [7] is one of the leading architectures in the history of Deep Convolutional Neural Network (DCNN). The author used backpropagation for handwritten zip code recognition. In this architecture, there are three hidden layers known as H1, H2, and H3. In the first hidden layer, H1, there are twelve groups of 64 units, which mean that there are twelve 8×8 feature maps. In the second hidden layer, H2, there are also twelve feature maps. In each feature map, there are sixteen units of 4×4 sizes. In the third layer, H3, there are 30 units and this layer is fully

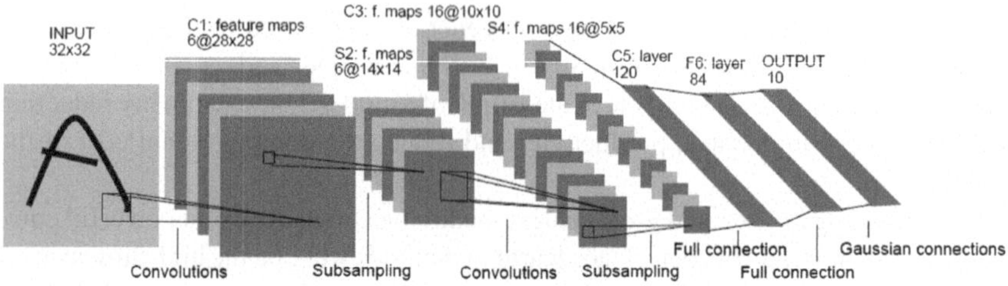

Fig. 1 Le-net-5 architecture used by Lecun et al. in [18]

connected to the H2 layer. The output layer contains 10 units and this layer is fully connected to H3. The input of this architecture is a 16 × 16 image.

Lenet-5 [18] architecture has seven layers as shown in Fig. 1 and is trained on MNIST [19] dataset. It takes the image of Input size is 32 × 32. Moreover, the first and third layers are convolutional. First layer takes the greyscale image sized 32 × 32 × 1 and produces the image of size 28 × 28. However, second and fourth are sub-sampling layers, fifth and sixth are fully connected layers and seventh layer is the output layer.

Lecun describes that the self-made features are not as helpful as the automatically learned features. In the publication, the gradient-based approach was used for loss minimization. The formula for calculation of output size is given as

$$n_{out} = \left[\frac{n_{in} + 2P - K}{S}\right],$$

where, n_{in} = number of input size features,
n_{out} = number of output feature,
K = convolution kernel size,
P = convolution padding size,
S = convolution stride size.

2.2 *AlexNet (2012)*

AlexNet [18] is a revolutionary signpost in the history of the deep convolutional neural network. AlexNet gained public attention when it emerged as the frontrunner of ILSVRC 2012. It achieved the top-5 error rate of 15.6% that was far better than ILSVRC 2011 winner on imagenet [20] which was 26.2%. It is the first CNN winner of ILSVRC challenge. The paper "ImageNet Classification with Deep Convolutional Neural Networks" by [8] described that AlexNet started the trend of CNN and left a remarkable achievement in the history of deep learning which brought many concepts into the picture. Authors in [8] used two GPUs (Graphics Processing Units) in this training, upper part with one GPU and the lower part with one GPU as depicted in

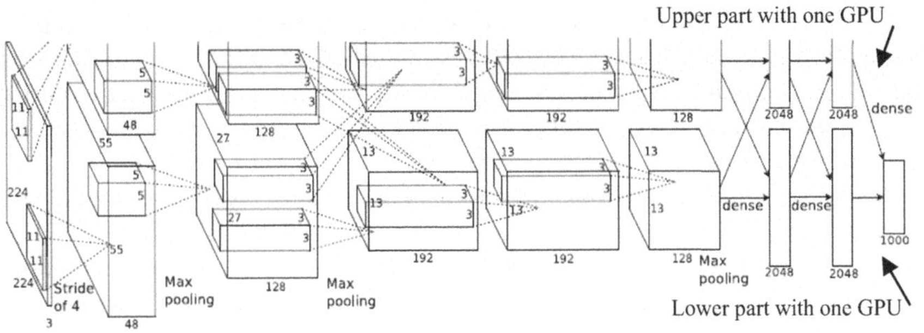

Fig. 2 The architecture used in "ImageNet classification with deep convolutional neural networks" [8]

Fig. 2. The key aspects of AlexNet are that it uses dropout and data augmentation to combat overfitting rather than regularization which decreases the top-5 error rate by 0.3%.

AlexNet takes the RGB image sized $256 \times 256 \times 3$ and extracts 224×224 patches for the first convolutional layer and then trains the model with these patches. Its first convolutional layer has 96 filters of size $11 \times 11 \times 3$ and the second convolutional layer processes the output of the previous layer, with 256 filters having size $5 \times 5 \times 48$. The third layer uses 384 filters of size $3 \times 3 \times 256$. The third layer is connected to the fourth layer of 384 filters of $3 \times 3 \times 192$. The fifth layer has 256 filters having size $3 \times 3 \times 192$. The sixth, seventh, and eighth layers are fully connected layers. The output layer that is the last layer contains 1000 nodes. In this model, ***ReLU*** function was used instead of logistic and tanh function that delivered a good performance.

2.3 ZFNet (2013)

After the AlexNet's win in 2012, it became a trend of CNN's entries in ILSVRC challenge. In 2013, Zeiler and Fergus presented a CNN model known as ZFNet [15] which gained 11.2% top-5 error rate and is better than the AlexNet model. In their paper titled "Visualization and Understanding CNN" [15], they mentioned that many researchers do not know the internal mechanism of the model. So, they also focused on the internal mechanisms of CNN although the interpretability of the model is still a challenge. Zeiler and Fergus used a Deconvolutional Network (DeConvNet) technique which is a visualization technique to deal with this issue. So, there is a deconvolutional layer attached to every convolution layer and it does the opposite task of the convolutional layer. DeConvNet [21] helps to understand the internal mechanism of CNN like examination of activation computed at each layer of ConvNet.

Generally, ZFNet is the modification of AlexNet by fine-tuning the hyper-parameters and making other minute changes. In the paper, Zeiler and Fergus, focused on the interpretability of the CNN model and described the feature map and activation at each layer adequately. AlexNet was trained on 13 million, and ZFNet was trained on 1 million images. AlexNet, as shown in Fig. 2, was trained on two GPUs and ZFNet was trained on one GPU. Hence, the sparse connection in AlexNet was replaced with the dense connection in ZFNet.

A lot of irrelevant information in the form of dead features was present at the first layer in AlexNet due to stride 4 and filter size 11 × 11, which was changed in ZFNet to filter size 7 × 7 and stride 2. Because of these changes made in ZFNet, there was a decrease in dead features as described in the paper "Visualization and Understanding CNN" by Zeiler and Fergus. The authors not only described the internal mechanism of CNN but also told us how to improve network architecture and describe a feature map at each layer in an effective way.

2.4 VGGNet (2013)

VGGNet-16 was the 1st runner up in ILSVRC 2014 with 7.3% top-5 error rate and GoogLeNet was the winner in ILSVRC 2014 with 6.7% top-5 error rate. VGGNet [22] has a simple architecture as shown in Table 1. VGGNet proved that even with simple architecture, good accuracy can be achieved. Small scale kernels show that good accuracy can also be achieved through them. ConvNet configuration in Table 1 shows different variants of VGGNet ranging from 11 layers to 19 layers. Also, ConvNet with lower depth is depicted on the left side. Depth increases from left to right and the number of filters used increases as we move downwards from 64 to 512. The configuration of ConvNet (Table 1) is represented as conv (size of the filter)—(no. of filters). For example, in conv3-64, 3 represents the size of the filter and 64 stands for the number of filters used.

VGGNet-16 has 13 convolutional layers with filter size 3 × 3 and the different number of filters with the range varying from 64 to 512. There are five pooling layers, in which max-pooling is used, and three fully connected layers are used with 4096 nodes and at the end, there is an output layer having 1000 nodes. ***ReLU*** activation function is used at every layer instead of ***tanh*** function. In convolutional layers, the processing is done with 3 × 3 filter having stride 1. The stride 2 is used where max-pooling is done with 2 × 2 window size. VGGNet has a large number of filters and parameters which are not so easy to handle.

VGGNet shows that depth also plays an essential role in gaining good results. It is a massive model with great depth having either 16 or 19 layers with respective computational cost (FLOPs) of 15.3 and 19.6 Billion. It gives us good generalization ability over a wide range of classes and describes the importance of depth of the model. As the author of the paper, "Very Deep Convolutional Networks for Large-Scale Image Recognition" [22], tell us that VGGNet not only proves itself right on ImageNet [20] but also works well on other image recognition datasets. The author describes the relationship between depth and accuracy. They mention how they managed to achieve accurate results with increased depth of the network.

Table 1 ConvNet configuration [22]

A	A-LRN	B	C	D	E
11 weights layers	11 weights layers	13 weights layers	16 weights layers	16 weights layers	19 weights layers
Input (224 × 224 RGB image)					
conv3-64	conv3-64	conv3-64	conv3-64	conv3-64	conv3-64
	LRN	conv3-64	conv3-64	conv3-64	conv3-64
Maxpool	Maxpool	Maxpool	Maxpool	Maxpool	Maxpool
conv3-128	conv3-128	conv3-128	conv3-128	conv3-128	conv3-128
		conv3-128	conv3-128	conv3-128	conv3-128
Maxpool	Maxpool	Maxpool	Maxpool	Maxpool	Maxpool
conv3-256	conv3-256	conv3-256	conv3-256	conv3-256	conv3-256
conv3-256	conv3-256	conv3-256	conv3-256	conv3-256	conv3-256
			conv1-256	conv3-256	conv3-256
					conv3-256
Maxpool	Maxpool	Maxpool	Maxpool	Maxpool	Maxpool
conv3-512	conv3-512	conv3-512	conv3-512	conv3-512	conv3-512
conv3-512	conv3-512	conv3-512	conv3-512	conv3-512	conv3-512
			conv1-512	conv3-512	conv3-512
					conv3-512
Maxpool	Maxpool	Maxpool	Maxpool	Maxpool	Maxpool
conv3-512	conv3-512	conv3-512	conv3-512	conv3-512	conv3-512
conv3-512	conv3-512	conv3-512	conv3-512	conv3-512	conv3-512
			conv1-512	conv3-512	conv3-512
					conv3-512
Maxpool	Maxpool	Maxpool	Maxpool	Maxpool	Maxpool
FC-4096	FC-4096	FC-4096	FC-4096	FC-4096	FC-4096
FC-4096	FC-4096	FC-4096	FC-4096	FC-4096	FC-4096
FC-4096	FC-4096	FC-4096	FC-4096	FC-4096	FC-4096
Soft-max	Soft-max	Soft-max	Soft-max	Soft-max	Soft-max

2.5 GoogLeNet (2014)

GoogLeNet or Inception Net was one of the earliest models which introduced blocks in the network. In GoogLeNet [23], they introduced inception building blocks as shown in Fig. 3. Later on, ResNet introduced residual blocks. These blocks work very well and provide a model with good generalization ability. In 2014, in the world's largest image recognition challenge ILSVRC, GoogLeNet performed very well with this new concept of introduction of blocks in the network. It attained the

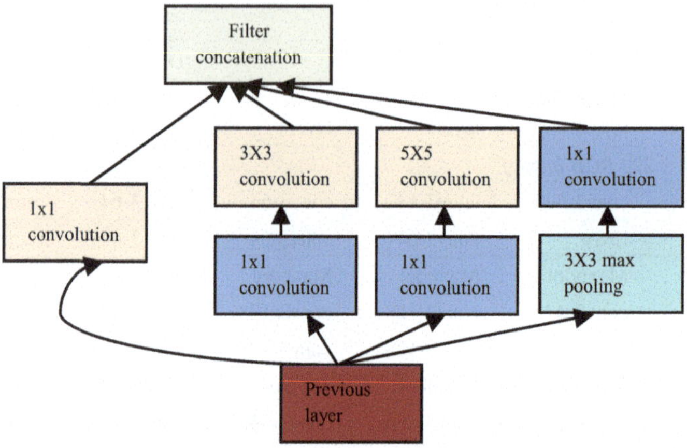

Fig. 3 Inception building block used in GoogLeNet [23]

top-5 error rate of 6.7% on the ImageNet dataset and emerged as the winner in the challenge.

Google also released the next versions of inception network which were inception-v2 [24], inception-v3 [24] and inception-v4. GoogLeNet comprises 9 linearly stacked inception modules. There are 27 layers in total, out of which, 5 are pooling layers. The total number of layers used for the structure of network is around 100 layers. In GoogLeNet [23], the author noticed that using average-pooling layers instead of fully connected layers improved top-1 accuracy by 0.6%. Dropout to combat overfitting is still necessary even after removing fully connected layers. As the position of an object in the image is significant, so they used the filter with multiple sizes of 1, 3, and 5, this makes the network wider. In this network, it appears that large stack of convolution operations is present and such a computation is expensive.

As the model upgraded frequently, new versions emerged. Inception-v2 and Inception-v3 came within a small-time gap in 2015. Inception-v2 performed well over ILSVRC challenge 2012 with 5.6%. top-5 error rate. Inception-v2 [24] incorporates the maximum features of GoogLeNet except for a few. In inception-v2, to reduce "representational bottleneck", filter banks were expanded in width. Other changes were introduced like factorization of 5×5 into 3×3 which led to reduction of expenses, complexity, and enhancement of performance. They also factorized the block of size $n \times n$ convolution in $1 \times n$ and $n \times 1$ like 5×5 convolution in 1×5 and 5×1. It emerged to be less expensive as compared to the VGGNet.

In inception-v3 [24], the focus was more on regularization and less on module modification. Moreover, inception-v3 gained good milestone by earning 3.58% top-5 error rate. On imagenet benchmark, this approach performed very well. Inception-v3 contained all the modifications of inception-v2. Moreover, in inception-v3, some modifications were done like usage of larger resolution input, and usage of RMSProp optimizer which effectively decrease the cost function.

We know that there is a lot of confusion among the reviewer about the versions of inception models after GoogLeNet. In inception-v3, a regularizing component was introduced in the loss formula to prevent the model from overfitting. As the focus was less on module modification, they also introduced batch normalization in the auxiliary classifiers of network. After inception-v3 and inception-v4, other versions were also introduced by Google and they performed well too. In the paper "Going Deeper with Convolutions" [23], the author described how they effectively utilized the resources within the network. In case of object detection work by GoogLeNet, their work is still competitive enough which actively tells us the greatness of inception module although they did not perform bounding box regression. Moreover, they also proved that in neural network, switching to the sparser network is also possible.

2.6 ResNet (2015)

ResNet is the unusual approach used in CNN. ResNet (2015) is the winner of ILSVRC 2015. Like GoogLeNet, it also introduces blocks in the network named as ResNet blocks and attained great attention. It also achieved the lowest top-5 error rate in ILSVRC 2015. In 2016, He et al. [25] presented a paper "Deep Residual Learning for Image Recognition [25]", introducing skip connections in CNN also known as residual learning. He et al. connected the output of one or more than one convolution layer to the original input. In plain deep models, when accuracy curve attains a saturation level, it suddenly falls because of difficulty in learning in these models. ResNet effectively addresses this problem of deep models.

ResNet-152 managed to achieve the top-5 error rate of 4.49% on ImageNet 2012 benchmark. Finally, in 2015, it led to attain the top-5 error rate of 3.57% and emerged as the winner of ILSVRC challenge 2015 with the right approach and introduction of the building blocks as shown in Fig. 4a and b.

The ResNet-152 architecture comprises 152 layers with 3×3 and 1×1 filters being used in these layers with residual learning as shown in Table 2. He et al. in the paper "Deep Residual Learning for Image Recognition" [25] discuss ResNet with the number of layers- 18, 34, 50, 101, 152. In 18 and 34 layered architectures, only 3×3 filters are used, whereas, in 50, 101 and 152 layered architectures, 3×3 and 1×1 filters are used. As per the paper [25], residual model is inspired by the approach used in VGGNet. As we move from left to right in the given Table 2, factors like depth, complexity, FLOPs (floating-point operations per second) and accuracy increases, whereas training and validation error decreases remarkably.

One interesting thing about ResNet is its complexity (11.3 billion) which is much less than VGGNet (15.3 billion for 16 layers/19.6 billion for 19 layers). The paper [25], effectively describe that training of deep residual network is easy to optimize than the plain model.

Moreover, it gives lesser training error and better accuracy than its counterpart plain network. Generic behavior of ResNet shows that it has better generalizing ability as it has won a lot of other image recognition challenges.

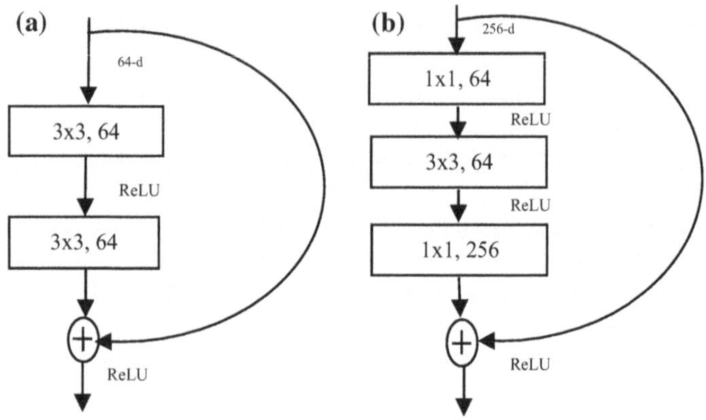

Fig. 4 **a** and **b** Residual blocks used, (left side) 5 (**a**) is for ResNet-18/34 and, (right side) 5 (**b**) for ResNet-50/101/152 [25]

2.7 SE-ResNet (2017)

In the paper "Squeeze-and-Excitation Networks" [26] Hu et al. presented a model with the number of fully connected layers and ResNet module, focusing on the reduction of computational cost and attainment of good accuracy level. In SE-ResNet network as shown in Fig. 5, two processes [27] work: one is squeeze process, and the other is the excitation process.

In this network, firstly, a transformation F_{tr} : X → U takes place then, in the squeeze process (F_{sq}), transformed features are passed to the squeeze process, which collaborates feature maps and produces a channel descriptor, and the descriptor produces the embedding of distributed channel-wise feature response with dimensions of $1 \times 1 \times$ C as depicted in Fig. 5.

After this collaboration of feature maps, the excitation process (F_{ex}) starts, it uses the self-gating mechanism, which uses the embedding of distributed channel-wise response, and it produces the modulation weight per channel as shown in Fig. 5 with a rectangle of different colors with dimensions of $1 \times 1 \times$ C. In the scaling process (F_{sc}), the weights generated by excitation process are applied to feature map U, to produce the final output ($\tilde{X}$) of Squeeze and Excitation block which can be used further for network layers. H' x W' and H x W are the spatial dimensions of feature maps across which F_{sq} and F_{ex} operations were performed, respectively, whereas C' and C are no filter used in these processes, respectively. SE-ResNet module as shown in Fig. 6a and b was used in this network model which has a global average-pooling layer for lowering the dimensions, two fully connected layers with two activation function.

Table 2 ResNet architectures having different depth and computation cost [25]

Layer name	Output size	18-layer	34-Layer	50-layer	101-layer	152-layer
conv1	112 × 112	7 × 7, 64, stride 2				
conv 2_x	56 × 56	3 × 3 max pool, stride 2				
		$\begin{bmatrix} 3\times3, 64 \\ 3\times3, 64 \end{bmatrix} \times 2$	$\begin{bmatrix} 3\times3, 64 \\ 3\times3, 64 \end{bmatrix} \times 3$	$\begin{bmatrix} 1\times1, 64 \\ 3\times3, 64 \\ 1\times1, 256 \end{bmatrix} \times 3$	$\begin{bmatrix} 1\times1, 64 \\ 3\times3, 64 \\ 1\times1, 256 \end{bmatrix} \times 3$	$\begin{bmatrix} 1\times1, 64 \\ 3\times3, 64 \\ 1\times1, 256 \end{bmatrix} \times 3$
conv 3_x	28 × 28	$\begin{bmatrix} 3\times3, 128 \\ 3\times3, 128 \end{bmatrix} \times 2$	$\begin{bmatrix} 3\times3, 128 \\ 3\times3, 128 \end{bmatrix} \times 4$	$\begin{bmatrix} 1\times1, 128 \\ 3\times3, 128 \\ 1\times1, 512 \end{bmatrix} \times 4$	$\begin{bmatrix} 1\times1, 128 \\ 3\times3, 128 \\ 1\times1, 512 \end{bmatrix} \times 4$	$\begin{bmatrix} 1\times1, 128 \\ 3\times3, 128 \\ 1\times1, 512 \end{bmatrix} \times 8$
conv 4_x	14 × 14	$\begin{bmatrix} 3\times3, 256 \\ 3\times3, 256 \end{bmatrix} \times 2$	$\begin{bmatrix} 3\times3, 256 \\ 3\times3, 256 \end{bmatrix} \times 6$	$\begin{bmatrix} 1\times1, 256 \\ 3\times3, 256 \\ 1\times1, 1024 \end{bmatrix} \times 6$	$\begin{bmatrix} 1\times1, 256 \\ 3\times3, 256 \\ 1\times1, 1024 \end{bmatrix} \times 23$	$\begin{bmatrix} 1\times1, 256 \\ 3\times3, 256 \\ 1\times1, 1024 \end{bmatrix} \times 36$
conv 5_x	7 × 7	$\begin{bmatrix} 3\times3, 512 \\ 3\times3, 512 \end{bmatrix} \times 2$	$\begin{bmatrix} 3\times3, 512 \\ 3\times3, 512 \end{bmatrix} \times 3$	$\begin{bmatrix} 1\times1, 512 \\ 3\times3, 512 \\ 1\times1, 2048 \end{bmatrix} \times 3$	$\begin{bmatrix} 1\times1, 512 \\ 3\times3, 512 \\ 1\times1, 2048 \end{bmatrix} \times 3$	$\begin{bmatrix} 1\times1, 512 \\ 3\times3, 512 \\ 1\times1, 2048 \end{bmatrix} \times 3$
	1 × 1	Average pool, 1000-d fc, soft-max				
FLOPs		1.8×10^9	3.6×10^9	3.8×10^9	7.6×10^9	11.3×10^9

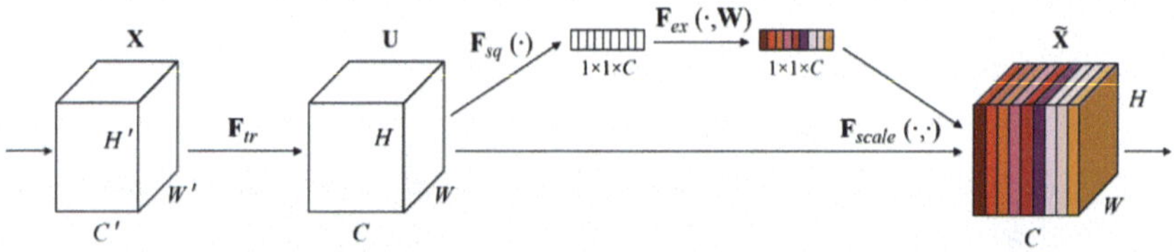

Fig. 5 Squeeze-and-excitation network [26]

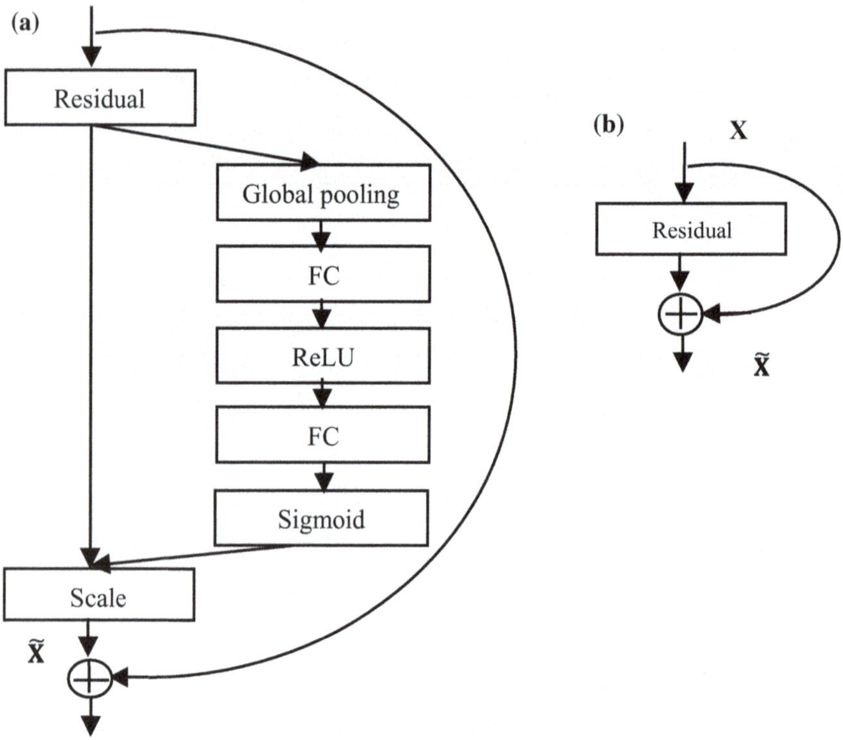

Fig. 6 **a** and **b** ResNet module (left side) and SE-ResNet module (right side) [26]

SE-ResNet network is the winner of ImageNet challenge 2017 with the top-5 error rate of 2.25%. SE-ResNet performs better than simple ResNet with reasonably high computational cost as it shows high generalization ability than the same ResNet. In the paper "Squeeze-and-Excitation Networks" [26], the author proves that combining SE-blocks to ResNet can improve accuracy with the reasonably high computational cost.

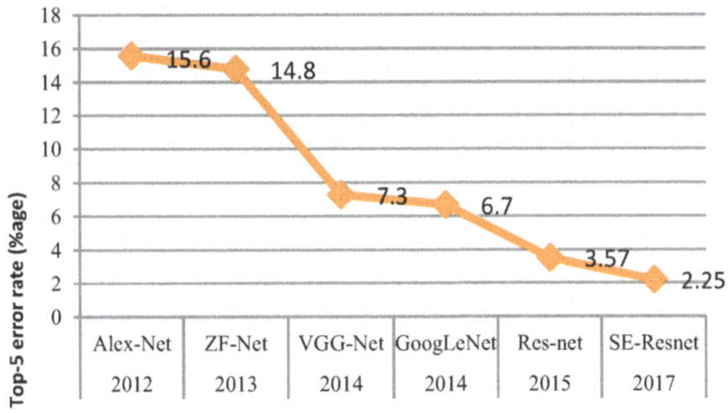

Fig. 7 Describing the top-5 error rate from ILSVRC (2012) to ILSVRC (2017)

3 Discussion

This paper describes the significant achievements in the world of Convolutional Neural Network concerning image recognition. We discussed the various models in the history of CNN in Image Classification. One model dealt with MNIST dataset and the rest focused on imagenet dataset challenge, i.e., ILSVRC. Nowadays, in the area of computer vision, image recognition and classification is a booming area with much undergoing research.

Figure 7 and Table 3 do not describe all the models in the area of image classification but cover all the major milestones achieved in this area. As we are discussing the ImageNet challenge, this paper covers all the CNN winners and runners-up over the past few years. Before AlexNet (2012), handpicked features were used which gave very high top-5 error rate as they have low generalization ability.

Figure 7 and Table 3 describe, how in ILSVRC challenge, top-5 error rate decreases from 15.3 to 2.25%. In Fig. 7, the x-axis shows top-5 error rate, whereas the y-axis shows the models which performed well in ILSVRC from 2012 to 2017. It describes that the same particular accuracy can be achieved with a lower depth. Moreover, in another case, the same particular accuracy can be achieved with the lower computational cost. Although computational cost increases from millions to billions, even depth increases but due to the high-performance computational resources available now, the desired output can be achieved in the image classification task.

4 Conclusion

In this paper, we discussed the achievements of CNN in the field of machine vision. We also mentioned the experimental results briefly in the form of graph and table. Efficient results of CNN on ImageNet describe its efficiency and capability in the area of deep learning. CNN has set remarkable changes in the history of image classification and made it an interesting field of computer vision (CV) or machine vision (MV). Introduction of building blocks such as residual and inception blocks in CNN is an excellent achievement in its history which gives unpredictable results. Moreover,

Table 3 Winners, runner ups and other models in ImageNet challenge (ILSVRC) on the basis of the top-5 error

CNN models	Developed by	Submitted in year	Place in ILSVRC	Top-5 error rate (ImageNet challenge)	Number of parameters	Computational cost (FLOPS)	Number of layers
Alex-Net [8]	Alex Krizhevsky et al.	2012	1st	15.3% (ImageNet' 12)	60 million	3300 million	8 layers
ZF-Net [15]	Matthew Zeiler et al.	2013	1st	14.8% (ImageNet' 13)	60 million	–	8 layers
GoogLe Net [23]	Szegedy et al.	2014	1st	6.7% (ImageNet' 14)	6.7 million	1502 million	22 layers
VGG-Net [22]	Simonyan et al.	2014	2nd	7.3% (ImageNet' 14)	138 million	15.3 billion	19 layers
Inception-v2 [24]	Szegedy, et al.	2015	–	5.6% (ImageNet' 12)	–	–	42 layers
Inception-v3 [24]	Szegedy et al.	2015	–	3.58% (ImageNet' 12)	21.8 million	–	48 layers
ResNet [25]	He et al.	2015	1st	4.49% (ImageNet' 12) 3.57% (ImageNet' 15)	60.2 million	11.3 billion	152 layers
SE-ResNet [26]	JieHu et al.	2017	1st	2.25% (ImageNet' 17)	–	11.32 billion	152 layers

how to fine-tune and optimize, how to deal with overfitting, and loss function, can be other classes of development in the field of CNN in image classification. Usage of good optimizer can enhance accuracy and generalizing ability of your model, reducing the error rate.

Availability and enhancement in hardware processing units are making image recognition a booming field of deep learning and may remain the same in the future as hardware capability increases. The number of parameters and computational cost used by these CNNs are still high. It is still a challenge to reduce the parameters and computational cost. As we discussed, most of the CNNs belong to supervised learning. So, shifting from supervised to unsupervised is still a problem. Unlike the human brain, which can easily classify without having a large amount of data, these CNNs require a huge dataset to have good generalization ability. As previously mentioned, there are many parameters in our CNNs but finding the parameter which could be used to tune to get definite results is still a challenge. So, the use of a suitable optimization algorithm is also a research problem.

In the end, we want to convey the notion that in the current era, deep learning or Artificial intelligence is solving a lot of challenges or problems by providing practical solutions. Being one of the fastest growing areas now, it will lead the technical world to a new era in the future.

References

1. Nvidia, What's the Difference between Artificial Intelligence, Machine Learning, and Deep Learning (2019), https://blogs.nvidia.com/blog/2016/07/29/whats-difference-artificial-intelligence-machine-learning-deep-learning-ai/. Accessed 25 April 2019
2. A. Karpathy, CS231n: convolutional neural networks for visual recognition (2016), http://cs231n.github.io/classification/
3. N. Dalal, B. Triggs, Histograms of oriented gradients for human detection, in *IEEE Computer Society Conference on Computer Vision and Pattern Recognition (CVPR)*, vol. 1 (IEEE, San Diego, USA 2005), pp. 886–893
4. D.G. Lowe, Distinctive image features from scale-invariant keypoints. Int. J. Comput. Vis. **60**(2), 91–110 (2004)
5. H. Bay, T. Tuytelaars, L. Van Gool, SURF: speeded up robust features, in *Computer Vision—ECCV 2006*, Lecture Notes in Computer Science, vol. 3951, ed. by A. Leonardis, H. Bischof, A. Pinz (Springer, Berlin, 2006)
6. O. Russakovsky*, J. Deng*, H. Su, J. Krause, S. Satheesh, S. Ma, Z. Huang, A. Karpathy, A. Khosla, M. Bernstein, A.C. Berg, L. Fei-Fei (* = equal contribution) ImageNet large scale visual recognition challenge. IJCV (2015)
7. Y. Lecun, B. Boser, J.S. Denker, D. Henderson, R.E. Howard, W. Hubbard, L.D. Jackel, Back-propagation applied to handwritten zip code recognition. Neural Comput. **1**(4), 541–551 (1989). Winter
8. A. Krizhevsky, I. Sutskever, G.E. Hinton, ImageNet classification with deep convolutional neural network, in *Proceedings of the 25th International Conference on Neural Information Processing Systems-Volume-1, NIPS* (2012), http://dl.acm.org/citation.cfm?id=2999134.2999257
9. M. Lin, Q. Chen, S. Yan, *Network in network* (2013). arXiv 1312.4400
10. Medium, Neural network fundamentals, https://medium.com/datadriveninvestor/neural-network-fundamentals-1956a3000c24. Accessed 25 April 2019

11. A. Gibansky, Convolutional neural networks, andrew.gibiansky.com/blog/machine-learning/convolutional-neural-networks/. Accessed 25 April 2019
12. G.E. Hinton, R.R Salakhutdinov, Reducing the dimensionality of data with neural networks. Science **313**(5786), 504–507 (2006)
13. M. Pak, S. Kim, A review of deep learning in image recognition, in *2017 4th International Conference on Computer Applications and Information Processing Technology (CAIPT)* (2017)
14. A. Razavian, H. Azizpour, J. Sullivan, S. Carlsson, CNN features off-the-shelf: an astounding baseline for recognition, in *Proceedings of the IEEE Conference on Computer Vision and Pattern Recognition Workshops* (2014)
15. M.D. Zeiler, R. Fergus, Visualizing and understanding convolutional networks. in *European Conference on Computer Vision* (Springer, Cham, 2014), pp. 818–833
16. A.A. Al-Saffar, H. Tao, M.A. Talab, Review of deep convolution neural network in image classification, in *2017 International Conference on Radar, Antenna, Microwave, Electronics, and Telecommunications (ICRAMET)* (IEEE, 2017), pp. 26–31
17. M. Sornam, K. Muthusubash, V. Vanitha, A survey on image classification and activity recognition using deep convolutional neural network architecture, in *2017 Ninth International Conference on Advanced Computing (ICoAC)* (IEEE, 2017), pp. 121–126
18. Y. Lecun, L. Bottou, Y. Bengio, P. Haffner, Gradient-based learning applied to document recognition, in *Proceeding of IEEE*, ed. by S. Haykin, B. Kosko (IEEE Press, 2001), pp. 306–351
19. [dataset] The Mnist Database, http://yann.Lecun.com/exdb/mnist/. Accessed 25 April 2019
20. J. Deng, W. Dong, R. Socher, L.-J. Li, K. Li, L. Fei-Fei, ImageNet: a large-scale hierarchical image database, in *CVPR09 2009*, http://www.image-net.org/papers/ImageNet_cvpr09.bib
21. Medium, ZFNet—Winner of ILSVRC 2013, https://medium.com/coinmonks/paper-review-of-ZFNet-the-winner-of-ilsvlc-2013-image-classification-d1a5a0c45103. Accessed 22 April 2019
22. K. Simonyan, A. Zisserman, Very deep convolutional networks for large-scale image recognition. in *ICLR* (2015)
23. C. Szegedy, W. Liu, Y. Jia, P. Sermanet, S. Reed, D. Anguelov, A. Rabinovich, Going deeper with convolutions, in *Proceedings of the IEEE Conference on Computer Vision and Pattern Recognition* (2015), pp. 1–9
24. C. Szegedy, V. Vanhoucke, S. Ioffe, J. Shlens, Z. Wojna, Rethinking the inception architecture for computer vision, in *Proceedings of the IEEE Conference on Computer Vision and Pattern Recognition* (2016), pp. 2818–2826
25. K. He, X. Zhang, S. Ren, J. Sun, Deep residual learning for image recognition, in *Proceedings of the IEEE Conference on Computer Vision and Pattern Recognition* (2016), pp. 770–778
26. J. Hu, L. Shen, G. Sun, Squeeze-and-excitation networks, in *Proceedings of the IEEE Conference on Computer Vision and Pattern Recognition* (2018), pp. 7132–7141
27. Towards data science, Squeeze-and-excitation-networks, https://towardsdatascience.com/squeeze-and-excitation-networks-9ef5e71eacd7. Accessed 25 April 2019

Prediction of Accuracy of High-Strength Concrete Using Data Mining Technique: A Review

Aman Kumar and Navdeep Mor

Abstract One of the most important strength parameters of concrete is compressive strength. This study primarily focused on the prediction of the accuracy of High-Strength Concrete (HSC) using data mining technique. HSC is the mixture of cement, water, blast furnace slag, coarse and fine aggregates, fly-ash, and admixtures. The design of the High-Strength Concrete is very complex due to intricacy in the materials. Compressive strength of concrete is a nonlinear function of its ingredients. Previous studies concluded that the major components responsible for the compressive strength of HSC are water-cement ratio and additive materials. The most commonly used computational methods by the researchers to predict the accuracy of HSC are: "Data-Mining (DM) techniques i.e. Artificial Neural Networks (ANN) and Support Vector Machines (SVM), and statistical models (Poisson Distribution, multiple additive regression trees and bagging regression trees)." This paper presented a review of previous research work covering the use of DM techniques for predication of optimized HSC.

Keywords Data mining · High-strength concrete · Prediction models · Comparison of models · Machine learning

1 Introduction of High-Strength Concrete

The concrete which has strength greater than 55 N/mm^2 is called High-Strength Concrete. HSC has been used for the last two decades in the construction industry because it provides high strength to the structure in earthquake-prone zones. HSC is prepared by mixing blast furnace slag, fly-ash, and other chemical additives such

A. Kumar (✉)
Structural Monitoring & Instrumentation Group, Department of Civil Engineering, Aimil Ltd., Chandigarh, India
e-mail: aman.civil16@nitttrchd.ac.in

N. Mor
Assistant Professor, Department of Civil Engineering, Guru Jambheshwar University of Science & Technology, Hissar, Haryana, India

M. Dutta et al. (eds.), *Proceedings of International Conference on IoT Inclusive Life (ICIIL 2019), NITTTR Chandigarh, India*, Lecture Notes in Networks and Systems 116,
https://doi.org/10.1007/978-981-15-3020-3_24

as superplasticizers. HSC is used for a structure such as tunnels, nuclear structures, highway bridges, and also in an aggressive environment. Chemical admixtures are used in the concrete to reduce the water content and at the same time to reduce the porosity of the hydrated cement. The addition of the chemical admixture in the concrete mainly depends on the temperature conditions, fineness, and cement chemistry, and, at the same time, it is important to decrease the water-cement (w/c) ratio as per temperature conditions. Cement replacement materials are also called as mineral admixtures, which act as pozzolanic materials and fine-fillers. The main function of the chemical admixture is to increase the micro-structure strength and density of cement paste. In concrete, Portland cement is the expensive component, so use of pozzolanic materials in construction reduces the overall costs as well as the emission of harmful gases. The main function of the additives is to improve concrete properties in terms of technical properties, e.g., workability, strength, and its durability. These additives further make concrete more complex and affect prediction accuracy. As per the literature review, the traditional model techniques only predict the behavior of concrete, especially, the compressive strength, but, these are time-consuming and unreliable.

1.1 Artificial Neuron Network (ANN)

Artificial Neuron Network is the computer designed intelligent system which stimulates the behavior of neuron and the human brain. The basic element of ANN is a processing element which is also known as node or neuron. These nodes comprise of very small amount of local memory data which is processed on the basis of mathematical operations. Further, these nodes are connected to the input and output layer through one or more hidden layers. The calculating power of ANN comes through these hidden layers when the data is processed through these hidden layers.

To assess the optimized prediction accuracy of the data mining techniques, this process is divided into three parts: (i) In the first part, we have to select the input parameters of the structures. In high-strength concrete, the parameters of the concrete are very complex in nature. In the ANN model of high-strength concrete, the authors select eight variables to produce the prediction models. (ii) The second part was to check the selected data mining models on the foundation of whole input data. (iii) "The performance of the model is based on their selection. Whether simpler models with eight variables and individual data sets will leads to better prediction accuracy".

An ANN model with possible eight variables is shown in Fig. 1. Here in this figure, "8-3-1" represents 8: Input Variables, 3: Hidden layers, and 1: Output data, i.e., compressive strength in MPa.

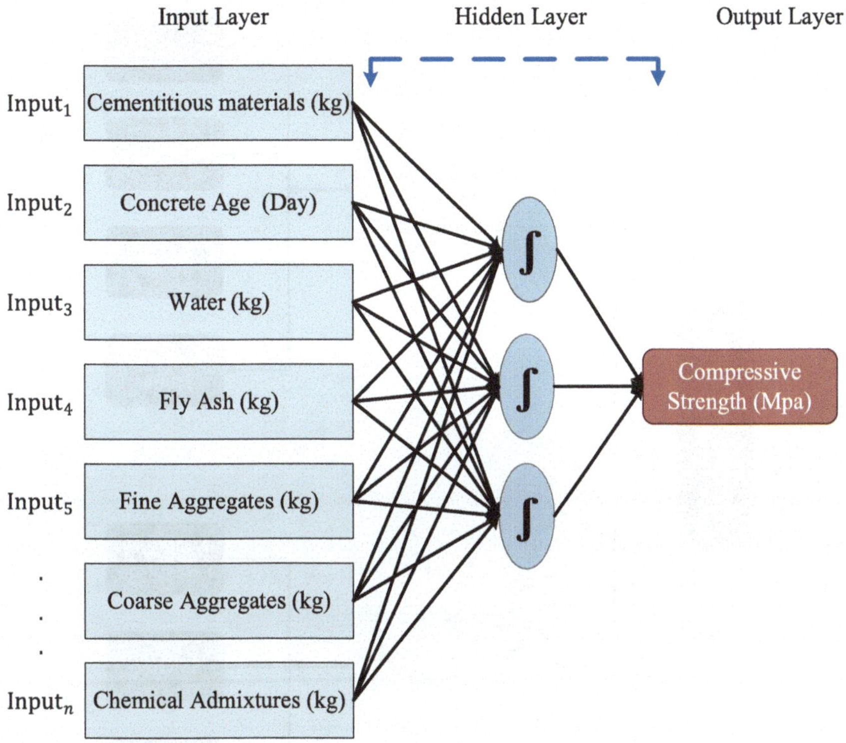

Fig. 1 Proposed structure of ANN Model used for high-strength concrete

1.2 Data Mining Techniques (DMT) for Estimation of HSC

The estimation of the strength of concrete using DM techniques is performed into two ways as shown in Fig. 2. The first method for prediction uses regression tree model which includes (a) "M5P, (b) M5-Rules, (c) REPTree (d) Decision Stump (DS) and the second one is using advanced prediction models which includes (a) Multilayer Perceptron, (b) Sequential Minimal Optimization Regression (SMOreg) and (c) Gaussian Regression."

The other techniques that can be used for estimation of accuracy of concrete are bagging and additive regression.

2 Literature Review

In the last few decades, there is a rapid change in the area of computing analysis software, which inspired the latest approaches to data analysis and its processing. In

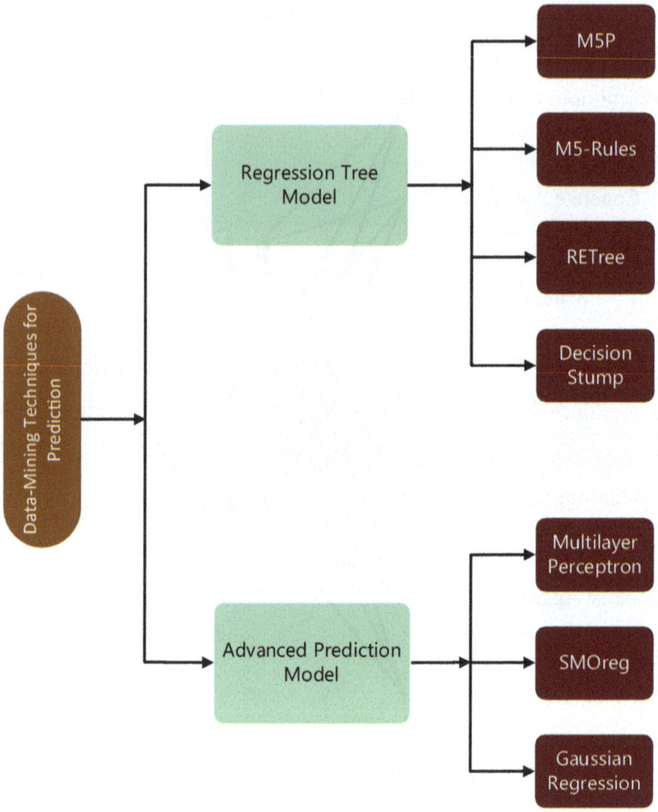

Fig. 2 Different types of DM techniques for prediction

terms of economy, among all the approaches that are being used to solve the engineering problems, the computer designed approaches are more useful and accurate. In recent years, ANN is useful to solve the many engineering problems in the field of concrete technology, geotechnical engineering, environmental engineering, and structural engineering.

2.1 *Regression Tree Models*

The regression techniques are mostly used in the data mining as the supervised techniques and mostly used in the numeric prediction. When we compared regression model techniques with other traditional techniques, its accuracy is less, but it is fast and easy to interpret. The review research is based on the four most used regression models.

Table 1 shows the details of different regression tree models used by the authors for the first time for determination of compressive strength of concrete along with their description.

Table 1 Description of regression tree models

Sr. no.	Technique	Author	Description
1.	M5P	Quinlan [1]	In this model, each branch stores a linear regression data which reaches to the leaf For breaking the data at each node or neuron, the standard deviation (SD) criteria are used The selection of the attributes directly affects the accuracy of the prediction model
2.	M5-Rules	Holmes et al. [2]	"This model works on the programming that uses divide and conquers method to generate the decision list for regression problems" This decision list is more compact and easier to understand This decision list works with both the nominal and continuous variables The M5-Rules works on M5P programming to build the model tree
3.	REPTree	Witten and Frank [3]	REPT stands for Reduced-Error Pruning Tree This tree model provides the fast build-up information that assembles within the model In this technique, the error is reduced by back-fitting technique In this model, the values taken by the system at the starting point of the run and after that this method are sorted automatically to calculate the right splits at each tree neuron
4.	Decision Stump	Iba and Langley [4]	This is also a machine learning tool, but, it has only one-level decision tree The internal neuron or node and intermediate connected nodes are called root node and terminal node, respectively Indecision stump, there is only one internal node which is intermediately connected to the nodes of the tree branches In this method, the attribute depends only on the single input This technique is mostly used to boost the prediction accuracy

2.2 *Advanced Prediction Models*

The advanced prediction models are of three types: (i) multilayer perceptron (ii) SMOreg and (iii) Gaussian Prediction model. Table 2 shows the details of different advanced prediction models used by the authors for the first time for determination of compressive strength of concrete along with their description.

Table 3 comprises of different DM techniques used by different authors, type of concrete, input variables, and the model accuracy in terms of R^2 in order to find the compressive strength (CS) of the concrete. It is observed that in previous studies, the main focus was on HSC with the addition of some minerals such as blast-furnace slag, superplasticizer, and fly-ash. It was found that most of the researchers used ANN technique for their analysis.

3 Conclusion and Future Directions

Numerous DM techniques used for the prediction of compressive strength of HSC have been explained and described in this study. It is concluded that among all data mining models, the decision stump can provide desirable results in terms of accuracy. Using REPTree and Gaussian Regression models, the overall value of the coefficient of determination (R^2) has been found 0.9217–0.9842, respectively. It is further observed that the Gaussian model gives the best accurate results if used individually and if we combine two models, the accuracy of the results may further increase.

The advanced prediction models (MP, SMOReg, and GR) have more accurate results as compared to "regression tree models" (RTM). In data mining, the most important step is to choose the more reliable method that can provide accurate results. When dealing with a huge amount of data, the regression tree models have more practical possibility. At last, it the study found that the previous studies used DM technique to predict the compressive strength of concrete only, but, in future, these techniques can be used to predict the tensile strength, durability, and slump values of the concrete as these methods give more precise and accurate results.

Table 2 Data Mining Advanced Prediction Models

Sr. No.	Technique	Author	Description
1.	Multilayer Perceptron	Caudil [5]	Artificial neuron network is a computerized programmed system which consists of simple and highly interconnected nodes to solve the particular problem An artificial neuron has mainly three layers which consists of "input layer," "hidden layer," and "output layer" and these are connected to each other with different nodes The weighting and bias values firstly are chosen randomly and then after that adjusted as per the training process
2.	SMOreg	Platt [6] & Cortes and Vapnik [8]	This is the sequential minimal optimization regression It is the supervised learning model and has been used in many tenacities such as regression and classification This technique is helpful to solve the large quadratic programming equations (QPE), by breaking the large QP into smaller ones This way the numerical optimization algorithms helps to solve problems
3.	Gaussian Regression	Bishop [7]	This model is a nonlinear prediction technique It is mostly used for the learning process and Bayesian regression for both the supervised and unsupervised learning works The main disadvantage of this model is that its formulation is based on the probability of the data taken by the system

Table 3 Prediction accuracy of the data from literature

Author	Data mining technique	R^2	Concrete type	Input data
Yeh [13]	ANN LR	0.914 0.574	HPC	Cement, water, Fa, FA, CA, BFS, SP and curing time
Gupta et al. [9]	NES	0.5776	HPC	Grade of concrete, maximum temperature, shape and size of specimen curing, period and technique
Zarandi et al. [12]	FPNN	0.8209	HPC	Cement, water, CA and FA, Silica Fume and SP
Yeh and Lien [14]	GOT ANN	0.8669 0.9338	HPC	Cement, water, BFS, CA and FA, Fa, SP and curing time
Deepa et al. [15]	ANN (MP) LR M5P Model	0.625 0.491 0.787	HPC	Cement, BFS, FA, and CA, Fa, SP, Curing age
Chou et al. [11]	ANN MR SVM MART BRT	0.9091 0.6112 0.8858 0.9108 0.8904	HPC	Cement, BFS, Fa, CA and FA, SP water and Curing Period
Atici [10]	ANN MR	0.9801 0.899	HSC	Concrete Mix, UPV and RH
Erdal et al. [16]	ANN ANN Gradient boosted ANN Wavelet gradient boosted ANN Bagged ANN wavelet bagged	0.9088 0.927 0.9528 0.9278 0.9397	HPC	Cement, Fa, BFS, FA and CA, Water and curing time
Orman et al. [17]	M5P Model M5-Rules REPTree MP SMOreg (SVM) GPR AR Bagging	0.9476 0.9482 0.9217 0.97 0.968 0.9843 0.9837 0.9816	HPC	Cement, Curing period, water, FA, LWA and Micro Air

Here, "R^2: Coefficient of Determination

ANN: Artificial Neuron Network, NES: Neural-Expert System, FPNN: Fuzzy Polynomial Neural Networks, GOT: genetic Operation Tree, MP: Multilayer Perceptron, MR: Multiple Regression, MART: Multiple additive regression tree, BRT: Bagging Regression tree, LR: Linear Regression, GPR: Gaussian Process Regression, AR: Additive Regression

HPC: High-Performance concrete, HSC: High-Strength Concrete

FA: Fine Aggregate, Fa: Fly-ash, CA Coarse Aggregate, BFS Blast Furnace Slag, SP: Superplasticizer, LWA: Light Weight Aggerate

UPV: Ultrasonic Pulse Velocity, R.H.: Rebound Hammer"

References

1. J.R. Quinlan, Learning with continuous classes, in *5th Proceeding of Joint Conference Artificial Intelligence* (World Science Press, Australian, 1992), pp. 343–348
2. G. Holmes, M. Hall, E. Frank, Generating rule sets from model trees. Lecture Notes of Computer Science (1999), pp. 1–12
3. I.H. Witten, E. Frank, *Practical machine learning tools and techniques* (Morgan Kaufman, San Francisco, 2005). 2nd Education
4. W. Iba, P. Langley, Induction of one-level decision tress, in *9th Proceeding of International Conference on Machine Learning* (Morgan Kaufman, San Francisco, 2005)
5. M. Caudill, Neural networks primer part I. AI Expert **2**(12), 46–52 (1987)
6. J.C. Platt, Sequential minimal optimizer: a fast algorithm for training support vector machines: Technical Report MSR-TR-98-14 (Microsoft Research, Redmond, WA, 1998)
7. C.M. Bishop, *Pattern recognition and machine learning* (Springer, New York, 2006)
8. C. Cortes, V. Vapnik, Support-vector networks. Mach. Learn. **20**(3), 273–297 (1995)
9. R. Gupta, A.K. Manish, A. Goel, Predication of concrete strength using neural-expert system. J. Mater. Civ. Eng. **18**(3), 462–466 (2006)
10. U. Atici, Prediction of the strength of mineral admixture concrete using multi-variable regression analysis and a artificial neural network. Expert Syst. Appl. **38**, 9609–9618 (2011)
11. J.S. Chou, C.K. Chiu, F. Mahmod, A.T. Ismail, Optimizing the prediction accuracy of the concrete compressive strength based on a comparison of data mining techniques. J. Comput. Civil Eng. **25**, 242–253 (2011)
12. M.H. Zarandi, T.B. Turksen, J. Sobhani, A.A. Ramezanianpour, Fuzzy polynomial neural networks for approximation of the compressive strength of concrete. Appl. Soft Comput. **8**, 488–498 (2008)
13. I.C. Yeh, Modelling of strength of high-performance concrete using artificial neural networks. Cem. Concr. Res. **28**(12), 1797–1808 (1998)
14. I.C. Yeh, L.C. Lien, knowledge discovery of concrete materials using genetic operations trees. Experts Syst. Appl. **39**, 5807–5812 (2009)
15. C. Deepa, S. Kumari, V.P. Sudha, Predication of the compressive strength of high-performance concrete mix using tree-based modeling. Int. J. Comput. Appl. **6**(5) (2010)
16. H.I. Erdal, O. Karakurt, E. Namli, High performance concrete compressive strength forecasting using ensemble models based on discrete wavelet transform. Eng. Appl. Artif. Intell. **26**, 1246–1254 (2013)
17. B.A. Orman, Q. Chen, R. Jin, Comparison of data mining techniques for predicting compressive strength of Environmentally friendly concrete. J. Comput. Civil Eng. **30**(6), 1–11 (2016)

AntMiner: Bridging the Gap Between Data Mining Classification Rule Discovery and Bio-Inspired Algorithms

Bhawna Jyoti and Aman Kumar Sharma

Abstract An exploratory study is presented which emphasize on how to bridge the gap between data mining classification rule discovery and bio-inspired algorithms. Real-world applications like expert system, facial recognition patterns mainly depends on the classification of structured and unstructured data. In this study, predictive accuracy values are calculated by implementing c-AntMiner2 and c-AntMinerpb (ant colony based algorithms) on six standard datasets. A comparative study of c-AntMiner2, c-AntMinerpb is presented against already existed rule induction algorithms (JRIP and PART) and shows a good performance of these bio-inspired algorithms over JRIP and PART.

Keywords Ant colony optimization · antMiner · Rule pruning · Rule induction · Classification rule discovery · Quality function

1 Introduction

Swarm intelligence and Bio-inspired algorithms are broadly used to solve many hard combinatorial problems and based on the self-organizing behaviour of social agents [1–6]. It acts as an interdisciplinary field and an active research area belongs to the category of building classification model from training and testing datasets [7–9]. In data mining, meaningful conclusions are drawn in the form of rules that provides a good strategy to discover hidden patterns in the form of knowledge, pattern processing and functional dependency analysis [10–12]. To do computational work on a huge amount of multidimensional data containing real-world datasets, research techniques are mainly concentrated on supervised classification. Classification is a data analysis task to build a model that describes data classes by identifying training examples and testing datasets [8, 13, 14]. In classification rule induction algorithms,

B. Jyoti (✉) · A. K. Sharma
Computer Science Department, HP University, Shimla, India
e-mail: bhawnashkl@gmail.com

A. K. Sharma
e-mail: sharmaas1@gmail.com

M. Dutta et al. (eds.), *Proceedings of International Conference on IoT Inclusive Life (ICIIL 2019), NITTTR Chandigarh, India*, Lecture Notes in Networks and Systems 116,
https://doi.org/10.1007/978-981-15-3020-3_25

a sequential approach is followed by Ant Colony (AC) procedure in which a single rule is created and included in discovered list of rules as well as the removal of the examples from training dataset [15–17] is simultaneously done that are used to create rules [4, 8, 18].

Ant Colony (Nature-inspired, self-organized) system is based on social co-operative agents(ants) and simulates the foraging behaviour of ants by adopting adaptation as well as cooperation mechanisms [19–22]. When designing specification of ant system, a problem dependent representation of probability describing heuristic function is needed so that ants can construct optimal solutions for the given problem [8, 13, 14, 23]. Support vector machines and neural networks have been successfully used to build classifier models and find their predictive accuracy but these models are difficult to interpret. Ant-Miner (AC based algorithm) has given outstanding performance to find rules in the field of data mining by building easy interpretable classifier models [24]. Parpinelli et al. [25] put their initial efforts to propose ant colony based classification rule discovery by using antMiner. AntMiner 2.0 and AntMiner 3.0 algorithms are based on density heuristic function for pheromone updation factor [1, 26]. Martens et al. [27] gave the idea of AntMiner+ which is based on MAX-MIN Ant system [28, 29] but has the limitation of not supporting continuous attributes directly and needs discretization of datasets as a pre-processing step [30–32].

The remainder of this paper is divided into the following sections: Sect. 2 outlines the AC based optimization and in Sect. 3, antMiner algorithm is discussed which is used to discover classification rules. The study of antMiner algorithm is experimentally proven in Sect. 4 and analysis of the experimental work is discussed in Sect. 5. Finally, Sect. 6 describes conclusion of the study as well as some directions of future research.

2 A Self-Reinforcing Ant System

This is a metaheuristic based optimization technique in which the term Meta means "in the upper level" and heuristic word is derived from the Greek verb "heuriskein" (to find). Ant colony based algorithms are population based, stigmergic (self-reinforcing system), robust, versatile and uses metaheuristic technique to solve hard combinatorial problems [18, 22, 23, 33]. Ant Colony behaves like a stigmergic system in which movements of ants takes place within as well as around their ant colony to communicate with each other for finding best solutions in large search space [1, 26, 34]. It was introduced as a self-organized and feedback directed mechanism based on real behaviour of ants by Dorigo [12, 28, 35] in 1990s.

When an ant goes from one place to other in the search of food, a chemical named pheromone is released and set path used by other ants because ants are basically blind in nature [10, 11, 36]. Specific heuristic information is used to update the associated concentration of pheromone trail released by ant [9, 18, 33]. A pruning technique is used to enhance the quality of rule by noisy information present in the candidate solution [36, 37]. The ants foraging behaviour is driven by probability function given below:

$$PF_{xy} = \frac{(Z_{xy})^{\alpha}(t_{xy})^{\beta}}{\sum_{x,y=1}^{n}(Z_{xy})^{\alpha}(t_{xy})^{\beta}}$$

where PF_{xy} denotes the functional probability showing ant movement; $(Z_{xy})^{\alpha}$ denotes pheromone concentration; $(Z_{xy})^{\beta}$ is the inverse of the displacement (x to y) and tells about the desirability to follow a particular path from x to y. A candidate solution will be formed by cooperating agents (ants) by using associated pheromone values to identify good optimal solution for a given problem [23, 29, 33]. A candidate solution will be formed by cooperating agents (ants) by using associated pheromone values to identify good optimal solution for a given problem. Good solutions are created by ants having the amount of concentration of pheromone with increased value while components those components will not be used whose pheromone concentration gradually decreased [9, 18].

3 Ant-Miner Description

The execution of steps in ant-Miner is described as follows:

- Input of all training examples is taken to create the rule.
- Discovered list is empty so that new rules can be added to it and initialized to **Ø**.
- While loop is executed until it considers maximum uncovered example in training set.
- The chemical pheromone is initialized with initial concentration.
- The heuristic function is calculated for a given problem.
- $\mathbf{R_{best}}$ is variable taken to store the best rule and initialized to **Ø.**
- **Ant** is initialized and antecedent part of the rule is constructed.
- Class of term of antecedent part is identified.
- Rule is discovered and pruning of rule is done to remove noisy data and improve the quality of rule.
- Quality function is used to calculate the quality of rule discovered and stored in $\mathbf{Quality_{current\ Rule}}$.
- Generated rule $\mathbf{R_{currentRule}}$ is compared with $\mathbf{R_{best}}$ and values are swapped by considering their quality function values.
- $\mathbf{R_{best}}$ is added to **listOfRulesDiscovered** and term (Example) is removed from the training set from which rule is discovered.
- This procedure continues until all terms converge to form discovered list of rules.

4 Computational Study

In the experimental setup, we used six datasets [21]. Ten-fold cross-validations are used by considering one test dataset and other nine as training datasets in each cross-validation. Parameters setting for computational study are done as follows (Table 1):

- Ant colony size is taken as {5, 10, 50, 100, 500}
- Evaporation factor is taken as {0.85}.
- Examples considered for coverage of experimentation are {2, 6, 14}
- Number of cycles repeated by ants (iterations) are 500.

To calculate the predictive accuracy of Ant miner algorithms, an open source framework based on java language known as Myra [3.1] is used. Weka tool is used for comparing the results with already existing rule induction algorithms JRIP and PART to calculate their predictive accuracy. IBM SPSS statistical open source software is used to find the statistical calculations. The following Table 2 shows precision accuracy measures of various algorithms against well known publicly available datasets and the most accurate results for a particular dataset obtained by a particular rule induction algorithm is formatted as bold characters.

Number of rules obtained in each dataset by rule induction algorithms is given in Table 3. The value of dataset having lowest entry of number of rules extracted is presented in bold face.

Table 1 Detailed description of datasets of LD, HC, HH, BT, BW, BI

Dataset	Continuous values, nominal values	Number of classes, size
liver disorder (LD)	(6, 0)	(2, 345)
heart-c (HC)	(6, 7)	(5, 303)
heart-h (HH)	(6, 7)	(5, 294)
breast tissue (BT)	(9, 0)	(6, 106)
breast-w (BW)	(30, 0)	(2, 569)
breast-i (BI)	(0, 9)	(2, 286)

Table 2 Average predictive accuracy of cant-Miner2, cAntMiner-pb along with its standard deviation

Dataset	c-AntMiner2	PART	JRIP	c-AntMinerpb
LD	66.56 ± 0.023	62.81 ± 3.40	66.33 ± 2.81	**66.72 ± 0.04**
HC	54.66 ± 0.23	53.73 ± 1.34	53.50 ± 1.52	**55.50 ± 0.37**
HH	63.71 ± 0.31	63.46 ± 1.58	63.93 ± 1.29	**64.76 ± 0.25**
BT	**69.61 ± 0.27**	64.36 ± 3.63	60.78 ± 3.34	67.23 ± 0.56
BW	93.68 ± 0.21	94.21 ± 1.02	93.76 ± 1.34	**94.28 ± 0.21**
BI	**76.21 ± 0.09**	68.89 ± 1.82	69.86 ± 2.12	72.42 ± 0.32

Table 3 Number of rules (model size) of various datasets

Dataset	JRIP	PART	c-AntMiner2	c-AntMinerpb
LD	8.66	22.12	**8.46**	11.87
HC	19.87	133.21	**7.01**	26.90
HH	16.02	72.90	**6.02**	22.51
BT	8.91	22.31	6.99	**6.34**
BW	13.12	11.68	9.78	**8.57**
BI	7.61	35.21	**4.51**	19.61

5 Comparative Analysis

In LD, HC, HH, BW datasets, c-AntMinerpb shows good results (66.72, 55.50, 64.76, 94.28), respectively, in terms of precision accuracy measure (shown in bold face in Table 2). In BT and BI datasets, c-AntMiner2 shows 69.61 and 76.21 precision accuracy rate. The following Fig. 1 shows a comparative outlook of various rule induction algorithms obtained in six publicly available datasets.

The Friedman test(non-parametric measure of repetitions (ANOVA)) is conducted. It provides ranks to each dataset for a particular rule induction algorithm. Under the null hypothesis, the Friedman statistics can be formulated as

$$(CHI)^2 = \frac{12N}{s(s+1)} \sum (Rj)2 - \frac{s(s+1)}{4}$$

where $(\mathrm{CHI})^2$ presents statistic distribution of data according to $\mathrm{k}-1$ degrees of freedom and $Rj = \frac{1}{N}\sum_i^j r$ is used to compare the average ranks of the algorithms.

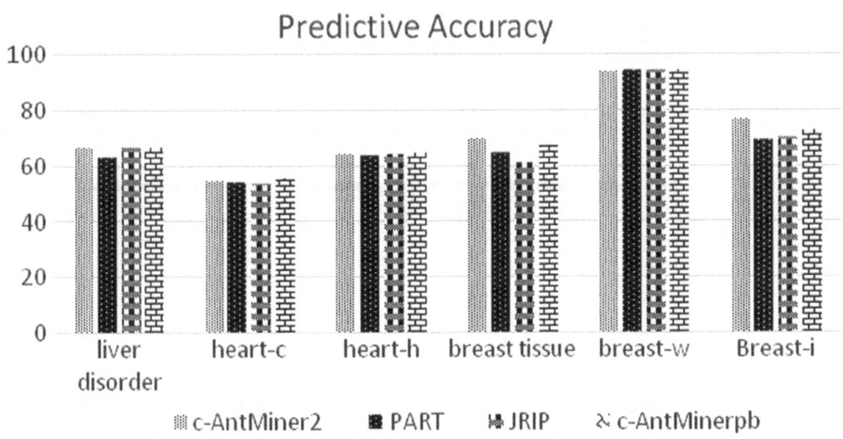

Fig. 1 A comparative study of accuracy measure of precision of c-AntMiner2, JRIP, PART and c-AntMinerpb algorithms

Here H_0: The probability distributions are identical and H_a: It is assumed that two of them have probability distribution differing in location (Fig. 2).

In this test mean, rank of c-AntMiner2 is 3.08, PART is 1.50, JRIP is 1.83, c-AntMinerpb is 3.58. The value of $(\text{CHI})^2$ is 11.411 and df is 3. By considering the statistical table for $(\text{CHI})^2$ where $\alpha = 0.05$, it is seen that it falls in rejection region, so the data provides sufficient evidence that at least two of the probability distributions differ in location. We conducted t-distribution test by considering population mean and considering the hypothesis H_0 that all cantMiner-2, JRIP, PART and c-AntMinerpb algorithms do not show significant performance differences (Fig. 3).

Under H_0, the test statistics is given by

$$t = \frac{x - \mu 0}{S/\sqrt{n}}$$

The statistical student t-test is conducted in IBM SPSS statistics 22 software for identifying the difference between two classifiers and we compared the result with

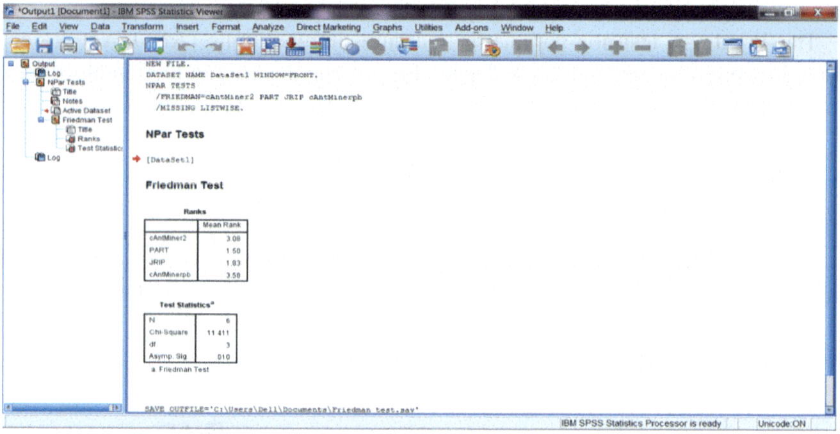

Fig. 2 Snapshot of Friedman test

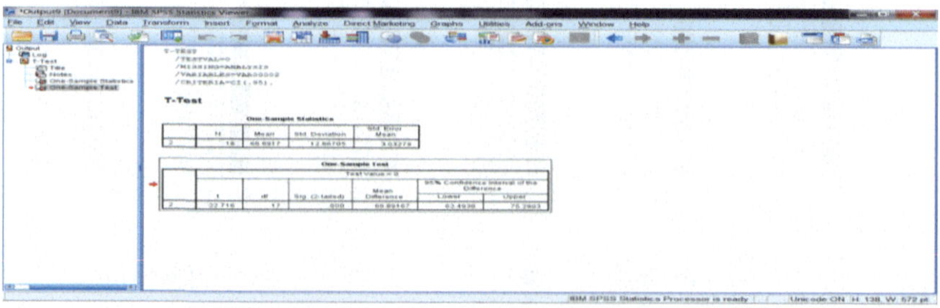

Fig. 3 Snapshot of T-test

standard value of t for 95% level of significance having mean difference of 68.89167 and found that t value is higher than given standard table value of t, so given hypothesis is rejected.

6 Conclusion and Future Work

Classification rule discovery is discussed based on AC procedure to find induction rules in knowledge based information systems. The main consideration is to use pruning techniques to simplify and enhance the quality of rule discovered by using quality function. We have implemented cant-Miner2 and cant-Minerpb involving 6 standard datasets and their predictive accuracy is calculated. Further, these algorithms are compared against PART and JRIP and it is observed that c-AntMiner2 and c-AntMinerpb have shown good results in classification rule discovery when compared to PART and JRIP. As a future research direction, global pruning procedure can be taken into consideration by evaluating the pruning of whole rule list rather than single created rule [34, 38, 39, 40]. Use of heuristic function can be done to sort the discovered rules by taking their threshold confidence level measures to guide the search effectively [24, 41].

References

1. B.C. Mohan, R. Baskaran, A survey: Ant Colony Optimization based recent research and implementation on several engineering domain. Expert Syst. Appl. **39**, 4618–4627 (2012)
2. W.J. Jiang, Y.H. Xu, Y.S. Xu, A novel data mining algorithm based on ant colony system, in *International Conference on Machine Learning and Cybernetics*, vol. 3 (IEEE, 2005), pp. 1919–1923
3. T. Stutzle, H. Hoos, M*AX*-M*IN* ant system. Futur. Generat. Comput. Syst. **16**(8), 889–914 (2000)
4. J. Demsar, Statistical comparisons of classifiers over multiple data sets. J. Mach. Learn. Res. **7**, 1–30 (2006)
5. A.A. Freitas, Understanding the crucial differences between classification and discovery of association rules—a position paper. SIGKDD Explor. **2**(1), 65–69 (2000)
6. Y. Chen, L. Chen, L. Tu, Parallel ant colony algorithm for mining classification rules, in *IEEE International Conference on Granular Computing*, 10 May 2006 (IEEE, 2006), pp. 85–90
7. S. Garcia, F. Herrera, An extension on statistical comparisons of classifiers over multiple data sets for all pairwise comparisons. J. Mach. Learn. Res. 2677–2694 (2008)
8. M. Dorigo, T. Stuzle, The ant colony optimization metaheuristic: algorithms, applications, and advances, in *Handbook of metaheuristics* (Springer, Boston, MA, 2003), pp. 250–285
9. M. Medland, F.E. Otero, A.A. Freitas, Improving the cAnt-Miner PB classification algorithm, in *International Conference on Swarm Intelligence,* 12 Sep 2012 (Springer, Berlin, 2012), pp. 73–84
10. F. Otero, A. Freitas, C. Johnson, Handling continuous attributes in ant colony classification algorithms, in *IEEE Symposium on Computational Intelligence and Data Mining (CIDM'09)*, March 2009, pp. 225–231

11. Q. Yang, W.N. Chen, Z. Yu, T. Gu, Y. Li, H. Zhang, J. Zhang, Adaptive multi-modal continuous ant colony optimization. IEEE Trans. Evol. Comput. **21**(2), 191–205 (2016)
12. K. Socha, M. Dorigo, Ant colony optimization for continuous domains. Eur. J. Oper. Res. **185**(3), 1155–1173 (2008)
13. C. Shah, A.G. Jivani, Comparison of data mining classification algorithms for breast cancer prediction, in *Fourth International Conference on Computing, Communications and Networking Technologies (ICCCNT)*, 4 July 2013 (IEEE, 2013), pp. 1–4
14. X.S. Yang, *Nature-Inspired Metaheuristic Algorithms*, Second edn (Luniver Press, 2010)
15. A. Amuthan, K.D. Thilak, Improved Ant colony algorithms for eliminating stagnation and local optimum problem—a survey, in *2017 International Conference on Technical Advancements in Computers and Communications (ICTACC)* (IEEE, 2017), pp. 97–101
16. A. Helal, J. Brookhouse, F.E. Otero, Archive-based pheromone model for discovering regression rules with Ant colony optimization, in *2018 IEEE Congress on Evolutionary Computation (CEC)* (IEEE, 2018), pp. 1–7
17. H.N. Al-Behadili, K.R. Ku-Mahamud, R. Sagban, Rule pruning techniques in the ant-miner classification algorithm and its variants: a review, in *2018 IEEE Symposium on Computer Applications & Industrial Electronics (ISCAIE)* 28 Apr 2018 (IEEE, 2018), pp. 78–84
18. M. Manfrin, M. Birattari, T. Stutzle, M. Dorigo, Parallel ant colony optimization for the traveling salesman problem, in: *International Workshop on Ant Colony Optimization and Swarm Intelligence*. 4 Sep 2006 (Springer, Berlin, 2006), pp. 224–234
19. K.M. Salama, A.M. Abdelbar, F.E. Otero, Investigating evaluation measures in ant colony algorithms for learning decision tree classifiers, in *IEEE Symposium Series on Computational Intelligence*, 7 Dec 2015 (IEEE, 2015), pp. 1146–1153
20. S.H. Ripon, Rule induction and prediction of chronic kidney disease using boosting classifiers, Ant-Miner and J48 Decision Tree. in *International Conference on Electrical, Computer and Communication Engineering (ECCE),* 7 Feb 2019 (IEEE, 2019), pp. 1–6
21. http://www.ics.uci.edu/~mlearn/MLRepository.html
22. V.K. Panchal, P. Singh, A. Narula, A. Mishra, Review on ant miners. in *2009 World Congress on Nature & Biologically Inspired Computing (NaBIC)*, 9 Dec 2009 (IEEE, 2009), pp. 1641–1644
23. M.N. Wahab, S. Nefti-Meziani, A. Atyabi, A comprehensive review of swarm optimization algorithms. PLoS ONE **10**(5) (2015)
24. K.M. Salama, F.E. Otero, Exploring different functions for heuristics, discretization, and rule quality evaluation in ant-miner, in *International Conference on Swarm Intelligence*. 12 Sep 2012. Springer, Berlin, pp. 344–345
25. R. Parpinelli, H. Lopes, A. Freitas, Data mining with an ant colony optimization algorithm. IEEE Trans. Evol. Comput. **6**, 321–332 (2002)
26. A. Helal, F.E. Otero, Automatic design of ant-miner mixed attributes for classification rule discovery, in *Proceedings of the Genetic and Evolutionary Computation Conference, July 1 2017* (ACM, 2017), pp. 433–440
27. D. Martens, M. De Backer, R. Haesen, J. Vanthienen, M. Snoeck, B. Baesens, Classification with ant colony optimization. IEEE Trans. Evolut. Comput. **11**, 651–65 (2007)
28. M. Dorigo, V. Maniezzo, A. Colorni, Ant system: optimization by a colony of cooperating agents. IEEE Trans. Syst. Man Cybern. B Cybern. **26**(1), 29–41 (1996)
29. M. Dorigo, V. Maniezzo, A. Colorni, The ant system: an autocatalytic optimizing process, Technical Report 91-016 (1991)
30. J. Smaldon, A. Freitas, A new version of the Ant-Miner algorithm discovering unordered rule sets, in *Proceedings Genetic and Evolutionary Computation Conference (GECCO)* (2006), pp. 43–50
31. S. Neelamegam, E. Ramaraj, Classification algorithm in data mining: An overview. Int. J. P2P Netw. Trends Technol. (IJPTT) **4**(8), 369–374 (2013)
32. G.M. Prabha, E. Balraj, A HM Ant Miner using evolutionary algorithm. Int. J. Innov. Res. Sci. Eng. Technol. **3**(3), 1687–1692
33. A. Colorni, M. Dorigo, V. Maniezzo, An investigation of some properties of an Ant Algorithm. in *Ppsn*, vol. 92, 28 Sep 1992 (1992)

34. B. Liu, H.A. Abbas, B. McKay, Classification rule discovery with ant colony optimization, in *IEEE/WIC International Conference on Intelligent Agent Technology* (IEEE, 2003), pp. 83–88
35. A. Colorni, M. Dorigo, F. Maffioli, V. Maniezzo, G.I. Righini, M. Trubian, Heuristics from nature for hard combinatorial optimization problems. Int. Trans. Oper. Res. **3**(1), 1–21 (1996)
36. F. Otero, A. Freitas, C. Johnson, cAnt-Miner: an ant colony classification algorithm to cope with continuous attributes, in *Proceedings of the 6th International Conference on Swarm Intelligence (ANTS 2008)*, Lecture Notes in Computer Science (2008)
37. F.B. Fakhar, A novel method for extracting classification rules based on Ant-Miner. J. Math. Comput. Sci. **8**, 377–386 (2014)
38. A. Chan, A. Freitas, A new classification-rule pruning procedure for an ant colony algorithm, in *Artificial Evolution, Lecture Notes in Computer Science*, vol. 3871 (2005), pp. 25–36
39. B. Liu, H.A. Abbass, B. McKay, Density-based heuristic for rule discovery with ant-miner, in *The 6th Australia-Japan joint Workshop on Intelligent and Evolutionary System,* 30 Nov 2002, vol. 184 (2002)
40. Z. Wang, B. Feng, Classification rule mining with an improved ant colony algorithm. in *Advances in Artificial Intelligence*, Lecture Notes in Computer Science, vol. 3339 (2004), pp. 357–367
41. J. Ranilla, O. Luaces, A. Bahamonde, A heuristic for learning decision trees and pruning them into classification rules. AI Commun. **16**(2), 71–87 (2003)

Fuzzy K-Medoid Clustering Strategy for Heterogeneous and Dynamic Data for IoT Scenario

Priya Dogra and Rakesh Kumar

Abstract IoT is a revolutionary vision pertaining to the fact that everything could be connected to the internet. With the increase in the number of internet users, IoT clients are also increasing. In IoT environment, the data generated is enormous. So, we need an efficient approach to manage such a gigantic multi-dimensional IoT data. It can only be done by applying some appropriate data mining algorithms. These algorithms organize and transform the data into a structured form. To generate such structured information, majorly adaptive clustering techniques are employed in data mining. Hence, in the proposed work, the authors focus on generating algorithm which enhances the performance and compares the proposed fuzzy k-medoid clustering with the existing clustering algorithm pertaining to IoT data collected in intelligent real-time traffic system. First, the data streams are uploaded and then adaptive k-means clustering is applied to classify the data. Then k-medoid clustering algorithm is applied to the same data streams, and equivalent distance matrix is generated based on the centroid. After that, the proposed fuzzy k-medoid clustering algorithm is applied. The proposed work gives better performance than the existing ones. So with this approach it is easier to manage a huge IoT data. One can easily extract the information when it is in groups. Conceivable result of this proposed approach is: generating algorithm which gives better performance and also there is a comparison of clustering techniques that helps in deciding better clustering scheme for the given IoT dataset which results in optimal performance.

Keywords IoT · Data mining · Clustering algorithms · Adaptive clustering and k-medoid clustering algorithm

P. Dogra (✉) · R. Kumar
NITTTR, Chandigarh, India
e-mail: priyadogra.cse@nitttrchd.ac.in

M. Dutta et al. (eds.), *Proceedings of International Conference on IoT Inclusive Life (ICIIL 2019), NITTTR Chandigarh, India*, Lecture Notes in Networks and Systems 116,
https://doi.org/10.1007/978-981-15-3020-3_26

1 Introduction

IoT is the network of devices (wired or wireless) where each device is connected to the other and it can exchange the data as well. It is a network of smart devices. It acts like an ecosystem where objects are connected through the internet. The "things" in IoT could be a person with a monitor or a physical device with in-built sensors which have the ability to collect and transmit data over a network without manual support or involvement [1, 2]. The embedded technology in this area helps them to interact with internal or the external environment, which in turn affects the decisions taken. IoT evolves a wide infrastructure for data, enables great services by intercommunication between the physical and virtual devices based on existing and evolving technologies [3]. Day-by-day IoT users as well as IoT devices are increasing. The data generated by the IoT devices are huge and gigantic.

Challenges and Open Research Issues in IoT [3, 4]
The various challenges in IoT are:
Security
Security is the essential pillar of IoT. IoT is developed on thousands of SNs and massive data is generated through communication system, so the data and its communication need to be secure. No one should be able to steal data easily.

Privacy
In most deployments, the data generated must be protected from unauthorized access. No one should be able to decipher the data in static form or while in communication. A Federal Trade Commission report entitled "Internet of Things: Privacy and Security in a Connected World" found that fewer than 10,000 households can generate 150 million discrete data points every day. This creates more entry points for hackers and leaves sensitive information vulnerable.

Lack of standards
In IoT, large numbers of heterogeneous devices are networked so it does not have a single platform for standardization. Numerous varying data storage and communication technologies/standards are used by these devices. Hence, to draw a common standard for such a scenario is extremely difficult. Further, to monitor and enforce such a standard if one exists is even more challenging.

Data management
The data generated through the IoT environment is huge and dynamic in nature. So, to extract useful data and manage it in such a big environment is too difficult. To handle this dynamism coupled with heterogeneity of data formats is a real issue. Data is the base of any IoT industry, so it has to be safe, secure and managed.

Motivation

IoT in the coming times is going to become a fundamental concept in internet technology because of its innovative vision of connecting devices and their communication via internet [4, 5]. With the increase of internet users IoT users are also increasing. Its rapid growth increases the risk of managing massive data. In IoT, the data is captured from different devices such as sensors, radio frequency identification (RFID), and ZigBee. Extracting specific information from such a wide-ranging data warehouse is difficult. The problem here generated is to manage such a huge data, so the data is to mine with a data mining approach which is employed in the proposed work to efficiently manage the data for storage and retrieval. Further, in data mining, clustering is the most suitable approach for this IoT environment. Because here the data generated is dynamic and heterogeneous in nature [6].

Need of Data Mining in IoT [7]

Data mining is a suitable approach which is used to manage the data. It efficiently manages the data in various forms, that is, it classifies the data, it generates clusters or groups to manage it, and it also creates patterns between the similarity data. The IoT data are different from each other where there is no possibility of matching so the pattern analysis is not being done in this area. Also, the classification is not suitable because it only classifies the data in different forms. So here we have chosen clustering as the key concept of our research because it is the most suitable in this environment. The clustering is a process of portioning the data in groups formed. It efficiently manages the data for storage and retrieval.

Applications in IoT [8]

Some applications are defined as:

Smart Home, Retail and cities
Ind. Internet
Connected Health and Farming in smart ways.

2 Existing Work

Following are the major work done in the area of IoT by other researchers in the past:

Data Stream Clustering [9–13]

Puschmann et al. [1] proposed a clustering scheme for dynamic IoT data streams. In this research paper, he focused on the state-of-the-art and discussed the benefits and drawbacks of different stream cluster algorithms. He analyzed stream data with concept and data drifts. He introduced the concept of adaptive clustering method which automatically computes the best clusters based on the data distribution. He compared the method against synthesized dataset and chose a silhouette coefficient to evaluate the experimental results. He gave various clustering techniques, namely such as a K-mean, K-mean++, and DBSCAN. Kumar [11] proposed a bunching plan to deal with enormous measure of information in IoT. The thing that continues going by in his brain is the tremendous measure of IoT applications. There he produces an information stream bunching plan which is unique in relation to the conventional information as just a single output is feasible for mining in light of its dynamic handling and also the stream keeps on developing. The enormous information stream created by IoT applications is considered of incredible significance, and for removing data, information mining calculations can be connected to the produced information stream. Information stream bunching, one of the critical procedures of information mining, is very useful in grouping the comparable information protests and identifying exceptions. The target of his examination is to survey the diverse information stream grouping calculations that are utilized to perform bunching on the information stream created by IoT applications. He utilized two bunching plans—K-medoid and DBSCAN [7, 8].

Data mining in IoT [6, 14–16]
Chen et al. [14] concentrated on the idea of information mining in IoT. Information mining includes finding novel, intriguing, and conceivably valuable examples from vast informational collections and applying calculations to the extraction of concealed data. The target of any information mining procedure is to construct a proficient prescient or distinct model of a lot of information that best fits or clarifies it, as well as ready to sum up to new information. They have applied the pattern mining technique to extract the information from a huge data but it was not that much efficient because it is difficult to find the pattern in such a huge environment. The internet of things idea emerges from the need to oversee, robotize, and investigate all gadgets, instruments, and sensors on the planet. Information mining advances are coordinated with IoT advances for basic leadership support and framework streamlining. Bin et al. [6] proposed four information digging models for the internet of things, which are multi-layer information mining model, dispersed information mining model, grid-based information mining model, and information mining model from multi-innovation combination viewpoint. Appropriate information mining model can take care of issues from saving information at various destinations. The internet of things will create expansive volumes of information. With respect to the few key issues in information mining of IoT are likewise talked about in their research papers.

As indicated by them, improvement heading of the up and coming era of internet, the internet of things draws in numerous considerations by industry world and scholastic circles. IoT data has numerous attributes, for example, disseminated capacity, mass time-related and position-related information, and restricted assets of hubs and so forth. This makes the issue of data mining in IoT.

IoT and Big Data [6, 8, 17–19]
Hongming et al. [8] proposed a framework that identifies the need of data mining in IoT, its processing, and mining areas of IoT big data. They categorized IoT data in cloud platforms where several conflicts are generated related to infrastructure resources, multitenant storage, and so on. They have also discussed the issues to implement IoT with cloud.

During the research they faced the challenge of handling these massive heterogeneous data in highly distributed environments, especially in cloud environments. They also gave a framework of IoT in which data processing process was given to show the overview of related studies on the view of application. They also discussed several open issues related to recent IoT developments and applications. Based on their framework of IoT data processing process, this research was divided into four modules. He also generated experimental results in which the comparison of implemented algorithms has been identified. Ortiz et al. [17] proposed a framework, that is, SIoT Architecture. According to them, with the increase of IoT users, number of devices used in IoT environment is also increasing and with this the data generated by these sensors has also increased. To handle SN, IoT data, devices was itself a big task. So they have generated a functional idea which created a group between internet of things (IoT) and SNs. In this structure, the data originate from IoT, and the SN conveys it to enable human-to-device communications. This paper investigates the novel approach to handle IoT data, indicated by SIoT. Hence, this paper initially addresses an entire view on SIoT and key viewpoints to imagine the genuine omnipresent figuring. They emphasized on the role of SN and also new challenges and open issues in SIoT.

Inferences drawn from literature review [20–24]
After the critical review, the following inferences are drawn:

- Most of the authors have proposed clustering algorithms to mine the data in IoT environment.
- K-medoid algorithm takes more time for the execution while random clustering is the fastest one as comparison to all other data clustering algorithms.
- Because of rapid growth in IoT world, it is too difficult to handle the data streams, so more robust algorithms will be needed to perform clustering.
- Too much data—IoT data is itself a largest type of data on the internet. It is difficult to manage such a huge data in an IoT environment.
- Data is not secure and safe.
- According to Gartner's research, processing large quantities of IoT data in real time will increase workload on data center. Service providers face difficulty of privacy, security, and reliability of data as well as devices.
- In IoT world tons of data are floating around, so there is always a risk of data to be stolen.
- The growth of IoT has led to the increase in the huge volume of real world streaming data.

3 Problem Formulation and Proposed Work

Problem Definition

Based upon the literature review, the following is the overall statement of the problem:

"To develop a scenario specific clustering technique to efficiently manage (storage/retrieval) information for a given IoT scenario and evaluate the performance parameters".

Proposed work

To meet the above objective, the following sub-objectives are proposed.

(i) To design a new method to calculate the distance between the clusters and the nearest threshold value calculated in the IoT data stream.
(ii) To compare the proposed scheme with the existing adaptive k-mean and k-medoid clustering algorithm and evaluate the performance.
(iii) To improve the precise rate and recall value in the clustering methods and reduce the error rate.

The proposed work is a combination of adaptive k-means clustering algorithm and k-medoid clustering algorithm to generate clusters in the dataset. The proposed research includes two sections:

Cluster formation and

Check the cluster quality by evaluating the performance parameters.

The algorithm which we have generated is **fuzzy k-medoid clustering algorithm.**

How it works and what are the steps included in it is explained below:

Fuzzy k-medoid clustering algorithm

Definition: This proposed method is inspired from agglomerative methods. The agglomerative clustering commerce comes with all instances as one cluster. Here merging approach is applied to evaluate the accuracy value and a grouped set of object or instance. Consequently, the offered method starts with large number of clusters as an input parameter, and the value of cluster centers is reduced or filtered during a loop. For calculating the cluster value, the fuzzy k-medoid algorithm has introduced. The cluster value is computed based on fuzzy clustering method, and then the nearest pair of group or clusters is determined and merged. In this proposal work, we calculated three types of clusters (Cluster 1, Cluster 2, and Cluster 3, that is, based on attributes clusters) which give high accurate value computed further and reduce the error rates in the traffic data stream. The Type 1 cluster is formed based on mean distance, the Type 2 cluster is formed based on the equidistance or centroid scheme, and Type 3 cluster is based on taking the base index from the attributes. Here the centroid acts as a base index.

In this method, the distance allocating membership function is assigned to individual data point that corresponds to each cluster or group center on the basis of distance between the group center and the data point.

The fuzzy k-medoid clustering algorithm has been explained by the distance membership function of each data point whose value shall be equal to 1. After that, each repetition of membership and group centers are altered according to the formula given by:

$$Uij = \frac{1}{\sum k = 1\left(\frac{ddji}{dik}\right)}(2/\mathrm{m}-1) \tag{1}$$

$$\mathrm{Vi} = \left(\sum\nolimits_{ii=1}^{nn} (\mu ij)mmxii\right) \Big/ \left(\sum\nolimits_{ii=1}^{n} (\mu ji)m\right), \forall \mathrm{jj} = 1,\ 2 \tag{2}$$

where “nn” is the number of data points;

U_{ij} is the equivalent matrix generated.

Here, Vjj defines the jth cluster, where “mm” is the fuzziness ID (index) and ddji defines the ED between iith and jjth cluster center.

The major aim of fuzzy k-medoid clustering algorithm is to minimize:

Jj (uu, vv) $= \sum\nolimits_{ii=1}^{nn} \sum\nolimits_{jj=1}^{cc} (\mu ij)mm||xxi - vvj||^2$, where $||xxi - vvj||^2$ is the Euclidean distance (ED) between the iith and jjth data center clusters.

Algorithm

Step 1: Suppose, XY = {xx1, xx2…..xxn} be the set of data-point and UV = {vv1,vv2……………vvc} be the set of data centers.
Step 2. Random selected ‘c’ as a cluster data point.
Step 3: Evaluate the Fuzzy Membership Uij using: -

$$\mu ij = \frac{1}{\sum k{=}1\left(\frac{ddji}{dik}\right)}(2/\mathrm{m}{-}1) \quad \text{………………………………………………………(i)}$$

Step 4: Calculate the Fuzzy K-medoid center as ‘vvi’ using:

$$\mathrm{vvi}{=}\left(\textstyle\sum_{ii=1}^{nn} \mu ij\right)\, mm\ xii) \,/\, \textstyle\sum_{ii=1}^{n}(\mu ji)m) \quad, \forall \mathrm{jj} = 1{,}2\text{……………………..(ii)}$$

Repeat step 2 and 3 until the reduce jj value is attained or $|| \mathrm{UU}^{(kk+1)} - \mathrm{UU}^{(kk)}|| < \beta\beta$.
Where ‘’kk’’ is the repetition step.
$\beta\beta$ is the finish standard value between {0,1}.
‘’UU’’ = $(\mu ij)\ n * c$ is the Fuzzy K-mediod membership matrix.
‘’JJ’’ is the OM (Objective Method).

Algorithm Steps

1. Suppose C1, C2, and C3 are sets of data clusters.
2. Randomly select the cluster point.
3. Implement a membership function, to calculate the distance, index, and nearest cluster points.
4. Rapidly, the data cluster center has been calculated, based on the nearest distance sum.

How the Proposed Scheme Works

There are some procedural steps of the algorithm which are explained below:

Step 1: Upload the dataset from UCI Machine Learning Repository site to download the traffic data stream dataset.

Step 2: Pre-process the dataset and extract the attributes in the column and row wise.

Step 3: Apply K-means clustering method to classify the data into cluster forms (Cluster 1 and Cluster 2).

Step 4: Then apply k-medoid clustering method to divide the data into closest clusters which is calculated in the index-based approach.

Step 5: Fuzzy k-medoid clustering: This approach divides the cluster into three forms (Cluster 1, Cluster 2, and Cluster 3). The Type1 cluster is formed based on mean distance, the Type2 cluster is formed based on the equi-distance or centroid scheme, and the Type3 cluster are based on taking the base index from the attributes. Here the centroid acts as a base index.

Step 6: Evaluate the performance parameters, namely precision, recall, and Silhouette coefficient and compared it to the existing one.

Advantages of the fuzzy k-medoid clustering scheme

It gives better consequences for overlapped data point. Comparatively, it is better than k-means algorithm. Unlike k-means where each data point must belong to one cluster center, here data point (Index) not only belongs to cluster center but it also assigned the cluster center as a member function.

4 Methodology of Proposed Novel Approach

Phase I: First of all, we collect the dataset from the UCI MACHINE LEARNING REPOSITORY SITE and download the IoT Traffic dataset.

Phase II: (Pre-processing phase): After uploading the IoT datasets, the next step is to calculate the intensity of the dataset, that is, the calculation of minimum and maximum value from the IoT traffic dataset.

Phase III: Apply the adaptive k-mean clustering algorithm. Divide the data into two types of cluster, that is, Cluster-1 and Cluster-2. For Cluster-1, there is an approach of vector quantization (VQ) which acts naturally in the algorithm and is famous for analyzing the cluster. Cluster-2 is generated by using the nearest distance based on max., min., and average values. The main objective of this approach is to divide the data into k-groups in which each group belongs to the cluster with the closest sum. The closest sum is calculated between the two clusters. The final group/cluster generated is known as segmented cluster.

Flowchart of the proposed work

The methodology adopted to carry out our proposed work is shown in the following flowchart (Fig. 1).

Phase IV: Implement a k-medoid clustering algorithm which is little changed from the k-means algorithm. They both are used to reduce the square error (SE), but the k-medoid method easily identifies the interference and then removes it. In this algorithm, they select the data points as medoid. K-medoid defines as the object of a cluster or group whose average distance is different to all the objects in the group.

Phase V: In this phase, the medoid value is allocated as individual data point that corresponds to each cluster or group center on the basis of which the distance between the group center and the data point is calculated by fuzzy k-medoid clustering algorithm.

Table 1 shows the performance parameters and generates a comparison between the proposed and existing work based on Silhouette metrics, precision, and recall rate.

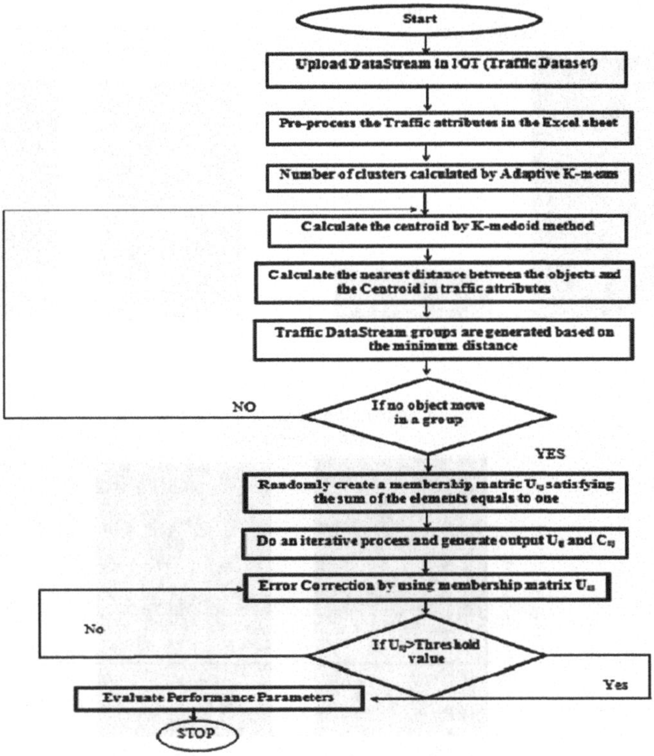

Fig. 1 Flowchart of the proposed work

Table 1 Comparison between proposed and existing work

Performance metrics	Values (proposed work)	Values (existing work)
Silhouette metrics	0.388	0.886
Recall	94.14	89.9
Precision	96.06	95.0

Fig. 2(i) enhances the performance metrics in Silhouette coefficient rate with Fuzzy k-medoid and existing clustering algorithm. Fuzzy k-medoid gives the high accuracy, that is, the minimum distance between the clusters. Fig. 2(ii) shows the comparison between proposed and existing work with adaptive K-mean clustering and k-medoid clustering algorithm. Recall value means the fraction of relevant objects, that is, the data streams (IoT) over the total amount of relevant stream. Figure 2(iii) shows the comparison between the proposed and existing work in precision value. The precision rate is the fraction of retrieved data streams that are relevant to the data

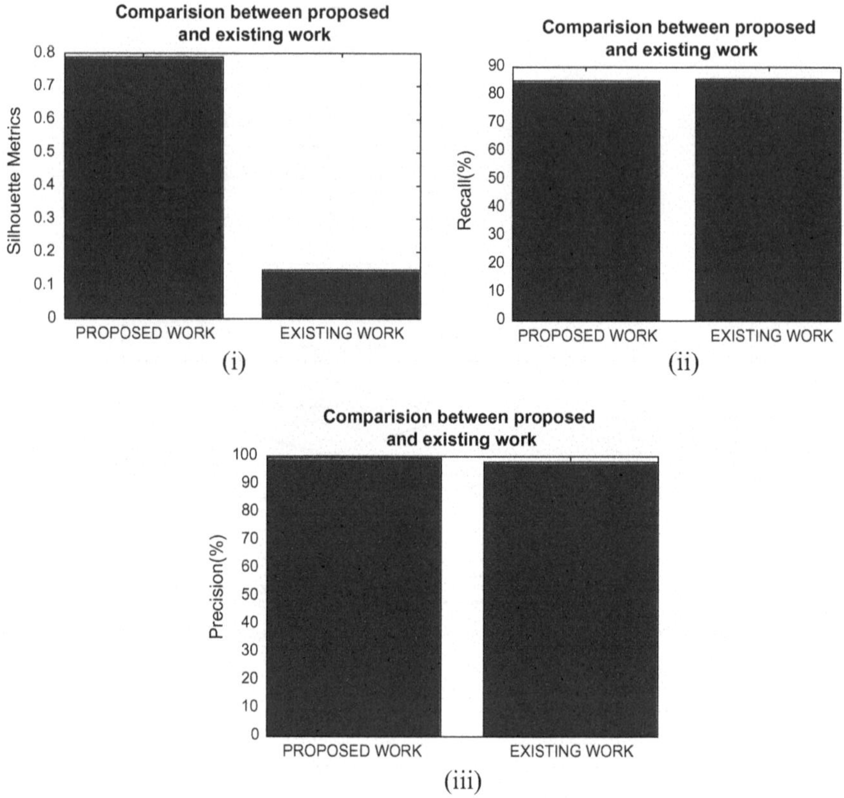

Fig. 2 (i) Comparison (proposed and existing work) in silhouette metrics; (ii) comparison between proposed and existing work (recall); and (iii) comparison between proposed and existing work (precision value)

stream. Precision rate takes all data retrieved from data stream in IoT into account, but it could also be calculated at a given rank measuring only the topmost consequences returned by the system.

5 Conclusion and Future Scope

The proposed work presents an efficient approach for managing IoT data streams using clustering techniques. The conclusion here has introduced a Fuzzy k-medoid clustering algorithm, that is designed to handle dynamic IoT data stream. The algorithm adapts to datasets of the underlying traffic data stream in IoT. The proposed methods are also able to determine the amount of classifiers which are searched inherently. In the data stream IoT depends on the data division and cluster improvement, that is, the quality of cluster. By working with the adaptive and K-medoid clustering algorithm we are able to search inherent classes from the data streams (IoT). The major issues found that too much data load on the data centers in IoT environment result in the largest type of data on the internet. It is difficult to manage such a huge data in an IoT environment. So we have evaluated a set of result experiments using manufacture data, and data taken from a traffic use case view where the traffic measurements done from the city areas. We improve the performance parameters which are Silhouette metrics, precision, and recall rate and compared them with the existing one. Internet of things network enables tracking of beneficial information about substance as they shift through the MATLAB platform. It defines important value for Internet things applications by giving an accurate knowledge of the recent IoT data-processing, which consequence in higher availability and reliable resource provision. Our main focus is to apply suitable clustering algorithm on situation-specific IoT data. This proposed strategy is supposed to reduce the complexity of IoT data streams collected in data center. Useful information can be also inferred from the given IoT data using data mining schemes.

With the more advancement in IoT applications there would be increase in amount as well as dimension of data stream. For handling these data stream, more reliable techniques are required to perform clustering methods. Several of research works are still going on this area and a lot of work will be required in the future also. It can implement an optimization method to improve the memory storage and processing time. Further, this approach helps in eliminating the replicated data streams and in efficiently using the storage capacity of data centre.

References

1. D. Puschmann, P. Barnaghi, R. Tafazolli, Adaptive clustering for IoT data streams. IEEE Internet Things **4**(1), 64–74 (2017)
2. J. Agrawal, S. Soni, S. Sharma, S. Agrawal, Modification of density based spatial clustering algorithm for large database using naive's bayes' theorem, in *Proceedings of the 4th International Conference on Communication Systems and Network Technologies,* vol. 1, April 2014, pp. 419–423
3. L. Zhou, M. Chen, B. Zheng, J. Cui, Green multimedia communications over internet of things, in *Proceedings of the IEEE International Conference on Communications*, vol. 2, Ottawa, Canada, June 2012, pp. 1948–1952
4. A. Zanella, N. Bui, A. Castellani, L. Vangelista, M. Zrorzi, Internet of things for smart cities. IEEE Internet Things J. **1**(1), 22–32 (2014)
5. R. Buyya, C.S. Yeo, S. Venugopal, Market oriented cloud computing: vision, hype, and reality for delivering IT services as computing utilities, in *Proceedings of the 10th IEEE International Conference on China*, vol. 2, no. 1, April 2008, pp. 5–13
6. S. Bin, L. Yuan, W. Xiaoyi, Research on data mining models for the internet, in *10th International Conference on Image Processing IEEE*, April 2010, pp. 127–132
7. C.-W. Tsai, C.-F. Lai, M.-C. Chiang, L.T. Yang, Data mining for internet of things: survey. IEEE Commun. Surv. Tutor. **16**(1), 77–97 (2014)
8. Hongming Cai, Xu Da Li, Xu Boyi, Cheng Xie, Shaojun Qin, Lihong Jiang, IoT-based configurable information service platform for product lifecycle management industrial informatics. IEEE Trans. Indus. Inf. **10**(2), 1558–1567 (2014)

9. R. Kataoka, N. Uchihira, Y. Ikawa, The evolutionary process of IT Concept words: a case study on bigdata, in *Proceedings of PICMET 16; Technology for Social Innovation*, vol. 2, pp. 1983–1992, March 2016
10. S. Nigam, S. Asthana, P. Gupta, IoT based intelligent billboard using data mining. in *1st International Conference on Innovation and Challenges in Cyber Security ICICCS*, August 2016, pp. 107–110
11. P. Kumar, Data stream clustering in internet of things. SSRG Int. J. Comput. Sci. Eng. (SSRG-IJCSE) **3**(8), 103–108 (2016)
12. J. Cooper, A. James, Challenges for database management in the internet of things. IETE Tech. **26**(5), 320–329 (2009)
13. M. Ester, H. Kriegel, J. Sander, X. Xu, A density-based algorithm for discovering clusters in large spatial databases with noise, in *Proceedings of the 2nd International Conference on Knowledge Discovery and Data Mining*, vol. 4, August 2016, pp. 226–231
14. F. Chen, P. Deng, J. Wan, D. Zhang, A.V. Vasilakos, X. Rong, Data mining for the internet of things: literature review and challenges. Int. J. Distrib. Sens. Netw. **11**(8) (2015). Article No. 431047
15. J.Y. Kim, H.-J. Lee, J.Y. Son, J.-H. Park, Smart home web objects–based IoT management model and methods for home data mining, in *IEICE APNOMS*, September 2015, pp. 327–331
16. Q. Li, P. Wang, W. Wang, H. Hu, Z. Li, J. Li, An efficient K-means clustering algorithm on map reduce, in *Proceedings of the 19th International Conference on Database Systems for Advanced Applications, Indonesia*, vol. 5, no. 2, pp. 357–371, April 2014
17. A.M. Ortiz, D. Hussein, S. Park, S.N. Han, N. Crespi, The cluster Between internet of things and social networks: review and research challenges. IEEE Internet Things J. **1**(3), 206–215 (2014)
18. Y. Chen, A.-X. Han, C.-H. Zhang, Research on data mining model in IoT, in *International Conference on Automation, Mechanical Control and Computational Engineering, AMCCE*, April 2015, pp. 1–7
19. N. Sharma, A. Bajpai, M.R. Litoriya, Comparison the various clustering algorithms of weka tools. Int. J. Emerg. Technol. Adv. Eng. **2**(5), 73–80 (2012)
20. J. Gubbi, R. Buyya, S. Marusic, M. Palaniswami, Internet of Things (IoT): A vision, architectural elements, and future directions. **29**(8), 1645–1660 (IEEE, 2013)
21. Q. Jing, A.V. Vasilakos, J. Wan, J. Lu, D. Qiu, Security of the internet of things: perspectives and challenges. Wireless Netw. **20**, 761–781 (2014)
22. A. Kumar, N.C. Narendra, U. Bellur, Uploading and replicating IoT data on distributed cloud storage, in *IEEE 9th Conference on Cloud Computing*, April 2016, pp. 670–677
23. D.C. Mocanu, E. Mocanu, P.H. Nguyen, M. Gibescu, A. Liotta, Big IoT data mining for real time energy disaggregation in buildings, in *IEEE International Conference on Systems, Man and Cybernetics*, vol. 4, no. 3, October 2016, pp. 3765–3769
24. W.T. Sung, J.H. Chen, M.H. Tsai, Applications of wireless sensor network for monitoring system based on IoT, in *IEEE International Conference on Systems, Man and Cybernetics*, October 2016, pp. 613–617

Leaf Disease Detection and Classification: A Comprehensive Survey

Manpreet Kaur and Rekha Bhatia

Abstract Agricultural crops are a pillar for Indian economy and source of livelihood. It is important to detect diseases of plant leaves in early stage, so as to get an increase in profit and crop yield. India is at a higher position in tomato production. Earlier method of naked eye observation is time-consuming and does not provide accurate results. In the present study various image processing techniques are used for leaf diseases detection and classification. The paper shows detailed survey for leaf diseases detection and classification methods. The summary of diseases and research work already done is tabulated to help the new researchers.

Keywords Leaf · Detection · Classification · Tomato · Deep learning

1 Introduction

Agriculture is the key necessity for a human being on the Earth. It is the backbone of Indian economy, which is highly dependent on the production and quality of the crops [29]. Tomato food crops are widely cultivated over the world. Tomato crops are more popular because of its nutrition power, such as vitamin C [19], vitamin E and beta-carotene which exist in tomatoes. They are also rich in mineral and potassium which are beneficial for good health. India is the chief producer and exporter of tomatoes and is at third position for production over the world. The country spans around 350,000 hectares and the production quantity is approximately 5,300,000 tons. Every year 10–30% crops are damaged due to diseases [31]. The leaf of tomato crop suffers from different diseases, mainly late blight, Septorial leaf spot [18], bacterial leaf spot, yellow leaf curl [32], leaf mold manual, tomato mosaic [31] and early blight [15, 18, 19]. Detection of diseases of leaf in early stage is important to decrease the usage of pesticide and upsurge the income of the farmer [14]. The traditional method such as naked eye observation method is very concentrated and

M. Kaur (✉)
M.Tech (CSE), Punjabi University Regional Centre for IT and Management, Mohali, India

R. Bhatia
Associate Professor, Punjabi University Regional Centre for IT and Management, Mohali, India

M. Dutta et al. (eds.), *Proceedings of International Conference on IoT Inclusive Life (ICIIL 2019), NITTTR Chandigarh, India*, Lecture Notes in Networks and Systems 116,
https://doi.org/10.1007/978-981-15-3020-3_27

takes a long time to perceive few diseases [1, 13, 19, 23, 29]. With less expenditure and time, the researchers develop the automatic method to observe the diseases in plant leaves. Various image processing-based machine learning and deep learning methods are used, such as color texture features that give us spatial arrangements of an image intensity and determine whether texture-based hue, saturation, and intensity (HSI) in color features of an image [1]. Supervised learning method in support vector machine uses hyperplane for classification. It is best suited for nonlinear problems [4, 5, 8, 19, 23]. The artificial neural network is a nonlinear complex data modeling, which detects the pattern based on input and output relationships [13, 21, 24], fuzzy system classifies images based on the wavelet color features [18, 21], convolution neural network is the advanced technique which contains a large number of layers and extraction of features from the images is more accurate than other models [29–31].

2 Leaf Diseases and Symptoms

Leaf quality describes the degree of superiority or the state which is free from imperfections. One of the main reasons of leaf diseases is on nursery stage. Lack of high-quality seeds of a plant in earlier stage suffers from major diseases and the climate is a major factor for healthy growth [7]. All plant leaves suffer from various diseases caused by viral, bacterial, deficiency, fungus, and so on. It is called infection agent or pathogens. The pathogens are divided into two categories: either saprophytes or autotrophies. Saprophytes are alive on dead tissues and autotrophies are alive on living tissues. An infectious agent is sprinkled from susceptible host to another [24] (Tables 1 and 2).

3 Literature Review

Pydipati et al. [1] developed a method for citrus leaf disease detection. Four different types of leaf are used for this study. The proposed method was based on an image processing technique and feature extraction was done by color co-occurrence methodology, which provides the unique feature of an image based on the color and texture. To evaluate the potential classification accuracy, the SAS discriminant analysis was used which provides 81% above using intensity feature, 95.8% on hue and saturation alone and 100% when HIS features used. An image texture feature of dataset was used to reduce hue and saturation. Two models emerged to perform this task for classification with 96% accuracy. Due to high accuracy, fewer number of data variables were used to minimize the computation time.

Phadikar and Sil [2] presented a method for disease recognition of rice leaf. The zooming algorithm was used for feature extraction. Only the infected part of the image was extracted. SOM neural network was used to classify the diseased images. The network was trained to extract feature and the two types of diseases images used

Table 1 All other leaf diseases and symptoms

Diseases	Leaf	Symptoms	Pathogens category
Angular spot	Cucumber	Water-soaked lesions	Bacterial
Anthracnose	Tobacco, watermelon	Irregular yellow spot	Fungi
Alternaria spot	Cotton	Dark brown to black circular leaf spots	Fungi
Brown spot	Rice, cucumber	Dark irregular brown spot	Fungi
Bacterial blight	Soybean, wheat, cotton	Yellowish-green Halo	Bacterial
Bacteria wilt	Cucumber	Yellow and brown at the margins	Bacterial
Blast leaf	Rice	Whitish to gray centre	Fungi
Canker	Citrus	Include round-to-irregular sunken, swollen, flattened, crack	Fungi, Bacterial
Cercospor	Cotton	Tan to brown spot	Fungi
Downey mildew	Cucumber, watermelon, soybean	Yellow to white patches	Fungi
Enation	Sugar beet	Crinkled texture, with small cracks	Virus
Faliar leaf	Cotton	Light yellow spot with Dark brown margin	Bacterial, fungi, virus
Frog eye	Tobacco, Soybean	Dark, water-soaked spots	Fungi
Greasy spot	Citrus	Yellowish brown blister	Fungi
Gall	Lime	Roundish rough-surfaced galls	Bacterial
Melanose	Citrus	Leaf becomes rough in texture	Fungi
Mosic virus	Okra	Pattern of light and dark green	Virus
Powdery mildew	Wheat	Grayish white or tan color spot	Fungi
Rust	Soybean, wheat	Pale leaf spots	Fungi
Stalk rot	Maize	Yellowing of the lower portions of the stalk and dull green leaf	Fungi
Yellow vein	Okra	Yellow-chlorotic flecks	Viral

Table 2 Tomato leaf diseases and symptoms

Tomato leaf diseases	Symptoms	Pathogens category
Bacterial spot	Water-soaked surrounded by yellow halo	Bacteria
Early blight	Dark spot ring around it yellow	Fungi
Late blight	Dark spot enlarge quickly	Fungi
Mosaic	Light and dark green pattern	Insect, Mite, Fungi
Powdery mildew	Grayish white color patches	Fungi
Potassium deficiency	Brown scorching and curling of leaf tips	Deficiency
Septoria spot	Yellow spots	Fungi
Veinal chlorosis	Leaf vein become yellow	Fungi, Bacteria
Yellow curl	Curly leaf and yellowish	Virus

for testing purpose are leaf blast and brown spot. Four different types of cases: RGB of the spots, Fourier transform of the spot, Random rotation of the spot was 50%, and Fourier alteration of the 50% rotation were developed for classification and achieved classification success of 92, 84, 82, and 70%, respectively.

Shen et al. [3] described a method for leaf spot disease detection which was based on the Otsu segmentation, and lesion segmentation was done using a Sobel operator that provides best results. The grading method was used to eliminate the bias of traditional classification method and also human-induced error. When the correct dataset was provided for detection, the method was more accurate and improved estimation credibility.

Tian et al. [4] presented a method founded on a support vector machine for wheat leaf disease recognition and classification. MCS which was based on two-stage offline SVM has three different feature levels. The lower level classifier used to improve correctness and provide stability for pattern recognition. The output of the proposed method corresponding to medal level categories describes the disease symptom of the crop as per the knowledge pathology of the plant. There was no method which combines the pattern recognition for crop diseases with biological crop. The proposed system may provide a way to take this knowledge for consideration.

Zhang and Zhang [5] presented a method in this paper based on SVM which was used to detect the diseases of cucumber leaf and to classify healthy and diseased leaf. To identify the leaf diseases two different kernel functions were used for two test sets. Clipping technology was used to extract spots from the complex background then wiping and threshold method was used for pre-processing an image. Other than two kernels, RBF kernel provides the best result and automatic identification was possible with this method.

Zhang and Meng [6] presented a scheme for detection of citrus canker disease of citrus leaf in live location. This was the good news for the farmer who used the system in the field. In this paper, the hierarchical detection strategy used to separate lesion images from the background. In previous studies, the leaf images were collected in the lab. Color and texture features were used to describe the feature of lesion images and local LBPH descriptor used spatial properties in each lesion zone of citrus leaf images. A modified version of AdaBoost algorithm was developed which provides the best features. All other feature machinists and classification techniques were examined and equated based on sample leaf of citrus disease in different environment. The proposed method provides best accuracy rate. Although the accuracy was the same as previous study who inspect the disease images of citrus leaf on the computer screen, but this method provides a higher potential than other systems.

Guru et al. [7] developed a model for detection of tobacco seedling leaf. The contrast stretching transformation was used to segment lesion on leaves and PNN system was used for classification. Two different features were considered for feature extraction. First-order statistical feature extracts the texture and grey co-occurrence matrix features. The proposed model provides good result.

Asraf et al. [8] proposed a classifier to categorize the oil palm leaf diseases which was based on SVM. The kernel function which was used in this paper had three types: linear, hard margin polynomial, and soft margin polynomial. The soft margin polynomial has best classification rate 95% than all other kernel. The study in this paper was based on evaluation of different kernel of SVM.

Husin et al. [9] presented a work based on DIP techniques for the detection of the leaf of chilli plant on whether it is healthy or not. The proposed method was effective and fastest, and the researcher considered it successful. Leaf recognition techniques were used to identify the diseased leaf of the chilli plant before a major damage. Using this method, usage of harmful chemical rate was reduced and the best quality production of chilli was increased.

Zhou et al. [10] presented a method based on the OCM which continuously monitors the field of sugar beet leaf diseases. Image registration algorithm was proposed in this work which provides a matching orientation code between two images. Quantized gradient angle of pixels in an image was calculated by operators such as Sobel, which calculate vertical and horizontal derivatives. OCM is better for further plant health analysis by accessing site-specific information of sugar beet. Luminance-chromaticity is used to separate healthier pixels from diseased rather than RGB. The RBF kernel of SVM was used for classification. The method provides correct results.

Hrishikesh et al. [11] developed an application to detect the leaf diseases detection in ten different plants. The algorithm was proposed and tested on these plant's leafs with less computational effort. An image was converted in HIS space and detects the most of green pixels in an image and removes it. Then color co-occurrence method was used for segmentation which is based on SGDM. Features from images were extracted and equated with the feature which is stored in feature collection. The

SVM classifier was used for classification. Accuracy was improved as compared to previous method.

Kutty et al. [12] describe the procedure for watermelon leaf diseases detection and classification based on digital image processing techniques. Based on RGB color component region of interest was identified from each leaf image and the mean value was calculated in this work. The crop images were inspected in SPSS software and error bar plot was used to test the performance. To reduce the noise median filter was used and classification was done using a toolbox of neural network pattern recognition. The proposed work achieved 75.95% accuracy.

Sannakki et al. [13] proposed a study based on computer vision which is used to classify the grape leaf diseases. Two categories of diseases of grape leaf were detected in this work. The developed algorithm was based on the feature extraction and classification. Feature extraction of an image done by color co-occurrence method extracts the texture of an image, and k-mean is used for clustering the lesion area of an image. The pattern recognition toolbox of the neural network in MATLAB® was used for classification. Training achieved 100% accuracy while hue features alone.

Keskar et al. [14] developed a system for citrus canker disease of citrus leaf detection and diagnosis with the assistant of an image processing. Features were extracted and stored in the database which was used by classifier. BPNN was used to classify the leaves. At testing time, features which were stored in database were compared with given feature of an image. As per the disease pesticides were used, and this information also is provided in this work.

Molina et al. [15] described the method of detection and classification of tomato leaf disease, early blight. Color descriptor was effective for visual feature representation related to the disease incidence. In this work, CSD was better to represent the color features and others were also applied for best classification results. CLD was not well and the DCT coefficients were not able to show all the variations of color which appeared at different scales. SCD privileges and Haar low-frequency coefficient were used for representation rather than fast color variations in the region marked as infected. The work had good classification rate.

Chaudhari et al. [16] developed a method for cotton leaf disease detection. A clustering algorithm for segmentation and DWT for feature extraction were implemented. The extracted features are reduced using PCA algorithm and neural network was used for classification. Accurate results were calculated to combine the method as compared to previous method for diseases detection and classification. By various learning rate of neural, the proposed method achieved 98% accuracy.

Shrivastava et al. [17] developed a method which is a fully automatic detection of soybean plant foliar diseases. In this paper, problems were highlighted related to soybean cultivation. The work deals with six foliar diseases. At severity level, diseases were classified into five categories in this work. New parameters were derived and developed, and indices like DLP, IPR, DSI were used for disease detection. The proposed method was implemented accurately and tested on field leaf dataset of soybean images. The proposed method was very simple to be used and this was the best news for farmers, and only mobile camera was used to click the image of leaf for testing. The methodology is fully automatic for RIO calculation and separation of background, to evaluate parameters, low in cost and widely usable in field conditions and simple method for classification was used (Table 3).

Table 3 Summarization for year 2006–2013

Article no.	Culture	Diseases	No. of images, size and format	Device used for dataset preparation	Accuracy	Technique applied
[1]	Citrus leaf	Greasy spot, melanose, scab	40 images of 480 × 640 JPEG	3CCD Camera (JAI, MV90)	96%	Color feature and discriminant analysis
[2]	Rice leaf	Leaf blast, brown spot	300 images	Nikon Cool Pix Py DC	N/A	Pattern recognition
[3]	Soybean	Gray leaf	N/A	N/A	N/A	Grading method
[4]	Wheat leaf	Leaf rust, puccinia triticinia, leaf blight, powdery mildew	800 images	N/A	N/A	SVM-based multiple classifier
[5]	Cucumber leaf	Downy mildew, mbrown spot, angular leaf spot	96 images	N/A	N/A	SVM
[6]	Citrus leaf	Canker	500 images	Sony, Canon and downloaded from internet	88%	Window union algorithm
[7]	Tobacco leaf	Anthracnose infection, frog eye spot	800 images of 1632 × 1224	Sony digital camera	75% and 73% according to diseases	Feature extraction and PNN classifier
[8]	Oil palm leaf	Nutrient diseases	300 images	N/A	95%	Kernel based SVM
[9]	Chilli plant leaf	Healthy and diseased leaf	107 images of 3872 × 2592	Web Cam images	N/A	Feature extraction
[10]	Sugar beet leaf	Leaf spot	640 × 480	CMOS camera	99.7%	RBF kernel of SVM for classification

(continued)

Table 3 (continued)

Article no.	Culture	Diseases	No. of images, size and format	Device used for dataset preparation	Accuracy	Technique applied
[11]	Ten different plants leaf	Bacterial brown spot, early scratch, yellow spot, brown spot, fungal disease	500 images	N/A	94.74%	Texture feature
[12]	Watermelon leaf	Anthracnose, downy mildew	200 images of 3872 × 2592 JPGE	Nikon D80	75.9%	Neural network pattern recognition
[13]	Grape leaf	Downy mildew, powdery mildew	300 × 300 Size	Nikon Cool PIX P510 and downloaded from internet	100%	Feed forward back propagation neural network
[14]	Citrus leaf	Canker	720 × 540 Size	Digital camera	N/A	Feature extraction and ANN classifier

Muthukannan et al. [18] proposed an approach which was based on the fuzzy rule system for classifying the unhealthy region of tomato leaf disease images. Color feature was used for feature extraction from an image and fuzzy inference system was used to classify the region in an image as healthy, more affected and less affected. The main purpose of this work is to classify the unhealthy region of tomato leaf and achieved 95% accuracy.

Mokhtar et al. [19] presented a work on automatic detection and classification of the two tomato leaf diseases based on SVM. For training and testing 200-image dataset was used. Different kernel was used to employ SVM as invmult kernel, Cauchy kernel and Laplacian. For parameter selection N-fold cross-validation and grid search techniques were used and the performance was also evaluated by using these techniques. The result of the proposed approach was based on three kernels, which shows very high accuracy combined and separately: 78, 100, and 98%, respectively.

Mondal et al. [20] presented a procedure for the detection and classification of YVMV disease in okra leaf. At the time of experiment if the size of the image dataset increases, the accuracy of classification also increased which was tracked from confusion matrix. From each leaf 23 features were extracted from a controlled feature set and the naïve-based classifier was used to classify the disease.

Billah et al. [21] presented a method of tea leaf diseases diagnosis which was based on neuro fuzzy inference system. Calculate the second-order statistical measure of co-variances for color feature extraction in tea leaf images which was based on different color band of wavelet frame transformation. Adaptive neuro fuzzy inference system was used in this work and very good results were obtained. Color wavelet is more accurate than other feature extraction and ANFIS had a higher accuracy rate.

He et al. [22] described the deep residual learning over all other models for CNN. This paper defines the error rate of all models and the residual network has less error rate as compared to other models. Residual net is easy to implement and has lower complexity.

Padol et al. [23] described the system for detection of grape leaf disease, which was based on SVM classifier. The pre-processing was done by thresholding, resizing, Gaussian filters. For segmentation k-mean clustering technique was used. In feature extraction phase, color and texture were extracted from the images. To detect which type of leaf disease the SVM technique for classification was used. The system shows 88.89% accuracy average for two types of diseases of grape leaf which is present in this work.

Pujari et al. [24] proposed an approach that mainly focused on plant disease detection. The developed algorithm was proved on five different types of plant diseases. Color features were extracted and feature reduction was done using threshold and delta method. For texture feature co-occurrence matrix developed. The advantage of this approach is that small number of features was developed for higher classification accuracy rate. ANN and SVM were used for classification of the diseased images, but SVM and K-NN were achieving high improvement in classification task over ANN.

Mohan et al. [25] presented a technique based on the image processing technique for paddy plant leaf disease detection. The SIFT method was used to extract the features from the disease-affected images. Then, using these features the SVM is used to recognize the images. The proposed work is mainly of three types of paddy leaf diseases. The experimental results shown by this model were suitable for disease detection and has 99.10% accuracy when SVM and 93.33 accuracy when K-NN was used.

Es-saady et al. [26] proposed an approach for detection and classification of six dissimilar types of plant leaf which is based on two serial SVM classifiers. For extraction of features of leaf diseases images by color, texture, shape features and detection of the plant leaf diseases, automatic detection system is developed. This work provides a decision support system which assists the farmers in earlier diagnosis phase.

Elangovan et al. [27] described the method which was created on image segmentation and SVM classifier for plant leaf disease detection. In this work, different techniques were used to segment an image, but the Otsu was the best method for binary image; in this process mean for each cluster was calculated, square the mean difference, multiply the number of pixels in each cluster, then the morphological method provides a better result for feature extraction and the SVM classifier was used for classification.

Rupa et al. [28] described a method for classification of various plant diseases which was based on digital image processing techniques. Distortion in images was removed by QAMS. Base study of defined method in this paragraph identify the histogram bins manually and spectral lights were used. Edges and reflection lights were removed using QAMS. An image was converted into the color channel and then pixel correction is applied. The CNN is used to separate the diseased region in images, and the SVM is used to detect the diseases. The work provides a better result for segmentation of leaf diseases and also provides a higher detection accuracy. Due to automatic identification bins in the histogram, the computational time was less.

Akila et al. [29] developed a model which was based on a deep leaning algorithm for diseases detection and classification of various plant leaves. In this paper various deep learning architectures with different feature extraction describe how the proposed detector of deep learning successfully detects different types of diseases in different plants. Faster R-CNN recently introduced region-based FCN and single short Multibox detector for detection and classification and also provided the information on which fertilizer is best for which type of disease.

Sabarinathan et al. [30] developed a model based on the CNN for identification of 50 various medical plants for classification and also showed the uses in medical by using MongoDB database. The pattern of the leaf was used for identification. The proposed method beats the best in classification such as SVM and on handcrafted feature of the medical plants datasets which was used in this work. The work was done in such a way so that researchers can train the network for classification of various other plants rather than medical plants and is called non-specific. This approach can be used in any other areas where classification is needed.

Tm et al. [31] described the modify model for tomato leaf disease detection created on the CNN and the modify LeNet model. The tomato leaf images were collected from the plant village dataset. The architecture of CNN which was used in this effort was a simple and has a minimum number of layers for classification. Different learning and optimization methods were used to experiment. The accuracy of modified model was 94–95% (Table 4).

4 Conclusion

In this paper, all techniques and concepts which are used by various researchers are highlighted for identification and classification of leaf diseases. The main goal of this paper is to develop the disease detection system at an early stage to increase the crop yield and profit. Massive number of applications is developed for leaf diseases detection and classification. This is the main reason of missing the solution for problematic issue. The researcher uses different methods with different tools for better result. The results are provided based on the accuracy and other parameter measurements. In this paper, a brief study of widespread detection and classification

Table 4 Summarization for year 2014–2018

Article no.	Culture	Diseases	No. of images, size and format	Device used for dataset preparation	Accuracy	Technique applied
[15]	Tomato leaf	Early blight	147 Images of 2592 × 1944	Tomato green house crop	99%	Color based classification
[16]	Cotton leaf	Cercospor spot, foliar leaf spot, atternaria spot, bacterial blight	256 × 256	Sony digital camera	98%	K-mean clustering and BPNN classification
[17]	Soybean leaf	Bacterial blight, downy mildew, brown spot, frog eye, sudden death syndrome, rust	1000 images of 1600 × 1200 JPEG	Camera of mobile phone (Samsung GT-S3770)	N/A	RIA, LCI, DSI technique
[18]	Tomato leaf	Late blight, early blight, Septoria leaf	N/A	N/A	95%	Fuzzy inference system
[19]	Tomato leaf	Powdery mildew, early blight	200 images	N/A	98-100%	SVM
[20]	Okra leaf	Yellow vein mosaic	N/A	Color camera	87%	Naïve based classification
[21]	Tea leaf	Diseased images	45	Digital camera	95.7%	Color wavelet feature extraction and neuro fuzzy

(continued)

Table 4 (continued)

Article no.	Culture	Diseases	No. of images, size and format	Device used for dataset preparation	Accuracy	Technique applied
[22]	Deep learning models	Check error rate	ImageNet dataset	Download from internet		Comparison between various deep learning techniques
[23]	Grape leaf	Downy mildew, powdery mildew	137 images JPEG	Digital camera and downloaded from internet	88.89%	K-mean clustering and SVM classifier
[24]	Various crop leaf	Fungal diseases, bacterial blight, viral diseases, deficiency diseases	9912 images	Sony DXC-3000A color camera	87%	SVM, ANN and feature reduction techniques
[25]	Paddy Plant leaf	Brown spot, leaf blast, bacterial blight	60 Images	N/A	91.1% SVM 93.3% KNN	SIFT feature transform and SVM, KNN classifier
[26]	Various crop leaf	Early blight, powdery mildew, thrips, tuta absolute, late blight, leaf miners	284 images	Digital camera and internet	87.7%	Serial combination of two classifiers
[27]	Plant leaf	Diseased leaf	N/A	N/A		K-Mean and SVM
[28]	Various plants leaf	Various plant diseases	N/A	N/A	91.9%	SVM and CNN
[29]	Various plant leaf	Various diseases	N/A	Camera and internet	N/A	CNN
[30]	50 Medical plants leaf	Leaf recognition	1500 images	Google images	98%	CNN and SVM Classification
[31]	Tomato leaf	Various diseases	18160 images JPEG	Plant village repository	94–95%	LeNet

techniques or procedures is provided. The keys are found in this literature which gives the idea for further research.

5 Future Scope

Deep learning is the best method which provides more accurate result over other detection and classification methods. It has a different standard model with large number of layers with different accuracy. One of the best models will be chosen for further research.

References

1. R. Pydipati, T.F. Burks, W.S. Lee, Identification of citrus disease using color texture features and discriminant analysis. Comput. Electron. Agric. **52**, 49–59 (2006)
2. S. Phadikar, J. Sil, Rice disease identification using pattern recognition techniques, in *2008 11th International Conference on Computer and Information Technology ICCIT 2008* (2008), pp. 420–423
3. W. Shen, Y. Wu, Z. Chen, H. Wei, Grading method of leaf spot disease based on image processing, in *2008 International Conference on Computer Science and Software Engineering CSSE 2008*, vol. 6 (2008), 491–494
4. Y. Tian, C. Zhao, S. Lu, X. Guo, SVM-based multiple classifier system for recognition of wheat leaf diseases, in *Proceedings of 2010 Conference on Dependable Computing* (2010), pp. 2–6
5. J. Zhang, W. Zhang, Support vector machine for recognition of cucumber leaf diseases, in *2010 2nd International Conference on Advanced Computer Control ICACC 2010*, vol. 5 (2010), pp. 264–266
6. M. Zhang, Q. Meng, Automatic citrus canker detection from leaf images captured in field. Pattern Recognit. Lett. **32**, 2036–2046 (2011)
7. D.S. Guru, P.B. Mallikarjuna, S. Manjunath, Segmentation and classification of tobacco seedling diseases (ACM, 2011), pp. 1–5
8. H.M. Asraf, M.T. Nooritawati, M.S.B.S. Rizam, A comparative study in kernel-based Support Vector Machine of oil palm leaves nutrient disease. Proc. Eng. **41**, 1353–1359 (2012)
9. Z.B. Husin, A.Y. Shakaff, A.H. Aziz, F.B. Farook, Feasibility study on plant chili disease detection using image processing techniques, in *2012 Third International Conference on Intelligent Systems Modelling and Simulation, ISMS 2012* (2012), pp. 291–296
10. R. Zhou, S. Kaneko, F. Tanaka, M. Kayamori, M. Shimizu, Matching-based cercospora leaf spot detection in sugar beet, in *2013 Proceedings of World Academy of Science, Engineering and Technology*, vol. 79 (2013), pp. 1603–1609
11. P.K. Hrishikesh, S.S. Lokhande, Detection of unhealthy region of plant leaves and classification of plant leaf diseases using texture features. CIGR J. **4**, 1777–1780 (2013)
12. S.B. Kutty et al., Classification of watermelon leaf diseases using neural network analysis, in *2013 IEEE Business Engineering and Industrial Applications Colloquium, BEIAC 2013* (2013), pp. 459–464
13. S.S. Sannakki, V.S. Rajpurohit, V.B. Nargund, P. Kulkarni, Diagnosis and classification of grape leaf diseases using neural networks (IEEE, 2013), pp. 3–7
14. P.V. Keskar, S.N. Masare, M.S. Kadam, S.U. Deoghare, Leaf disease detection and diagnosis. Int. J. Emerg. Trends Electric. Electron. **2**, 104–127 (2013)

15. J.F. Molina, R. Gil, C. Bojacá, F. Gómez, H. Franco, Automatic detection of early blight infection on tomato crops using a color based classification strategy (IEEE 2014), pp. 1–5
16. V. Chaudhari, C.Y. Patil, Disease detection of cotton leaves using advanced image processing. Int. J. Adv. Comput. Res. **4**, 653–659 (2014)
17. S. Shrivastava, S.K. Singh, D.S. Hooda, Color sensing and image processing-based automatic soybean plant foliar disease severity detection and estimation. Multimed. Tools Appl. **74**, 11467–11484 (2014). Springer
18. K. Muthukannan, P. Latha, Fuzzy inference system based unhealthy region classification in plant leaf image. Int. J. Comput. Inf. Eng. **8**(11), 2103–2107 (2014)
19. U. Mokhtar, M.A. Ali, A.E. Hassenian, H. Hefny, Tomato leaves diseases detection approach based on support vector machines (IEEE 2015), pp. 246–250
20. D. Mondal, A. Chakraborty, D.K. Kole, D.D. Majumder, Detection and classification technique of Yellow Vein Mosaic Virus disease in okra leaf images using leaf vein extraction and Naive Bayesian classifier, in *International Conference on Soft Computing Techniques and Implementations, ICSCTI 2015* (2015), pp. 166–171
21. M. Billah, M. Badrul, A. Hanifa, M. Ruhul, Adaptive neuro fuzzy inference system based tea leaf disease recognition using color wavelet features. Commun. Appl. Electron. **3**, 1–4 (2015)
22. K. He, X. Zhang, S. Ren, J. Sun, Deep residual learning for image recognition, in *Proceedings of the IEEE Conference on Computer Vision and Pattern Recognition*, 2016–Decem (2015), pp. 770–778
23. P.B. Padol, A.A. Yadav, SVM classifier based grape leaf disease detection, in *2016 Conference on advances in signal processing, CASP 2016* (2016), pp. 175–179
24. D. Pujari, R. Yakkundimath, A.S. Byadgi, SVM and ANN based classification of plant diseases using feature reduction technique. Int. J. Interact. Multimed. Artif. Intell. **3**, 6 (2016)
25. K. Jagan, M. Balasubramanian, S. Palanivel, Detection and recognition of diseases from paddy plant leaf images. Int. J. Comput. Appl. **144**, 34–41 (2016)
26. Y. Es-Saady, I. El Massi, M. El Yassa, D. Mammass, A. Benazoun, Automatic recognition of plant leaves diseases based on serial combination of two SVM classifiers, in *Proceeding of 2016 International Conference on Electrical and Information Technologies ICEIT 2016* (2016), pp. 561–566
27. K. Elangovan, Plant disease classification using image segmentation and SVM techniques. Int. J. Comput. Intell. Res. **13**, 1821–1828 (2017)
28. M. Rupa, S. Pathur Nisha, A. Geetha, P. Shanthi, A classification approach for predicting plant leaf diseases in digital image processing. Int. J. Comput. Sci. Mob. Comput. **6**, 48–56 (2017)
29. M. Akila, Detection and classification of Plant Leaf diseases by using Deep Learning Algorithm. Int. J. Eng. Res. Technol. **6**, 2–7 (2018)
30. C. Sabarinathan, A. Hota, A. Raj, V.K. Dubey, V. Ethirajulu, Medicinal plant leaf recognition and show medicinal uses using convolutional neural network. Int. J. Glob. Eng. **1**, 120–127 (2018)
31. P. Tm, A. Pranathi, K. Saiashritha, N.B. Chittaragi, S.G. Koolagudi, Tomato leaf disease detection using convolutional neural networks. in *2018 11th International Conference on Contemporary Computing IC3 2018* (2018), pp. 2–4
32. https://en.wikipedia.org/wiki/List_of_tomato_diseases. Accessed 25 July 2019

Performance Evaluation of Different Classification Factors for Early Diagnosis of Alzheimer's Disease

Agha Alfi Mirza, Maitreyee Dutta, Siddheshwari Mishra and Agha Urfi Mirza

Abstract In India, approximately 4 million people are being suffered from some form of dementia. 44 million people are affected with dementia worldwide, so it becomes the global health crisis that must be resolved. Identification of pre-MCI and MCI patients at higher level of risk before conversion to AD is very effective for patient treatment. Before the content-based image retrieval (CBIR) system only clinical inputs were taken with some relevant data from genetic analysis where the diagnosis is totally dependent on knowledge and experience of doctor, which can be biased also. Multi-modality, network structure, and measures of classification play very important role to predict AD and its prodromal stages. Neuroimaging scanners like MRI, PET, CT scan; biomarkers like CSF, cerebral glucose, tau proteins, amyloid precursor proteins, apolipoproteins E (APOE); and clinical scores like ADAS-Cog, MMSE are being used as multi-modality inputs to predict disease and its prodromal stages. These imaging techniques along with clinical inputs and biomarkers have their own level of mechanisms to refine classification technique. AI techniques like machine learning, deep learning, and artificial neural network play an important role to diagnose and predict the AD and its prodromal stages. KNN, SVM, RF, naïve Bayes classifier, and CNN are techniques which are used for feature selection as well as for classification. Different combinations of these techniques are being used for optimal prediction. Somebody used them as multi-stage classifier and others used them as multi-view classifier.

A. A. Mirza (✉) · M. Dutta · S. Mishra
NITTTR, Chandigarh, India
e-mail: alfimirza@gmail.com

M. Dutta
e-mail: d_maitreyee@yahoo.co.in

S. Mishra
e-mail: rakeshismishra@gmail.com

A. U. Mirza
Al Musanna College of Technology, Muladdah, Sultanate of Oman
e-mail: urfi@act.edu.om

M. Dutta et al. (eds.), *Proceedings of International Conference on IoT Inclusive Life (ICIIL 2019), NITTTR Chandigarh, India*, Lecture Notes in Networks and Systems 116,
https://doi.org/10.1007/978-981-15-3020-3_28

Keywords Alzheimer's disease · Pre-MCI detection · AD prodromal stages · Support vector machine · Principal component analysis

1 Introduction

Early detection of AD progress detection is very effective in improving life for Alzheimer's disease patients. MCI is an intermediate stage between cognitive decline and AD [1]. Different methodologies have been proposed to predict AD, its prodromal stages, and conversion to AD. Biomarkers such as GM, WM, CSF, hippocampus size, and cortical thickness are extracted from structural MRI [2], cerebral metabolic rate for glucose (CMRgl) for glucose consumption evaluation, which is measured through fluorodeoxyglucose positron emission tomography (FDG-PET), hyper-phosphorylated tau (p-tau) measurement through neurofibrillary tangle (NFT) and PET imaging, and integrated biomarker methodology used for biomarker extraction [3].

Multi-modality through MRI, PET, and CT-scan are very effective for selecting features but it further increases dimensionality and integrates incomplete data which needs pre-processing. Multi-hypergraph learning plays an important role to deal such conditions. 90.29% accuracy for AD versus NC classification and 74.68% accuracy for pMCI versus sMCI classification were achieved [4]. Clinical scores like MMSE and ADAS-Cog are being used as multi-modality inputs to predict disease and its stage [5, 6]. Principal component analysis is very effective for dimensionality reduction but it needs improvisation for predicting the classes in input dataset [7, 8]. Multi-view and multi-template learning is used to reduce the dimensionality [9]. Machine learning methodologies such as SVM, KNN, RF, and naïve Bayes classifier are used to classify patterns in input dataset.

2 Literature Review

2.1 Issues of Dimensionality Reduction

High-dimensional data is a major concern of classification [7]. However, the classification task becomes more complex because of high dimensionality present in samples taken from MRI, PET, and CT scan [10, 11]. To get the lower dimensionality that contains geometric and topological characteristic from the original dataset, manifold learning is proved as the most powerful tool [12, 13]. The major drawback of this method is that the results from this method are not easy to interpret and investigate. Furthermore, the sparse regression models are very effective in handling high-dimensional data [8]. Various methods are proposed to reduce dimensionality but they are limitedly applicable to the underlying methods due to interpretational requirements. Sparse regression methods with support of regularization are highly effective classification technique. Deep learning is proved as the most effective technique in various applications like image processing, medical image analysis, and speech recognition. Convolution neural network is a very effective tool to capture whole brain modeling in a cost-effective way. Various fast convolution neural network-based networks like AlexNet and GoogleNet are used to effectively extract features and reduce multi-dimensionality in less amount of computing time.

PCA is a linearly associated orthogonal transformation-based machine learning technique that reduces high-dimensional data into lower dimensions [14]. Thus first principal component (PC 1) reduces the dimensionality but it does not provide any discriminative information for classification. Second principal component (PC 2) is used for classification purpose with feature selection step (Table 1).

2.2 Issues of Multi-modality

Although the use of multi-modality yields excellent results but the main disadvantage of using this is the presence of incomplete data in input dataset. Liu et al. [4] suggested a novel multi-modality-based approach to handle AD. According to them there was an ambiguity and insufficiency in ADNI database while evaluating AD using multi-modality. Integration of multi-modality is a challenge, because more than half of ADNI AD subjects have no fluorodeoxyglucose PET and CSF data. Previous studies proposed multi-hypergraph learning method to establish the relationships among subjects. They performed this for better multi-modality integration and got good results for different types of AD stages conversion.

Moradi et al. [21] presented an integrated biomarker learning and classification technique in which they first applied semi-supervised learning on MRI biomarkers and then it was integrated with clinical measures using a supervised learning approach. The proposed experiment demonstrates the MCI-to-AD conversion on ADNI dataset.

Two approaches are focused while using modality: (1) single modality, or (2) multi-modality with separate results for each modality. Hinrichs et al. [22] proposed an integrated multi-modality-based feature selection method to select optimal feature and then apply multi-kernel learning framework. The primary objective of MKL is to perform learning and classification simultaneously. They created multi-modal disease marker which successfully predicted the conversion from prodromal stages of AD. Hojjati et al. [23] proposed the performance of sMRI and fMRI and according to them there was unavailability of integration of both these mechanisms. These techniques were implemented and tested for all types of conversion (Table 2).

Table 1 Summary and critical evaluation of **dimensionality reduction**

Ref. no.	First author, year	Methodology	Model type	Remarks	Limitations
[15]	Shlens (2005)	Principal component analysis (PCA)	Linear	PCA is better when number of samples per class is less	It is a non-parametric analysis
		Linear discriminant analysis (LDA)	Linear	LDA is better with large dataset having multiple classes	LDA weakness: Number of projection directions is lower than the class number
[16]	Weinberger et al. (2006)	Isomap	Manifold	It preserves the pairwise geodesic distances between feature vectors	Prone to topological instabilities
		Kernel PCA	Nonlinear	Kernel extension can be applied to inner product-based algorithms	Kernel PCA is limited to NLDR
[17]	Saul et al. (2000)	LLE	Manifold	Local linear structure preservation	It is limited to simple datasets like Swiss roll dataset
[18]	Niyogi et al. (2003)	Eigenmaps based on Laplacian	Manifold	It preserves the local similarity structure in feature space	It suffers from over fitting on the manifold
[19]	Yan et al. (2007)	Marginal Fisher analysis (MFA)	Linear	Local manifold structure present	Kernel MFA is better than traditional MFA as it is linear in nature
[20]	Chawla et al. (2002)	SMOTE	Nonlinear	Resampling is used to remove the class imbalance	Nearest biasing is possible as it uses k-nearest neighbors to generate new synthetic minority class members

Table 2 Summary and critical evaluation of **multimodality**

Ref. no.	Author(s), year	Modality	Technique	Accuracy	Limitations
[24]	Chaves et al. (2010)	SPECT	Apriori	95.87%	Missing value handling did not mention
[25]	Chaves et al. (2011)	SPECT	Apriori	94.87%	Uncertain missing values in input data
[26]	Zhang et al. (2011)	MRI + FDG − PET + CSF	SVM	93.2%	Unhandled missing values and class imbalance
[27]	Chaves et al. (2012)	FDG-PET + PiB − PET	Apriori	94.74%	Unproven data
[28]	Chaves et al. (2012)	SPECT PET	Apriori	92.78%	Unproven data
[29]	Chaves et al. (2013)	SPECT PET	Apriori	SPECT: 96.91% PET: 92%	Pathologically unproven data and unhandled missing values
[30]	Meenakshi et al. (2014)	PET	Apriori	91.33%	No dataset details, unproven missing values

2.3 Role of Network Structure and Classification Techniques

Liu et al. [1] demonstrated a technique to predict AD using MRI-based images by extracting features from whole brain using hierarchical network structure. The AAL atlas was proposed to break whole brain into small regions. They used spatial correlations rather than selecting all the features for improving the classification accuracy. The AD classification was based on the F-score method which was used to select features from raw feature spaces. Finally, a multiple kernel classification algorithm (MKBoost) is used to improve the performance of final classification. Using this technique an accuracy of 89.63% and an AUC of 0.907 for AD/MCI classification was achieved.

Liu et al. [31] suggested an hierarchical networks-based AD classification approach of brain regions using edge features as well as node features. They used multiple kernel classifiers to combine edge feature and node feature. Evaluation was based on ADNI dataset of 710 subjects' images.

Liu et al. [32] proposed a unique relational-based multi-template learning methodology for AD and MCI. Previous methodologies performed concatenation and average on multiple purpose of features extracted from multiple sources of templates which basically ignore the relevant feature in brains. Each brain MRI image registered separately onto different pre-selected templates. They developed the novel feature selection algorithms which extracted multiple templates from feature vector, sparse feature selection, and ensemble-based classification. They evaluated the technique on 459 subjects from the ADNI database.

Li et al. [14] suggested a technique to check progressive stages of Alzheimer's disease in patients using multi-modality. Images were taken from different sources including MRI, PET, and CSF and further analyzed by PCA to transform these input features into linearly uncorrelated variables. To remove the typical cause of overfitting, a dropout technique using multi-task deep learning was applied. The AD/MCI diagnosis was conducted on ADNI dataset, and AD diagnosis was very effective using dropout technique which improves the accuracy by 5.9%.

2.4 Optimization of Classification Methodologies

Kruthika et al. [33] suggested methods to diagnose AD in early stage because AD is treatable if it could be diagnosed in early stage. They used multi-stage classifier technique to classify Alzheimer's disease using naïve Bayes classifier, KNN, and SVM. Finally, the PSO technique is applied to individual feature vectors to extract useful features from it.

Zeng et al. [34] presented a simple model to diagnose the AD prophase called MCI to detect AD in earlier phase. The classification was undergone through preprocessing, PCA, and the SVM. To give more strong support to classification a variant of particle swarm optimization algorithm was used.

Liu et al. [35] proposed an ensemble learning framework of multi-view technique for dementia diagnosis using multi-modality. This framework is based on ANN which not only solves feature fusion learning problem but also prediction problem simultaneously. Finally, multi-modality added to refine the classification.

Basaia et al. [36] suggested a single cross-sectional brain scan method with deep learning-based technique to predict Alzheimer's disease and conversion from cMCI to AD. CNN is applied on 407 HC, 418 AD, 280 cMCI, and 533 sMCI subjects. 99% accuracy was achieved using ADNI dataset only, and 98% with ADNI + non-ADNI dataset.

2.5 Classification Using Region of Interest (RoI)

RoI-based classification becomes most popular approaches for AD diagnosis. Each RoI can be analyzed through one of the two methods: (1) through single RoI methods or (2) through multiple RoI methods [37]. AD, MCI, and HC subjects' discrimination is done using hippocampus volume, extracted from a segmentation method. They combined hippocampus visual features for AD diagnosis [1]. Other methods of multiple RoI integration for different subjects into 137 RoIs are then analyzed individually and all the individual RoIs are used to extract the useful features from them [37].

2.6 Integration of Biomarkers

The voxel of a brain gives features which work as a biomarker. GM, WM, CSF, size of the hippocampus, and cortical thickness are features extracted from structural MRI [2]. Active brain neurons blood cells move at higher rate than inactive brain cells. Nowadays, the activity of a brain at rest can be evaluated through rsfMRI, and these measures are very useful to check the correlation between two regions of a brain.

Alzheimer's disease patients have lower cerebral glucose consumption at cortex regions. CMRgl evaluates this glucose consumption, which is measured through FDGPET. Patient's tau proteins remain separated to microtubules which make it unstable in Alzheimer's disease [38]. The unbound tau proteins which are also called hyper-phosphorylated tau (p-tau) bind together to form neurofibrillary tangle (NFT) [39]. NFT can be observed through biopsy, and with the advent of recent technologies, it can be evaluated using T807, THK-5117, and PBB3 tracers (Table 3).

2.7 Integration of Clinical Scores

Li et al. [5] suggested a sparse tensor regression model with clinical assessments. They applied multiple clinical scores simultaneously with feature selection from different subjects. MMSE clinical scores were fused with training dataset of pre-processed MRI data. This training data is then converted to sparse multivariate tensor regression and optimal feature vector selected from this correlated and complementary information. Finally, they implemented this using ADNI dataset to show the outcome and performance of their methodology.

Huang et al. [6] presented the comparative study to evaluate the sensitivity measure in AD progression. They performed analysis on MCI population with hippocampal volume, CSF β-amyloid, and apolipoprotein (APOEε4) as a core constituent of brain matter. The proposed composite score was very effective in accordance to existing clinical endpoints and could be easily implemented and standardize (Table 4).

3 Conclusion

Different CBIR-based methodologies are adapted to predict the Alzheimer's disease and its prodromal stages. Classification gives more accurate result when it uses multi-modality. Incompleteness, present in data while performing multi-modality,

Table 3 Summary and critical evaluation of **biomarkers**

Ref. no.	First author, year	PET ligand	CSF assay (Aβ1-42)	Cases	Concordant
[40]	Landau et al. (2013)	Florbetapir	Luminex	Total = 374	322 (86.1%)
[41]	Palmqvist et al. (2014)	Flutemetamol	ELISA	Total original cohort = 118 Total validation cohort = 38	109 (92.4%) 37 (97.4%)
[42]	Fagan et al. (2007)	PiB	ELISA	Total = 50	50 (100%)
[43]	Forsberg et al. (2008)	PiB	ELISA	Total = 16	16 (100%)
[44]	Fagan et al. (2009)	PiB	ELISA	Total = 189	157 (83%)
[45]	Jagust et al. (2009)	PiB	Luminex	Total = 55	50 (91%)
[46]	Tolboom et al. (2009)	PiB	ELISA	Total = 37	31 (84%)
[47]	Degerman et al. (2010)	PiB	ELISA	Total = 10 (probable AD = 10)	10 (100%)
[48]	Weigand et al. (2011)	PiB	Luminex	Total = 41 (Controls = 11, MCI = 34, AD = 10)	41 (100%)
[49]	Zwan et al. (2014)	PiB	ELISA	Total = 136	114 (84%)
Total				1064	937 (88.0%)

is further refined by using multi-hypergraph learning or multi-view ensemble learning schemes. Clinical diagnosis improves the classification but 100% pathological verification was not found in any of the literature. Classification accuracy is also controlled through hierarchical or multi-level structural execution of classifiers. Deep neural network and restricted Boltzmann machine are very effective techniques in selecting different features in classification and also executing big data in a very efficient manner.

Table 4 Summary and critical evaluation of **clinical scores** on occupational exposure

Ref. no.	First author, year	Outcome: assessment criteria	Results • Male • Female • Both	Bias score
[50]	Geller et al. (2007)	• Medical inputs • Clinical inputs • HIS, MMSE, expert	• – • – • 2.10 (0.20, 23.60)	10
[51]	Crowe et al. (2010)	• Population-based study • Clinical inputs • ADRDA (possible AD)/NINCDS	• 1.80 (0.64, 5.05) • 1.08 (0.57, 2.07) • 1.38 (0.88, 2.26)	12
[52]	Sorahan and Mohammed (2014)	• Death certificate • Mentioned AD • ICD-10, ICD-9	• – • – • 0.73 (0.33, 1.61)	12
[53]	Tseng et al. (2014)	• Population study • Clinical inputs • MMSE (dementia including two-thirds of AD cases)	• – • – • 3.40 (1.30, 8.90)	11
[54]	Schouten et al. (2015)	• Death certificate • Death causes • ICD-10, ICD-9	• 0.91 (0.39, 2.12) • – • –	12
[55]	Poulsen et al. (2017)	• Medical • Clinical inputs • ICD-10, ICD-8	• 1.13 (0.70, 1.82) • – • –	12

References

1. J. Liu, M. Li, F. Wu, Y. Pan, J. Wang, Classification of Alzheimer's disease using whole brain hierarchical network. IEEE/ACM Trans. Comput. Biol. Bioinform. **15**(2), 624–632 (2018)
2. R. Chaves, J. Ramirez, J.M. Gorriz, Integrating discretization and association rule-based classification for Alzheimer's disease diagnosis. Expert Syst. Appl. **40**(5), 1571–1578 (2013)
3. E. Moradi, A. Pepe, C. Gaser, H. Huttunen, J. Tohka, Machine learning framework for early MRI-based Alzheimer's conversion. NeuroImage **104**, 398–412 (2015)
4. M. Liu, Y. Gao, P. Yap, D. Shen, Multi-Hypergraph learning for incomplete multimodality data. IEEE J. Biomed. Health Inf. **22**(4), 1197–1208 (2018)
5. Z. Li, H. Suk, D. Shen, L. Li, Sparse multi-response tensor regression for Alzheimer's disease study with multivariate clinical assessments. IEEE Trans. Med. Imaging **35**(8), 1927–1936 (2016)
6. Y. Huang, K. Ito, C.B. Billing, R.J. Anziano, Development of a straightforward and sensitive scale for MCI and early AD clinical trials. Alzheimer's Dement. **11**(4), 404–414 (2015)
7. P. Cao, X. Liu, J. Yang, D. Zhao, M. Huang, J. Zhang, O. Zaiane, Nonlinearity-aware based dimensionality reduction and over-sampling for AD/MCI classification from MRI measures. Comput. Biol. Med. 21–37 (2017)
8. H. Suk, S. Lee, D. Shen, Deep ensemble learning of sparse regression models for brain disease diagnosis. Med. Image Anal. **37**, 101–113 (2017)
9. M. Liu, D. Zhang, E. Adeli, D. Shen, Inherent structure-based multiview learning with multi-template feature representation for Alzheimer's disease diagnosis. IEEE Trans. Biomed. Eng. **63**(7), 1473–1482 (2016)

10. Z. Liu, T. Xu, C. Ma, C. Gao, H. Yang, Alzheimer's disease diagnosis via interested structure selection in MRIs, in *International Conference on Natural Computation*, June 2018
11. H. Suk, S. Lee, D. Shen, Hierarchical feature representation and multimodal fusion with deep learning for AD/MCI diagnosis. NeuroImage (2014)
12. X. Zhu, H. Suk, D. Shen, Matrix-similarity based loss function and feature selection for Alzheimer's disease diagnosis, in *IEEE Conference on Computer Vision and Pattern Recognition*, September 2014
13. G. Fiscon, E. Weitschek, G. Felici, P. Bertolazzi, S. De Salvo, P. Bramanti, M.C. De Cola, Alzheimer's disease patients classification through EEG signals processing, in *IEEE Symposium on Computational Intelligence and Data Mining (CIDM)*, December 2014
14. F. Li, L. Tran, K. Thung, S. Ji, D. Shen, J. Li, A robust deep model for improved classification of AD/MCI patients. IEEE J. Biomed. Health Inf. **19**(5), 1610–1616 (2015)
15. J. Shlens, A tutorial on principal component analysis. Systems Neurobiology Laboratory, University of California at San Diego, vol. 82 (2005)
16. L.K. Saul, K.Q. Weinberger, J.H. Ham, F. Sha, D.D. Lee, Spectral methods for dimensionality reduction. Semisupervised Learn. 293–308 (2006)
17. S.T. Roweis, L.K. Saul, Nonlinear dimensionality reduction by locally linear embedding. Science **290**(5500), 2323–2326 (2000)
18. M. Belkin, P. Niyogi, Laplacian eigenmaps for dimensionality reduction and data representation. Neural Comput. **15**(6), 1373–1396 (2003)
19. S. Yan, D. Xu, B. Zhang, H.-J. Zhang, Q. Yang, S. Lin, Graph embedding and extensions: a general framework for dimensionality reduction. IEEE Trans. Pattern Anal. Mach. Intell. **29**(1), 40–51 (2007)
20. N.V. Chawla, K.W. Bowyer, L.O. Hall, W.P. Kegelmeyer, SMOTE: synthetic minority oversampling technique, J. Artif. Intell. Res. **16**(1), 321–357 (2002)
21. E. Moradi, A. Pepe, C. Gaser, H. Huttunena, J. Tohka, Machine learning framework for early MRI-based Alzheimer's conversion prediction in MCI subjects. Neuroimage **104** (2015)
22. C. Hinrichs, V. Singh, G. Xu, S.C. Johnson, Predictive markers for AD in a multi-modality framework: an analysis of MCI progression in the ADNI population. NeuroImage **55**(2), 574–589 (2010)
23. S.H. Hojjati, A. Ebrahimzadeh, A. Khazaee, A. Babajani-Feremi, Predicting conversion from MCI to AD by integrating rs-fMRI and structural MRI. Comput. Biol. Med. 30–39 (2018)
24. R. Chaves, J. Ramírez, J.M. Gorriz, M. Lopez, D. Salas-Gonzalez, I.A. Illán, F. Segovia, P. Padilla, Effective diagnosis of Alzheimer's disease by means of association rules, in *Hybrid Artificial Intelligence Systems*, vol. 1 (Springer, 2010), pp. 452– 459
25. R. Chaves, J.M. Górriz, J. Ramírez, I.A. Illán, D. Salas-Gonzalez, M. Gómez-Río, Efficient mining of association rules for the early diagnosis of Alzheimer's disease. Phys. Med. Biol. **56**(18), 6047 (2011)
26. D. Zhang, Y. Wang, L. Zhou, H. Yuan, D. Shen, Multimodal classification of Alzheimer's disease and mild cognitive impairment. NeuroImage **56**(18) (2011)
27. R. Chaves, J. Ramírez, J.M. Gorriz, C.G. Puntonet, Association rule based feature selection method for Alzheimer's disease diagnosis. Expert Syst. Appl. **39**(14), 11766–11774 (2012)
28. R. Chaves, J. Ramírez, J.M. Gorriz, I.A. Illan, Functional brain image classification using association rules defined over discriminant regions. Pattern Recogn. Lett. **33**(12), 1666–1672 (2012)
29. R. Chaves, J. Ramírez, J.M. Gorriz, Integrating discretization and association rule-based classification for Alzheimer's disease diagnosis. ACM Digit. Libr. **40**(5), 1571–1578 (2013)
30. A. Veeramuthu, S. Meenakshi, P.S. Manjusha, A new approach for Alzheimer's disease diagnosis by using association rule over PET images. Int. J. Comput. Appl. **91**(9), 9–14 (2014)
31. J. Liu, J. Wang, B. Hu, F. Wu, Y. Pan, Alzheimer's disease classification based on individual hierarchical networks constructed with 3-D texture features. IEEE Trans. Nanobiosci. **16**(6), 428–437 (2017)
32. M. Liu, D. Zhang, D. Shen, Relationship induced multi-template learning for diagnosis of Alzheimer's disease and mild cognitive impairment. IEEE Trans. Med. Imaging **35**(6), 1463–1474 (2016)

33. K.R. Kruthika, H.D. Maheshappa, Multistage classifier-based approach for Alzheimer's disease prediction and retrieval. Inf. Med. Unlocked (2018)
34. N. Zeng, H. Qiu, Z. Wang, W. Liu, H. Zhang, Y. Li, A new switching-delayed-PSO-based optimized SVM algorithm for diagnosis of Alzheimer's disease. Neurocomputing **320**, 195–202 (2018)
35. J. Liu, S. Shang, K. Zheng, J. Wen, Multi-view ensemble learning for dementia diagnosis from neuroimaging: An artificial neural network approach. Neurocomputing **195**, 112–116 (2016)
36. S. Basaia, F. Aqosta, L. Waqner, E. Canu, G. Maqnani, R. Santangelo, M. Filippi, Automated classification of Alzheimer's disease and mild cognitive impairment using a single MRI and deep neural networks. NeuroImage (Clinical) (2018)
37. J. Liu, J. Wang, Z. Tang, B. Hu, F. Wu, Y. Pan, Improving Alzheimer's disease classification by combining multiple measures. IEEE/ACM Trans. Comput. Biol. Bioınform. **15**(5), 1649–1659 (2018)
38. S.S. Beagum, A.A. Almas, S. Sheeja, Alzheimer's disease, bio-markers, and the role of classification techniques in early diagnosis from neuro-images - An analysis, in *IEEE International Conference on Computational Intelligence and Computing Research*, December 2016
39. H. Jalilian, S.H. Teshnizi, M. Röösli, M. Neghab, Occupational exposure to extremely low frequency magnetic fields and risk of Alzheimer disease: a systematic review and meta-analysis. Neurotoxicology **69**, 242–252 (2018)
40. S.M. Landau, M. Lu, A.D. Joshi, M. Pontecorvo, M.A. Mintun, J.Q. Trojanowski, L.M. Shaw, W.J. Jagust, Comparing positron emission tomography imaging and cerebrospinal fluid measurements of β-amyloid. Alzheimer's Dis. Neuroimaging Initiative **74**(6), 826–836 (2013)
41. S. Palmqvist, H. Zetterberg, K. Blennow, S. Vestberg, U. Andreasson, D.J. Brooks, R. Owenius, D. Hägerström, P. Wollmer, L. Minthon, O. Hansson, Accuracy of brain amyloid detection in clinical practice using cerebrospinal fluid β-amyloid 42: a cross-validation study against amyloid positron emission tomography. Jama Neurol. **71**(10) (2014)
42. A.M. Fagan, C.M. Roe, C. Xiong, M.A. Mintun, J.C. Morris, D.M. Holtzman, Cerebrospinal fluid tau/beta-amyloid(42) ratio as a prediction of cognitive decline in nondemented older adults. Arch. Neurol. **64**, 343–349 (2007)
43. A. Forsberg, H. Engler, O. Almkvist, G. Blomquist, G. Hagman, A. Wall, A. Ringheim, B. Långström, A. Nordberg, PET imaging of amyloid deposition in patients with mild cognitive impairment. Neurobiol. Aging **29**, 1456–1465 (2008)
44. A.M. Fagan, M.A. Mintun, A.R. Shah, P. Aldea, C.M. Roe, R.H. Mach, D. Marcus, J.C. Morris, D.M. Holtzman, Cerebrospinal fluid tau and ptau(181) increase with cortical amyloid deposition in cognitively normal individuals: implications for future clinical trials of Alzheimer's disease. EMBO Mol. Med. **1**, 371–380 (2009)
45. W.J. Jagust, S.M. Landau, L.M. Shaw, J.Q.Trojanowski, R.A. Koeppe, E.M. Reiman, N.L. Foster, R.C. Petersen, M.W. Weiner, J.C. Price, C.A. Mathis, Relationships between biomarkers in aging and dementia. Neurology **73**, 1193–1199 (2009)
46. N. Tolboom, W.M. van der Flier, M. Yaqub, R. Boellaard, N.A. Verwey, M.A. Blankenstein, A.D. Windhorst, P. Scheltens, A.A. Lammertsma, B.N. van Berckel, Relationship of cerebrospinal fluid markers to 11C-PiB and 18F-FDDNP binding. J. Nucl. Med. **50**, 1464–1470 (2009)
47. M.D. Gunnarsson, M. Lindau, A. Wall, K. Blennow, T. Darreh-Shori, S. Basu, A. Nordberg, A. Larsson, L. Lannfelt, H. Basun, L. Kilander, Pittsburgh compound-B and Alzheimer's disease biomarkers in CSF, plasma and urine: An exploratory study. Dement. Geriatr. Cogn. Disord. **29**, 204–212 (2010)
48. S.D. Weigand, P. Vemuri, H.J. Wiste, M.L. Senjem, V.S. Pankratz, P.S. Aisen, M.W. Weiner, R.C. Petersen, L.M. Shaw, J.Q. Trojanowski, D.S. Knopman, C.R. Jack, Jr, Transforming cerebrospinal fluid Abeta42 measures into calculated Pittsburgh compound B units of brain Abeta amyloid. Alzheimers Dement. **7**(2), 133–141 (2011)
49. M. Zwan, A. van Harten, R. Ossenkoppele, F. Bouwman, C. Teunissen, S. Adriaanse, A. Lammertsma, P. Scheltens, B. van Berckel, W. van der Flier, Concordance between cerebrospinal fluid biomarkers and [11C]PIB PET in a memory clinic cohort. J. Alzheimers Dis. **41**(3), 801–807 (2014)

50. A. Seidler, P. Geller, A. Nienhaus, T. Bernhardt, I. Ruppe, S. Eggert, M. Hietanen, T. Kauppinen, L. Frolich, Occupational exposure to low frequency magnetic fields and dementia: a case-control study. Occup. Environ. Med. **64**(2), 108–114 (2007)
51. R. Andel, M. Crowe, M. Feychting, N.L. Pedersen, L. Fratiglioni, B. Johansson, M. Gatz, Work-related exposure to extremely low-frequency magnetic fields and dementia: results from the population-based study of dementia in Swedish twins. J. Gerontol. Ser. A. Biol. Sci. Med. Sci. **65**(11), 1220– 1227 (2010)
52. T. Sorahan, N. Mohammed, Neurodegenerative disease and magnetic field exposure in UK electricity supply workers. Occup. Med. (Oxford,England) **64**(6), 454–460 (2014)
53. Z. Davanipour, C.-C. Tseng, P.-J. Lee, K.S. Markides, E. Sobel, Severe cognitive dysfunction and occupational extremely low frequency magnetic field exposure among elderly Mexican Americans. Br. J. Med. Med. Res. **4**(8), 1641 (2014)
54. T. Koeman, L.J. Schouten, P.A. van den Brandt, P. Slottje, A. Huss, S. Peters, H. Kromhout, R. Vermeulen, Occupational exposures and risk of dementia-related mortality in the prospective Netherlands cohort study. Am. J. Ind. Med. **58**(6), 625–635 (2015)
55. C. Pedersen, A.H. Poulsen, N.H. Rod, P. Frei, J. Hansen, K. Grell, O. Raaschou-Nielsen, J. Schüz, C. Johansen, Occupational exposure to extremely low- frequency magnetic fields and risk for central nervous system disease: an update of a Danish cohort study among utility workers. Int. Arch. Occup. Environ. Health 1–10 (2017)

Real-Time Multi-cue Object Tracking: Benchmark

Ashish Kumar, Gurjit Singh Walia and Kapil Sharma

Abstract Object tracking is imperative research domain of computer vision. In the last decade, a lot of progress has been made in terms of tracking approach and the datasets. However, to estimate the performance of trackers in real time it is essential to have a dataset that can have videos of adequate length with various attributed challenges. In order to bridge the gap, we have proposed a fully annotated datasets with video sequences of adequate length for object tracking. Also, the existing tracking of the state-of-the-art is reviewed on the proposed dataset and the performance is evaluated in real-time.

1 Introduction

Object tracking is rapidly evolving and widely explored area in video surveillance, human interactions, robot surveillance, and augmented reality. Object tracking is to estimate the location of the target in the first frame either manually or through detection algorithm. After localization, the track of the target is kept in subsequent frames. In the last 10 decades, object tracking has been investigated a lot but still there is scope of improvement in terms of available datasets. Object tracking encompasses either environmental challenges or hardware challenges. Environmental challenges include illumination variation, fast motion, motion blur, scale variation, in-plane and out-of-plane rotation, full occlusion, out-of-view, partial occlusion, and background clutters. On the other hand, hardware challenges refer to the problems that arise due to poor camera quality and precision. This has developed poor dataset in which

A. Kumar (✉) · K. Sharma
Delhi Technological University, New Delhi, India
e-mail: ashish.gupta14d@gmail.com

K. Sharma
e-mail: kapil@ieee.org

G. S. Walia
Ministry of Defence, DRDO, New Delhi, India
e-mail: gurjit.walia@gmail.com

M. Dutta et al. (eds.), *Proceedings of International Conference on IoT Inclusive Life (ICIIL 2019), NITTTR Chandigarh, India*, Lecture Notes in Networks and Systems 116,
https://doi.org/10.1007/978-981-15-3020-3_29

tracking the target is a challenge. So, the evaluation of the dataset is a must by analyzing the performance and accuracy of the state-of-the-art. In order to evaluate the existing state-of-the-art, there is a requirement of dataset that includes diverse dynamic challenges in one video sequences.

Recently, many datasets are proposed, viz., CAVIAR [1], PETS, object tracking benchmark (OTB) [2], and Templecolor (TC128) [3] to evaluate the performance of trackers. However, most of these datasets include the small target and static background. These datasets do not include the real-time scenarios which may actually happen during tracking. The real-time scenario includes the environmental variations due to the presence of sunlight, fog, smoke fumes, and so on. There is also need to include the videos in the dataset in which the target is stationary and camera is moving. The trackers are run on video sequences in order to evaluate their performance in dynamic environmental variations.

Object tracking approach can be divided either as generative approach or discriminative approach. Generative approach-based trackers search for the most analogous area to the target while discriminative trackers consider tracking as a binary classification problem. Generative trackers include DFT [4] and ASLA [5], while discriminative trackers include UGF [6], WMIL [7], and STRUCK [8]. These trackers are not able to handle the long-term challenge. Also, the recent particle filter-based trackers [9, 10], correlation filter-based trackers [11–13], and deep learning trackers [14–16] performance degraded in long videos. In order to compare various state-of-the-arts, we have evaluated the performance of 10 publicly available trackers on our datasets having adequate length video sequences. The main contributions of the paper are as follows:

1. We have developed a fully annotated dataset with 12 video sequences of adequate length. The ground truth values are provided in both .txt as well as .mat files.
2. An extensive review of dataset is done by evaluating 10 trackers on the video sequences.
3. For performance evaluation, precision versus CLE threshold plot and success versus overlap threshold plot [17] are considered as performance metrics.

The remainder of the paper is organized as follows: The review of closely related works is described in Sect. 2. In Sect. 3, experimental analysis of the dataset on state-of-the-art is discussed. Also, the performance metrics and evaluation results are analyzed. Finally, the concluding remarks are given in Sect. 4.

2 Related Works

In this section, the existing tracking approaches and the object tracking datasets are reviewed. Recent object tracking algorithms considered either generative approach or discriminative approach. ASLA [5], IDCT [18], L1-APG [19] trackers exploit generative model while CT [20], MIL [6], and WMIL [7] exploit discriminative

Table 1 Evaluated state-of-the-art

Tracker	Extracted features	Approach
ASLA [5]	Image patch	Generative
WMIL [7]	Random Haar-like features	Discriminative
CT [20]	Haar-like features	Discriminative
DFT [4]	Multi-layer distribution field	Generative
IDCT [18]	RGB histogram	Discriminative
MIL [6]	Haar-like features	Discriminative
MTT [22]	Holistic image intensity	Generative
SCM [23]	Holistic image intensity and histograms	Hybrid
STRUCK [8]	Haar-like features	Discriminative
L1-APG [19]	Holistic sparse representation	Generative

approach. Table 1 tabulates the details of various state-of-the-art trackers, considered approach, and the features extracted for target representation.

Recently, many object tracking datasets have been proposed. CAVIAR, PETS, VOT 2014, VOT 2015, OTB-100, and UAV [21] are few examples of object tracking datasets. CAVIAR and PETS datasets are quite old datasets. Targeted object in these datasets is quite small with static background and hence, not suitable to assess the performance of the trackers. VOT (2014, 2015, 2016, 2017), OTB 50, OTB 100, and TC128 datasets contain many common video sequences. Number of distinct videos in these datasets is quite small. UAV dataset includes substantial large number of aerial video sequences. Most of the videos have small target size and fast abrupt motion. It is difficult to test the performance of tracking algorithms on this dataset. This dataset is suitable to test the performance of trackers under low-altitude target tracking.

Attributes (attrb) in video sequence: Evaluating the performance of trackers is a challenging task. A video sequence is attributed by many tough environmental challenges. We have considered eight challenges in the proposed datasets which often exist in the videos. Table 2 tabulates the description of various attributes analyzed in a video sequence.

The details of the video sequences and the attributed challenge are given in Table 3.

Figure 1 includes the first frame of the test video sequence. In order to highlight the target, target is surrounded with red color rectangular bounding box. The following section details about the experimental analysis of tracking algorithms on test video sequence.

Table 2 Various considered attributes in video sequences of dataset

Attrb	Description
IV	Change in illumination: significant change in illumination in the target region
FOC	Full occlusion: target is totally occluded
SV	Variation in scale: ratio of at least one frame and subsequent frame is out of range
BC	Background clutters: analogous color, texture, shape background near the target
FM	Fast motion: background motion between the subsequent frames is more than 20 px
POC	Partial occlusion: target is partly occluded
CM	Camera motion: camera moves abruptly
MB	Motion blur

Table 3 Video sequence and their attributed challenge

Video	Attributes	Total frames
Auto	SV, FM, CM	125
Bounce	FM, MB	304
Boy1	IV, SV, CM	334
Cyclist	IV, CM, FM	358
Wagon	SV, FM, CM	186
Field	IV, SV, BC, POC	486
Person	IV, BC	219
Man	SV, CM	452
Stairs	SV, BC, POC	287
Stone	SV, FM	188
Walk	SV, FM, FOC, POC	172
Badminton	BC, FM, CM	357

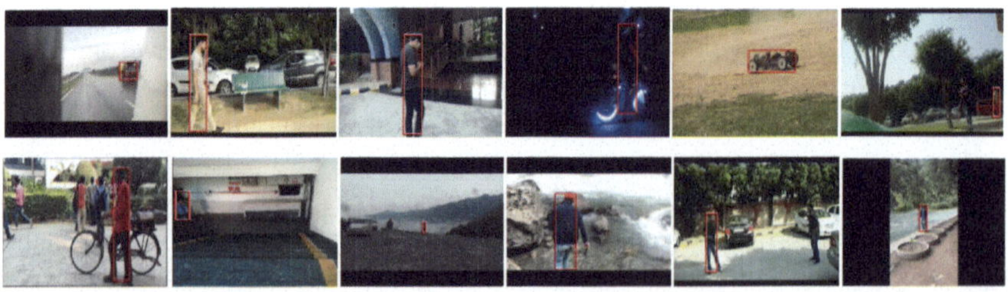

Fig. 1 First frame of video sequences and the target is surrounded with red color rectangular bounding box

3 Experimental Analysis

Two evaluation metrics, namely, precision plot and success plot, are exploited to determine the performance of tracking algorithms. Precision plot defines the average precision of the tracker taken over various overlap thresholds and are depicted in Fig. 2. Average centre location error (CLE) defines the sum of overall frames for one sequence over all the frames. Precision plot considers the location error thresholds only and ignores the overlap information. Figure 3 illustrates the success plot which considers the success rate versus overlap threshold. Overlap threshold defines ratio as the number of successfully tracked frames w.r.t. to ground truth. For each tracker, author-release codes are used for evaluation. Matlab2015b on 2.4 GHz machine with 6 GB RAM is used for running the trackers.

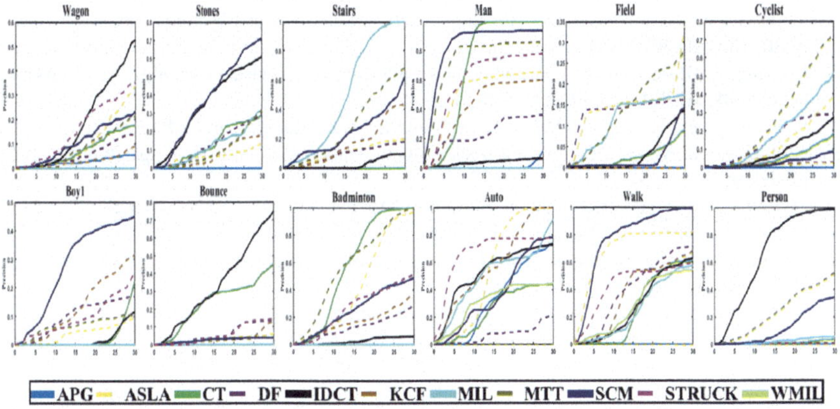

Fig. 2 Precision plot (precision vs. CLE threshold plot)

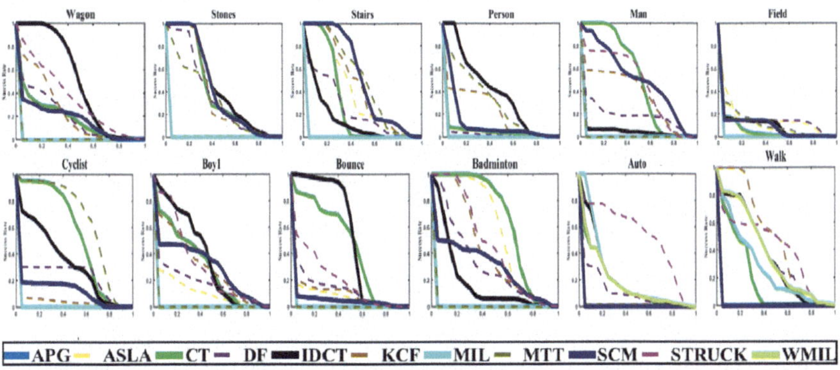

Fig. 3 Success plot (success vs. overlap threshold plot)

4 Conclusion

In this manuscript, object tracking dataset has been proposed for evaluation of various state-of-the-arts. The dataset includes video sequences encompassing various environmental challenges. The video sequences are of adequate length to determine the real-time performance of various trackers. Experimental outcomes infer that the proposed dataset is appropriate to estimate the performance of existing tracking solutions.

References

1. R.B. Fisher, The PETS04 surveillance ground-truth data sets, in *PETS* (2004)
2. Y. Wu, J. Lim, M.H. Yang, Object tracking benchmark. IEEE Trans. Pattern Anal. Mach. Intell. **37**(9), 1834–1848 (2015)
3. P. Liang, E. Blasch, H. Ling, Encoding color information for visual tracking: algorithms and benchmark. IEEE Image Process. (2015), pp. 1–14
4. L. Sevilla-Lara, E. Learned-Miller, Distribution fields for tracking, in *CVPR* (2012)
5. X. Jia, H. Lu, M.-H. Yang. Visual tracking via adaptive structural local sparse appearance model, in *CVPR* (2012)
6. G.S. Walia, H. Ahuja, A. Kumar, N. Bansal, K. Sharma, Unified graph-based multicue feature fusion for robust visual tracking. IEEE Trans. Cybern. (2019)
7. K. Zhang, H. Song, Real-time visual tracking via online weighted multiple instance learning. Pattern Recogn. **46**(1), 397–411 (2013)
8. S. Hare, A. Saffari, P.H.S. Torr, Struck: structured output tracking with kernels, in *ICCV* (2011)
9. G.S. Walia, A. Kumar, A. Saxena, K. Sharma, K. Singh, Robust object tracking with crow search optimized multi-cue particle filter. J. Pattern Anal. Appl. 1–17 (2019)
10. G.S. Walia, R. Kapoor, Robust object tracking based upon adaptive multi-cue integration for video surveillance. Multim. Tools Appl. **75**(23), 15821–15847 (2016)
11. M. Danelljan, G. Bhat, F. Shahbaz Khan, M. Felsberg, Eco: Efficient convolution operators for tracking., in *CVPR* (2017)
12. T. Zhang, C. Xu, M.H. Yang, Multi-task correlation particle fillter for robust object tracking, in *CVPR* (2017)
13. T. Zhang, C. Xu, M.H. Yang, Learning multi-task correlation particle filters for visual tracking. IEEE Trans. Pattern Anal. Mach. Intell. **41**(2), 365–378 (2018)
14. H. Hu, B. Ma, J. Shen, H. Sun, L. Shao, F. Porikli, Robust object tracking using manifold regularized convolutional neural networks. IEEE Trans. Multim. **21**(2), 510–521 (2018)
15. Y. Song, C. Ma, X. Wu, L. Gong, L. Bao, W. Zuo, … M.H. Yang, Vital: visual tracking via adversarial learning, in *Proceedings of the IEEE Conference on Computer Vision and Pattern Recognition* (2018), pp. 8990–8999
16. J. Gao, T. Zhang, C. Xu, Smart: joint sampling and regression for visual tracking. IEEE Trans. Image Process. (2019)
17. Y. Wu, J. Lim, M.H. Yang, Online object tracking: a benchmark, in *Proceedings of the IEEE Conference on Computer Vision and Pattern Recognition* (2013), pp. 2411–2418
18. A. Asvadi, H. Mahdavinataj, M. Karami, Y. Baleghi, Incremental discriminative color object tracking, in *International Symposium on Artificial Intelligence and Signal Processing* (Springer, Cham 2013), pp. 71–81
19. C. Bao, Y. Wu, H. Ling, H. Ji, Real time robust L1 tracker using accelerated proximal gradient approach, in *CVPR* (2012)
20. K. Zhang, L. Zhang, M.-H. Yang, Real-time compressive tracking, in *ECCV* (2012)

21. M. Mueller, N. Smith, B. Ghanem, A benchmark and simulator for uav tracking, in *European Conference on Computer Vision* (Springer, Cham, 2016), pp. 445–461
22. T. Zhang, B. Ghanem, S. Liu, N. Ahuja, Robust visual tracking via multi-task sparse learning, in *CVPR* (2012)
23. W. Zhong, H. Lu, M.-H. Yang, Robust object tracking via sparsity-based collaborative model, in *CVPR* (2012)

IoT Enabling Technologies

IoT-Driven Real-Time Monitoring of Air Pollution with Dynamic Google Mapping

Nikhil and Milanpreet Kaur

Abstract For the estimation of air quality index of an area, monitoring the air quality only at some locations is not enough. Hence the monitoring of air quality at multiple locations is required to get the fair idea about pollution at different areas. This paper basically represents the monitoring of air pollution caused by the integrity of residential and industrial areas located at different places. An IoT-enabled gas-sensing module is used, which can either be connected on a vehicle or can be placed at an open area for getting real time and continuous values of air pollutants present in the air. The use of open source database management enables the system to collect and send the data of gas-sensing module on web server. Also the use of dynamic Google mapping technique helps to locate the areas with poor air quality. These areas can be identified by the pollution control boards and immediate steps can be taken to reduce the emission of polluted gases.

Keywords IoT · Air quality index · Google mapping · Microcontroller

1 Introduction

The major intimidation to human health is poor quality of air for breathing. Burning of fossil fuels, industrialization, ejections from motor vehicles and many other factors can be one of the crucial sources of air pollution. Air pollutants not only engender impaired health conditions but also affect the regular climatic cycle. Air pollution leads to the death of as many as nearly 5.5 million people according to the report of World Health Organization (WHO) every year. The death rate is more in metro cities since metro cities have more exposure to smoke from industries and plenty of vehicles which add harmful gases into atmosphere. More than 140 million people in

Nikhil (✉)
National Institute of Technical Teachers Training and Research, Chandigarh, India
e-mail: nikhil.elect@nitttrchd.ac.in

M. Kaur
Indian Institute of Technology, Ropar, India

M. Dutta et al. (eds.), *Proceedings of International Conference on IoT Inclusive Life (ICIIL 2019), NITTTR Chandigarh, India*, Lecture Notes in Networks and Systems 116,
https://doi.org/10.1007/978-981-15-3020-3_30

the world breathe air which is 10 times above the WHO safe limit. The situation is alarming right now and will worsen with the increasing population.

In order to ensure the accessibility of uncontaminated and clean air, its quality should be examined and managed carefully on a regular basis. The traditional methods of monitoring air pollution include the installation of a system at one place of the city and giving air quality index as the average for the whole city. But the air quality and its contaminants can vary from place to place and its detection at a certain place does not assure the results for the other place which is far away. Moreover, earlier systems used a number of sensors based on electrochemical principle to detect different contaminations, which resulted in high cost and high complexity. This paper focuses on the concept of air pollution monitoring at different areas by placing the sensing module on a vehicle. The location of vehicle using Google mapping technique will provide the information of polluted area through wireless technologies. Also the system uses metal oxide semiconductor-based (MOS) sensors which are cheap and more efficient.

2 Literature Survey

The author proposed a system for measuring the air quality which is based on real-time statistics data using different sensors as presented in paper [1]. The system has lamp posts placed along roads for transmitting information of air quality to server as well as mobile phones.

In [2, 9] the authors presented a bicycle-born device for collecting the data of air quality near roadways. In this method a module was connected on the bicycle widely used in European and Asian inner city transportation system composed of GPS, Bluetooth, exhaust gas sensors.

The paper [3] describes the three-phase air pollution monitoring system. The monitoring of air pollution is done with the help of android application called IoT Mobair for accessing the relevant data from cloud. This android application along with data logging is used to predict Air Quality Index (AQI) level.

A novel system for visualization of weather data such as parallel coordinates, polar coordinates with enhanced parallel coordinates for analyzing the air pollution problem in Hong-Kong is described in [4]. This interesting pattern is developed for pollution checking from any coordinate and the system is available in public domain on the official website ENVF and HKUST (Hong Kong University of Science and Technology).

The author describes a modular sensor system which is adopted for monitoring the pollution at low concentration and high efficiency in [5]. This modular sensor system has six plugs to drive sensing modules which has multiple number of wireless sensors for monitoring real-time air pollution.

The author described in paper [6] the low-cost sensing system which includes smart phones, mobile apps and portable mobile sensor units for pollution sensing of metropolitan area. This system named as Haze Watch has three contributions for better understanding the impact of air pollution in urban areas as follows:

- The prototype and comparison of this system in between multiple devices for the collection of air pollution real-time data with high spatial density
- The mobile app is developed for estimation and better visualization of pollution along with exposure customized to everyone.

To validate the system, various trails are conducted to demonstrate the better accuracy and efficiency of the system as compared to previous system for monitoring air pollution.

The author presented an algorithm for detecting vehicular air pollution in paper [7] by extracting the concentration of pollution. When windows of vehicle are opened, the sensed data is converged known as crowd source-based air quality monitoring system. The real-time air quality data is collected from different cities for three months for sampling the data of air quality.

In paper [8] the author describes the use of internet of things for monitoring of toxicity of gases in air. The web page is created with the use of JavaScript and HTML for enabling the device to publish the data of air quality monitoring on cloud along with Google mapping. The use of GPS sensor for finding the exact location helped in controlling the pollution of the smart city.

The author explained the functionality of internet service protocol MQTT in paper [10]. This paper explains the basic concept of working of MQTT. In MQTT, a publisher/subscriber model is used in which a publisher publishing messages and users subscribing to topics considering as message subject. Subscriber can subscribe to a certain topic to obtain certain messages. The broker is the interlink between the publisher and the client and control the distribution of information.

One of the main problems is ground-level ozone which is adversely affected by pollution in urban cities as presented in paper [11]. The detection of gases with the help of MQ-7 sensor is done which are emitted from vehicles and the data is monitored using TCP/IP protocol and sent to GPS for the better localization of vehicle.

The author reviewed methods for estimating CO_2 gas from the field burning on global scale in paper [12]. The estimation of CO_2 during the period of 2002–2012 has been presented and compared with the GFED 3.1 database extracted from global fire emission 3.1 versions.

3 System Architecture

The proposed system consists of a sensing module which has different types of gas sensors. This module is interfaced with internet of things using any local area network. The sensing node of system receives the data of air quality. GPS module provides the location of polluted area. The data of air quality and location of polluted area is given to the web server for online monitoring via computer/mobile app (Fig. 1).

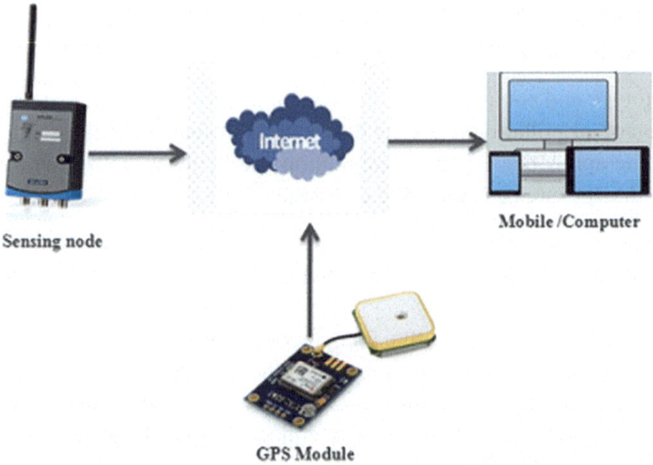

Fig. 1 System architecture

The controller at the sensing node is equipped with different sensors used to detect air quality of an area under observation. The MQ-7, MQ-135, MHZ14, DHT22 and GPS modules are connected with the controller for sensing purpose. ESP8266 Wi-Fi module is also interfaced with controller for transmission of data of air quality. The MQ7 gas sensor is used to detect the carbon monoxide in air. MQ-135 is used to check the toxicity level of gases such as NO_2, SO_2, NH_3 and smoke in air. The MHZ14 gas sensor is used to check the CO_2 content in air which is infrared based, long life and good selectivity. ESP8266 Wi-Fi module is a low-cost transceiver which is used to provide internet connection to the controller using AT commands. The wireless communication between the controller and the server is done with ESP8266 module using any third-party server. Ublox NEO-6 M GPS (global positioning system) based on satellite navigation system is used to track the location of module mounted on vehicle, which gives the information of polluted area.

4 Hardware Implementation

Figure 2 shows the schematic block diagram of the whole setup. The data from different sensors used to detect the contaminants in air are collected by the controller in the form of parts per million (PPM), where PPM is given by:

$$PPM = \frac{R_s}{R_o} \tag{1}$$

$$R_s = \left(\frac{V_{cc}}{V_{rl}} - 1\right) \times R_l \tag{2}$$

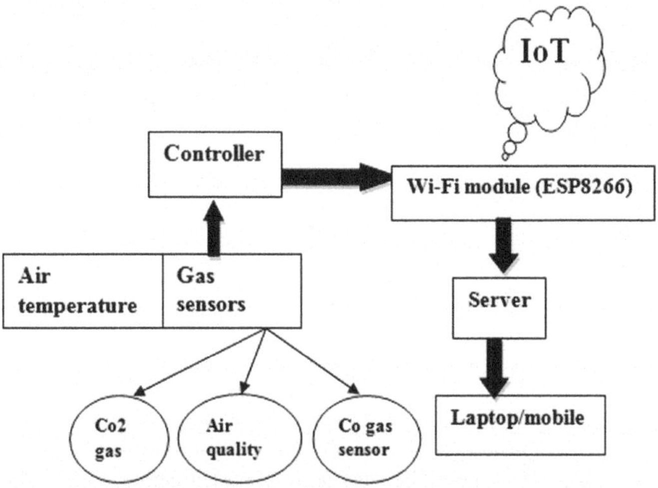

Fig. 2 Block diagram of setup

R_s is sensor resistance at various concentration of gases, R_o is sensor resistance in the clean air, V_{cc} is 5 V, R_l is the load resistor connected to sensor and V_{rl} is analog voltage of sensor. But the term PPM is not user-friendly as it is not understandable by every person. Therefore government agencies use a new term, that is, Air Quality Index (AQI) for better communication with the public to represent the pollutants in air of a specific area.

PPM can be calculated by taking the average of air pollutant value (PPM) measured over a long period of time. The controller is given with the data from different sensors in the form of PPM. It has to be converted in the form of AQI for better communication. AQI can be calculated by the equation as

$$AQI = \frac{(I_{high} - I_{low})}{(C_{high} - C_{low})} \times (C - C_{low}) + I_{low} \tag{3}$$

where C is pollutant concentration, C_{low} is concentration (PPM) breakpoint, that is, $\leq C$, C_{high} is concentration (PPM) breakpoint, that is, $\geq C$, I_{low} is the index breakpoint related to C_{low}, I_{high} is the index breakpoint related to C_{high}.

So AQI reflects the national air quality standards with which real-time data of monitored air is compared to define the quality of air. Table 1 shows the standards of air quality index (AQI) and the concentration of different pollutants associated with it.

Figures 3 and 4 show the flowchart for the proposed methodology and the actual hardware setup used to measure the AQI of an area, respectively. From Fig. 3 we can see that first the hardware set which is mounted on the vehicle initializes all the sensors connected to it. They all work in a synchronized manner to collect the data of air quality in the form of PPM. Data received from sensors is pressed and calibrated

Table 1 AQI standards and associated pollutant concentration

AQI category (range)	NO_2 (24-h)	O_3 (8-h)	CO (8-h)	CO_2 (24-h)	NH_3 (24-h)
Good (0–50)	0–40	0–50	0–1.0	0–40	0–200
Satisfactory (51–100)	41–80	51–100	1.1–2.0	41–80	201–400
Moderately polluted (101–200)	81–180	101–168	2.1–10	81–380	401–800
Poor (201–300)	181–280	169–208	10–17	381–800	801–1200
Very poor (301–400)	281–400	209–748*	17–24	801–1600	1200–1800
Severe (401–500)	400+	748+	34+	1600+	1800+

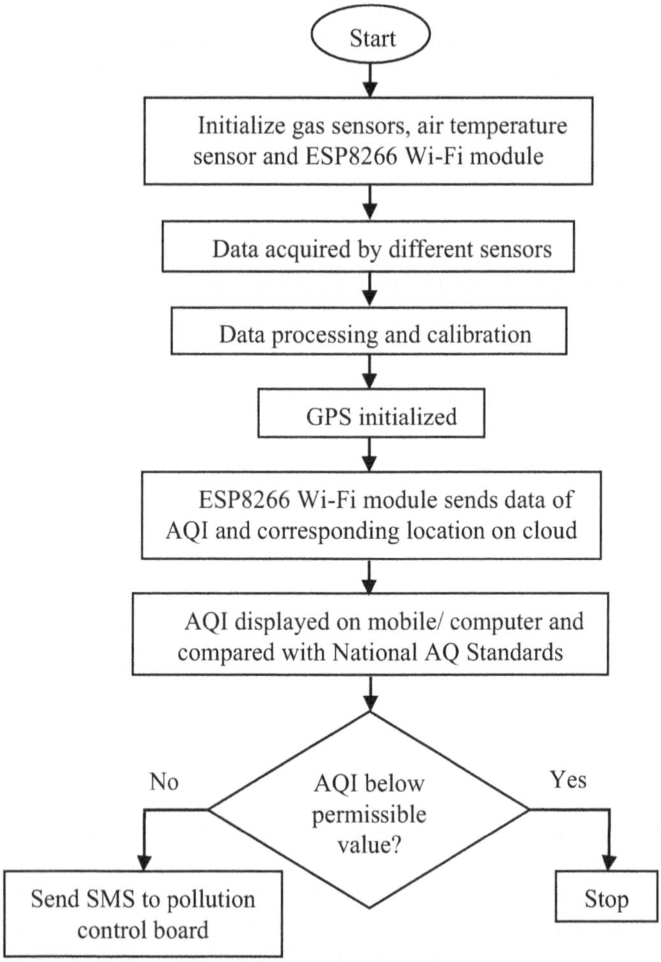

Fig. 3 Flowchart for the proposed methodology

Fig. 4 Hardware setup

in the form AQI for public interest. Now the GPS is initialized to locate the area under monitoring, using Google mapping technique. Now ESP8266 Wi-Fi module sends data of AQI and corresponding location on cloud using IoT. By using mobile app or a computer, health of air of any area can be monitored. If the air quality does not match with the prescribed quality standards, an SMS will be sent to the pollution control boards, so that preventive measures can be taken in time.

5 Result and Discussion

An experiment is performed in the laboratory to check the air quality by using the setup shown in Fig. 4. Figures 5, 6, and 7 show the experiment results of MQ7, MQ135 and MHZ14 sensor on the serial monitor. The data for air quality in PPM, humidity and air temperature on serial monitor can be shown. This data is processed and calibrated to calculate AQI and is given to cloud using IoT. An app is developed for a mobile phone/computer to monitor the AQI on a regular interval of time. In Fig. 8 exact location of the area under monitoring is tracked using Google mapping technique. Figure 9 represents the air quality index of the laboratory where the experiment has been performed.

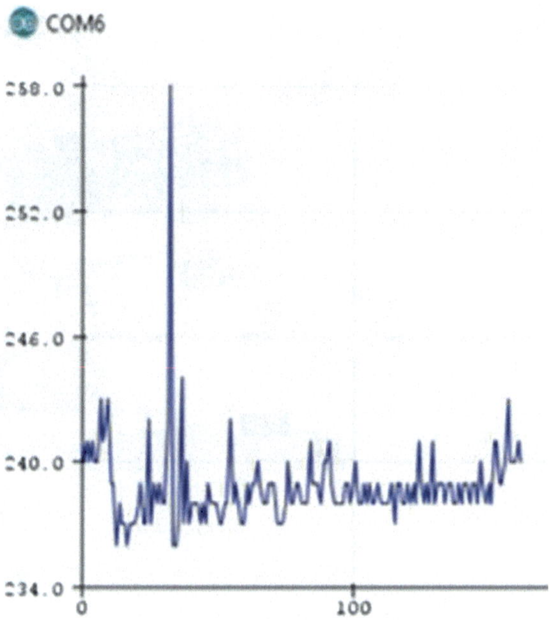

Fig. 5 Serial monitor data for MQ7

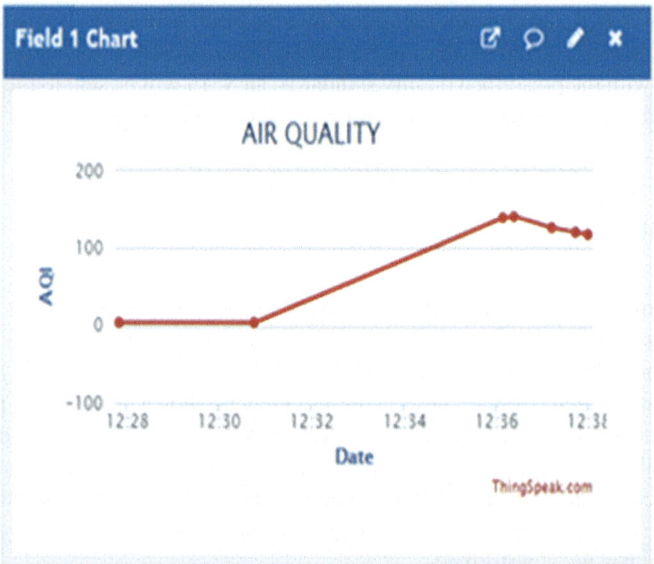

Fig. 6 Data on serial monitor for MQ135

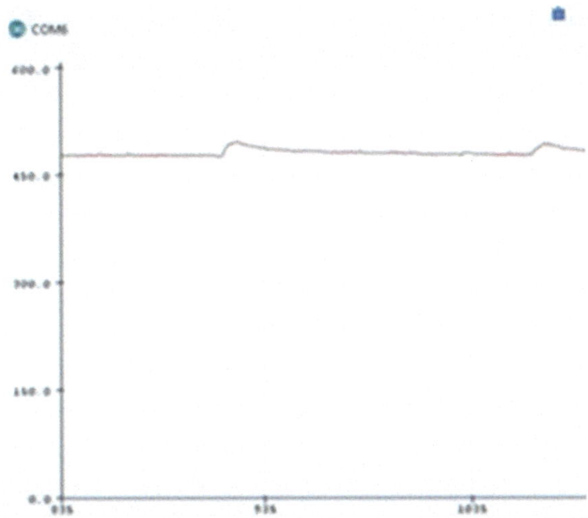

Fig. 7 Data on serial monitor for MHZ14

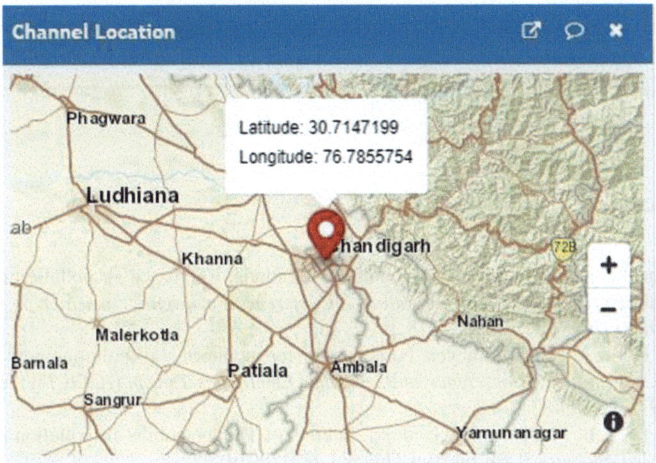

Fig. 8 GPS location on Google map

6 Conclusion

It can be seen from Fig. 9 that the AQI is coming out to be 17. By comparing the experiment value with national air quality standards in Table 1, it can be said that the AQI of laboratory is in the range of 0–51; therefore it can be referred as good and appropriate for breathing. Therefore, it can be concluded that the proposed setup is appropriate for monitoring the air quality of an area. It can be mounted on any

Fig. 9 AQI of the laboratory

vehicle to keep a continuous check on the air quality of different locations for a given area, so that better estimate of AQI can be made.

References

1. S. Muthunkumar, W.S. Mary, S. Jayanthi, R. Kiruthgia, IOT based air pollution monitoring and control system, in *2018 International Conference on inventive research in computing applications (ICIRCA)*,12 July 2018, pp. 2–18
2. X. Liu, B. Li, A. Jiang, A bicycle-borne sensor for monitoring air pollution near roadways, in *2015 International Conference on Consumer Electronics-Taiwan (ICCE-TW)*, June 2015. ISBN: 978-1-4799-8745-0,8
3. S. Dhingra, R.B. Madda, A.H. Gandomi, Internet of Things mobile–air pollution monitoring system (IoT-Mobair). IEEE Internet Things J. **6**(3) (2019)
4. H. Qu, W.-Y. Chan, A. Xu, K.-L. Chung, Visual analysis of the air pollution problem in Hong Kong. IEEE Trans. Vis. Comput. Graph. **13**(6) (2007)
5. W.-Y. Yi, K.-S. Leung, Y. Leung, Modular Sensor System (MSS) for urban air pollution monitoring, in *2016 International Conference on Sensors* (2016)
6. K. Hu, V. Sivaraman, B.G. Luxan, A. Rahman, Design and evaluation of a metropolitan air pollution sensing system. IEEE Sens. J. **16**(5) (2016)
7. J. Huang, N. Duan, P. Ji, C. Ma, F. Hu, Y. Ding, A crowdsource-based sensing system for monitoring fine-grained air quality in urban environments. IEEE Internet Things J. **6**(2) (2019)
8. G. Spandana, R. Shanmughasundram, Design and development of air pollution monitoring system for smart cities, in *Proceedings of the Second International Conference on Intelligent Computing and Control Systems (ICICCS 2018)*, 15 June 2018. ISBN:978-1-5386-2842-3
9. A. El Fazziki, D. Benslimane, A. Sadiq, J. Ouarzazi, An agent based traffic regulation system for the roadside air quality control, in *2017 IEEE. Translations and Content Mining are Permitted for Academic Research*, vol. 5, November 2016

10. D. Soni, A. Makana, A survey on MQTT: a protocol of Internet of Things, in *International conference on Telecommunications, Power Analyzing and Computing Techniques* (2017), pp. 252–258
11. M. Rathod, R. Gite, A. Pawar, An Air Pollutant Vehicle Tracker system using Gas sensor and GPRS, in *IEEE International Conference on Electronics, Communication and Aerospace Technology* (2017)
12. W. Takeuchi, A. Sekiyama, R. Imasu, Estimation of global carbon emissions from wild fires in forests and croplands, in *IEEE International conference on Geoscience & Remote Sensing Synopsium (IGARSS)* (2013), pp. 1805–1808

Android Things: A Comprehensive Solution from Things to Smart Display and Speaker

Rohit Roy, Sayantika Dutta, Sagnick Biswas and Jyoti Sekhar Banerjee

Abstract Currently, it is one of the biggest concerns of research for making the smart world to control and operate each and every device embedded in various systems through the Internet. Many IoT platforms like IBM Watson IoT platform, IoT platform provided by Amazon Web Service, Google cloud platform are commonly used by the new generation developers. Android Things is one of those platforms developed by Google, which could also be used for developing IoT-enabled systems. It is the new Operating System (OS) developed by Google for building professional IoT projects and android apps. It is thought that Android Things would be the next technological revolution, as it can open a new field of interest for android developers. Google being the technological boon of the era gives us more reliability and sustainability. In this paper, the authors prepare a detailed review report on Android Things. Though it a very trendy topic, still a very few research findings are available presently.

1 Introduction

Establish a wireless connection of different objects and to control those objects remotely is not new to us, but in recent years, it demands to develop further [1, 2]. As the circuits are getting miniaturized and demand low power consumption [3], hence, the Internet of Things (IoT) is the best solution to address all of the problems. According to many analysts, IoT is the most innovative technology of the present

R. Roy · S. Dutta · S. Biswas · J. S. Banerjee (✉)
Department of ECE, Bengal Institute of Technology, Kolkata 700150, India
e-mail: tojyoti2001@yahoo.co.in

R. Roy
e-mail: rohit.roy7728@gmail.com

S. Dutta
e-mail: 2017nehagrowup123@gmail.com

S. Biswas
e-mail: sagnickbiswas98@gmail.com

M. Dutta et al. (eds.), *Proceedings of International Conference on IoT Inclusive Life (ICIIL 2019), NITTTR Chandigarh, India*, Lecture Notes in Networks and Systems 116,
https://doi.org/10.1007/978-981-15-3020-3_31

decade. It also promises that in the future, it can modify our old habits and would have a massive impact on society. Already IoT has started spreading its roots in various sectors like Industry [4–6], Healthcare [7, 8], Smart city, Agriculture [9], etc. On the other hand, Google developed an operating system (OS) called ‘Android Things’ to contribute to the market of IoT. This OS has been developed with the motto to build IoT projects and to develop professional applications with the help of Android and trusted platforms; therefore, implementing IoT projects with Android is considered as the main aim of Android Things OS. Google also claimed that Android Things could be the next technological revolution after IoT. The name ‘Android Things’ has been derived from the name ‘Android’, but there is a difference between Android and Android Things basically, the latter is the modified version of the former [10]. Hence, it becomes an advantage of Android developers who work with Android Studio and other applications to build IoT projects comfortably.

For communicating with other peripherals, a new library has been developed by Google called Things Support Library, which is not present in Android SDK. Low PAN protocol can be used to connect other external peripherals.

In [4], the authors have proposed a remote sensor network deploying for a wide area using cloud computing and IoT. The wireless [10, 12, 13, 14] sensor network helps the Food Reserve Agency for timely action, in analytics [15, 16] and real-time reporting from all the storehouses around Zambia. In [17], the authors have framed a model which ensures the security [5, 11, 18] of children and women around the world. To our best knowledge, there is no published work so far on review of Android Things. Authors are also claiming that very few research findings are available presently in the said topic.

The letter has been framed as follows. Specifications are being defined in Sect. 2. Section 3 includes the functions of Android Things in IoT. Section 4 outlines the difference between Android and Android Things accompanied by the Protocols in Sect. 5. The advantages of using Android Things over other OS are present in Sect. 6, The disadvantages of using Android Things as an OS and applications of using Android Things as an OS are described in Sects. 7 and 8. Section 9 consists of additional applications of using Android Things and accompanied by conclusions in Sect. 10.

2 Specifications

Google has developed a new operating system, Android Things, which is previously known as Brillo that helps us to build professional projects using other various renowned platforms. It is the upgraded system of Android, and hence, we can efficiently utilize our basic Android knowledge to apply smart IoT-based applications. So, we can operate with the similar precious Software Development Kit, which is being used in Android apps to develop Android Things app.

We can connect GPIO pins with the help of a significant class, ‘Peripheral Manager Service’ provided by the Android Things SDK [19]. Some handful of actions can be done using this particular class, such as obtaining the pins list, pin state, and assigning the pin state.

Renowned and trusted Google services like Firebase, the Google cloud platform, and tensor flow can be easily merged with the Android Things apps. In Android Things, as it boots right into the apps, no further browsers or launchers are needed. It helps us by decreasing boot time and lesser memory footprints. This particular platform was constructed to be dependable by supplying strong and robust protection to fight infectious bugs. Every developed application that is made with android things is provided with a free automatic system update lasting for three years.

2.1 Layer Structure of Android Things OS

Mainly The Intel Edison, the NXP Pico i.MX6UL and raspberry Pi 3 are being supported by android Things [20]. The main characteristics feature in these boards are Intel and ARM combination-based CPUs, which support 32-bits as well as 64-bits. A RAM of a minimum of 512 Mb is required to work smoothly with Android Things. Moreover, Wi-Fi and Bluetooth must be supported by these boards.

In Android Things OS structure, Linux kernel is used as the fundamental element (see Fig. 1). Linux is an operating system which supports virtual memory. This results in Android Things, which requires a processor that has virtual memory. Mostly microcontroller is being used in several IoT products having minimal memory and flash storage. Hence, Android Things prefers to work with ARM processors from

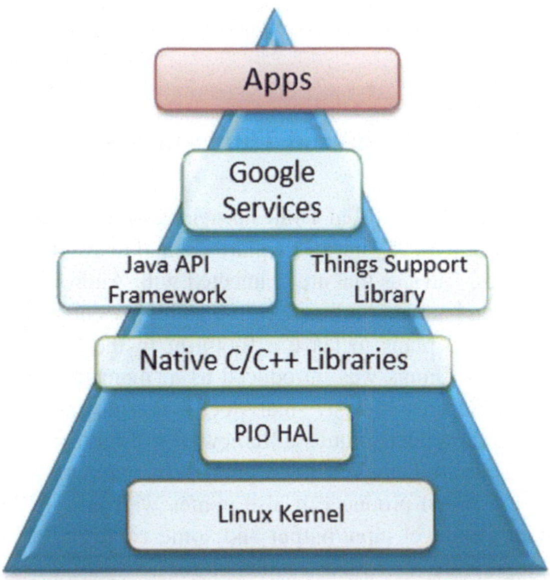

Fig. 1 Android things OS structure

Table 1 Different supported platforms and their specifications

Platform	Raspberry Pi 3	Intel Edison	NXP Pico i.MX6UL
CPU	Broadcom BCM 2837, 1.2 GHz Quad-core ARM cortex A53	Two atom silver mount cores, (500 MHz) One Intel Quark core (100 MHz)	Single-core ARM cortex A7 (528 MHz)
Memory	1 GB RAM	1 GB RAM	512 MB RAM
Storage	Micro SD card slot	SD card connector (4 GB EMMC)	SD card connector (4 GB EMMC)
USB	4 × USB 2.0 host	OTG with micro USB type AB connector	Standard USB 2.0 host connector
Display	HDMI Raspberry Pi touch display	HDMI	A touch and RGB TTL display
Interfaces	UART, I2C, SPI, PWM, GPIO	USB, GPIO, SPI, PWM, I2C	USB, I2C, UART, GPIO, PWM
Networking	10/100 Ethernet Wi-Fi 802.11n (2.4 GHz) Bluetooth 4.1	Bluetooth 4.0	10/100 Ethernet Wi-Fi 802.11bgn Bluetooth 4.0
Size	85 mm × 56 mm	35.5 mm × 25 mm × 3.9 mm	14 mm × 14 mm, 9 mm × 9 mm

Cortex-A range rather than any microcontroller from Cortex-M range. A detailed information is being given in Table 1.

With an open to all products, Google is aiming at low power devices, which has restricted storage workload and finite processing.

3 Roles of Android Things in IoT

In a nutshell, Android Things provides us with a well-modified operating system to function with any hardware platform and to efficiently perform on any low powered devices. Devices that are connected with Android Things are ensured with Google certified back-end infrastructure and highly enhanced security updates.

Google plans to leave a footprint in the IoT world through Android Things. Android Things was introduced as a 'managed Operating system' for smart IoT-based devices such as smart locks, thermostats, and much more. To build IoT-enabled products requisite frameworks of both software and hardware are provided by Android Things. It has a good pace compared to other OS in the development of IoT-based products. It also comes with new APIs and library to quickly work with low-level input/output and some common elements like display controllers, temperature sensors, and much more.

Now we would understand the importance of Android Things in IoT by having an overview of women security system using IoT and Android Things.

3.1 Women Security System Using IoT and Android Things

Women Security system ensures the safety of women, especially the late-night workers [17]. Mainly a band, loaded with the proximity sensors, heart rate sensors, motion sensors, GPS tracker, and GSM module, is provided to track the victim and ask for help to the pre-defined numbers. The basic approach can be divided into two sections. The mechanism of the transmitter part consists of activation of the device by just triggering the panic button (see Fig. 2a). Next, the GPS module tracks the location of the victim and the GSM module conveys both the location and a distress message to some pre-set contacts.

The receiver part is mainly used for retrieving the messages. We can use Raspberry Pi or typically a laptop to process and display the messages and the location of the user (see Fig. 2b).

Thus, women safety is ensured through the simplified women security system using IoT and Android Things. Particularly, Android things have been used in this project for some situations, where we are stuck in a place with the poor network. Android Things plays a vital role here. It supports both Bluetooth and Bluetooth low energy APIs, which provides significantly low power consumption, unlike Android.

4 Difference Between Android and Android Things

Though Android [21] and Android Things have the same thing, i.e., Linux at the core but still there exists some difference between these two like the former does not support APIs, whereas the latter supports. The structure layer of Android Things is much more compact and therefore, different from Android OS. Another difference is that the notification bar is not supported in Android Things app, which is supported in Android OS; for this reason, notifications cannot be triggered from the app. It is

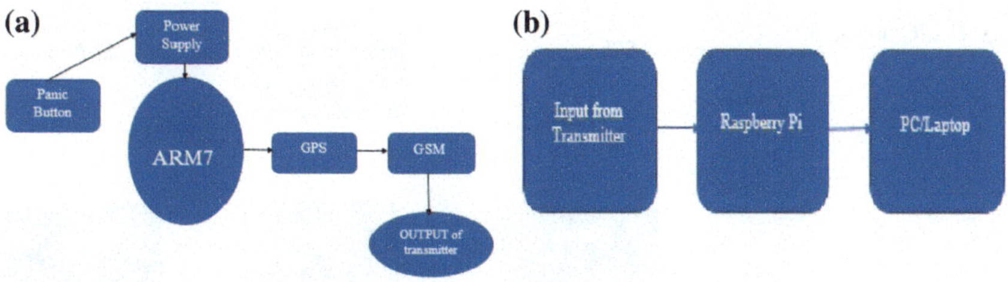

Fig. 2 **a** Transmitter block diagram. **b** Receiver block diagram

known to everyone that Android OS supports themes and styles, but it is not the case for Android Things app. Android apps require permissions from the user at the time of install, whereas Android Things apps do not have this feature, but it is present in Android 6 (API level 23).

5 Protocols

It can be challenging for IoT to create powerful connections between the gadgets. While being omnipresent, Wi-Fi and Bluetooth work pretty well in various situations; on the other hand, it also endures several restrictions. When a massive number of devices are involved in a single network, it results in a limited power supply. Different new communication technologies [17, 23, 24] are introduced to mitigate power consumption related problems and arrange the system in such a way which could solve the mentioned problems. Some of the few introduced protocols [25, 26] using Android Things are I2C, MQTT, CoAP, and LOWPAN.

5.1 I2C Protocol Overview

I2C, in other words, Inter-Integrated Circuit is a communicational protocol for transferring and receiving data serially using two wires. Philips developed this protocol in the 1980s to exchange vital data within various integrated circuits. As I2C protocol was improved a few times, there exist various protocols obtained from it. Though this updated protocol is almost equivalent to each other, the well-known is the SMBUS protocol by Intel. As we learned previously, this protocol uses two wires, namely, SCL, that is the serial clock wire and SDA, which is the serial data wire. Furthermore, this particular settlement uses two individual nodes, which are a Master node that triggers the clock signal and a Slave node which uses this generated clock signal from the master node to concur its work. The typical structure of this protocol is a master and one or more slaves connected to it (see Fig. 3). Also, there exist a handful of constructions, including more master and lesser slaves.

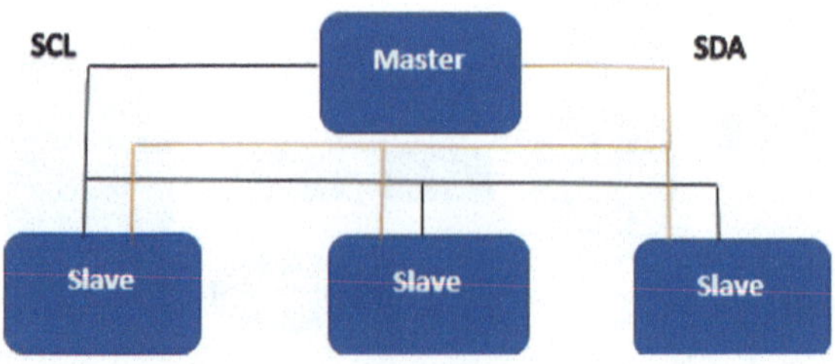

Fig. 3 I2C protocol structure

By observing this figure, we can easily realize the fundamental concept which is used to communicate [20, 28, 29, 30] the sensors with the Android Things board. Let us assume the Android Things board as the master that produces the clock signal while the sensors are the slaves that utilize the triggered clock. As there are several slaves connected to one master, every slave is assigned with a unique address that helps to detect the slaves individually. To understand communication through data among master and slaves, we have to follow these essential ways:

- The master initiates a start condition by triggering the clock and notifying all the slaves that the communication is going to start.
- The master conveys the slave a unique address with read (R) or write (W) command.
- The slave having the ID which is identical with the sent address, answers with an ACK signal.
- Data starts to trade between the master and the slave.
- Finally, when the communication is done, the master conveys a signal to terminate the transmission.

Since the protocol is straightforward and one can work with ease here, it is a broadly suggested and adopted protocol. As the crucial drawback of this I2C protocol is the speed, it connects with low-speed devices like various sensors and converters and much more.

5.2 *MQTT Protocol Overview*

Message Queue Telemetry Transport is a light-weighted messaging protocol. It was constructed around 1999 as an open settlement to mainly operate with Machine to Machine (M2M) data transferring in IoT. It is easy to use protocol with a small overhead. MQTT is a suitable approach for M2M communication with networks bandwidth restrictions, remote connections, and small code footprint.

We can separate the components into three parts according to their works in the message exchanging process (see Fig. 4):

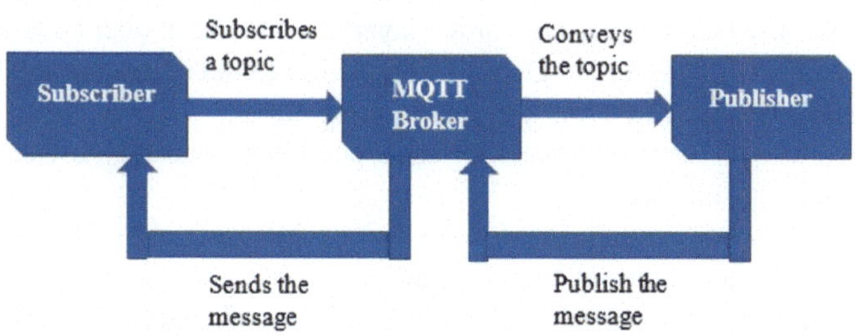

Fig. 4 MQTT protocol architecture

- Publishers: From this device, the message is generated conveying important information. Hence, we can consider this device as a source of information. Let us assume the IoT board as the publisher which retrieves the data from the sensors and passes it on.
- MQTT Broker: This is the crucial device that initiates the message flow within the publisher and the subscriber. It plays the role of a conveyor that collects data from the publishers and passes it on to the subscribers.
- Subscriber: In this device, the message or data is collected from the publisher through the brokers. Here we can assume that the Android Things board is acting as the subscriber.

A topic plays an essential role between the publisher and subscriber as a virtual channel to filter the requirements of a subscriber.

The open to all nature of this protocol and reliable yet straightforward features are exaggerating the pace of adopting the MQTT protocol. Moreover, various IoT cloud platforms have already declared it as their standard protocol in exchanging data from IoT boards.

Some common yet worldwide uses of MQTT protocol are as follows:

- Facebook messenger uses it for their online chat purpose.
- Amazon IoT is used combined with MQTT in Amazon Web Services.
- Several IoT platforms communicate between millions of products using MQTT to fulfill the requirements of the M2M communication.

5.3 CoAP Protocol Overview

CoAP stands for Constrained Application Protocol [31]. It is a web transfer protocol to work with constrained nodes and networks. CoAP was developed for M2M applications, namely, smart energy and automatic buildings based on the Request-Response model within endpoints. Interaction is done by asynchronous message exchange between the client and the server.

We can make two different layers out of this CoAP protocol that is Messages and Request/Response (see Fig. 5). The former layer is mainly built over UDP as it has a light mechanism which offers us great reliability. The Request/Response layer deals with the interaction based on the asynchronous messages.

CoAP works with four individual message types: Confirmable, Non-Confirmable, Acknowledgement, and Reset.

Each CoAP data or message is provided with its different unique ID, which helps us to recognize the duplicity of any message.

A Confirmable message is the reliable one while exchanging data. This strong message is received by using a particular command (CON) (see Fig. 6a). Opting to use these types of messages, clients can get the assurance that it would surely go to the server directly. A confirmable of reliable message is conveyed over and over

Fig. 5 CoAP protocol architecture

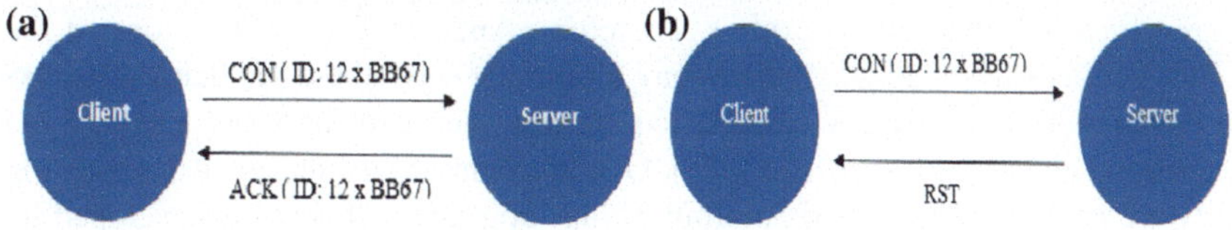

Fig. 6 **a** Conveying an acknowledge message. **b** Conveying a reset message

again unless a command of acknowledge message (ACK) obtained. The IDs of these two messages are equal to each other.

If the server faces complications to manage the incoming request, it conveys back a Reset message (RST) (see Fig. 6b).

The other message category left is the Non-confirmable message (NON) (see Fig. 7). Even being unreliable, these messages also have their unique ID.

CoAP plays a vital role to work smoothly with low-power sensors. Reducing the overheads and several complications it helps to make the working easier by resolving

Fig. 7 Conveying non-confirmable message

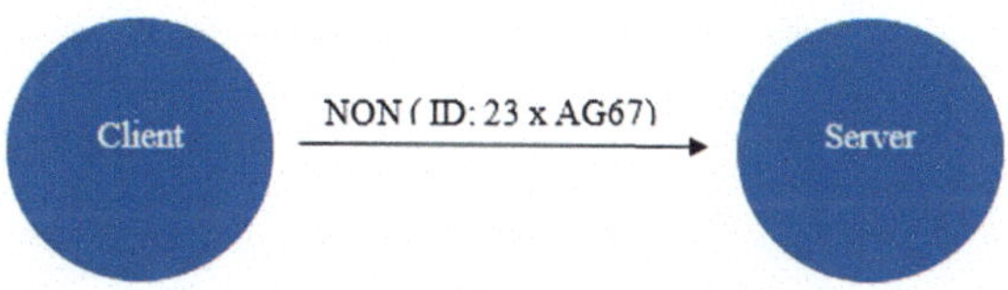

with various components. It also supports proxy and caching abilities to help in secure messaging with M2M communication.

5.4 Low PAN Protocol Overview

The Low-power Wireless Personal Area Networks API provides the developers to work with IP-based, low power, lossy networks at a smooth and fast pace [32]. It was developed to work with tiny gadgets having restricted processing capabilities that transfer information with the help of an Internet protocol. This protocol helps in peer–to-peer communication between various devices. Low PAN is produced as a segment of Android Things developer preview.

6 Advantages of Using Android Things Over Other OS

Android Things has initially been intended to be used on a wide range of hardware. Different IoT products or devices have a different way of communication techniques with other devices, and uses different protocols and hence consist of many security gaps which help hackers to exploit. Therefore, Android Things OS is one of the robust services from Google, which may ultimately solve these problems. This OS can even work with the devices that require low power to work and have limited processing and storage spaces [33, 21]. One of the main advantages is that existing software can be used to build Android Things app. Since this OS is developed by Google so it can be integrated with all the Google services like FireBase, Tensor Flow, and Google cloud platform with ease. This OS is speedy and does not require launchers as it directly boots into the application; this helps in reducing the boot time and memory occupancy. Another significant advantage is that it can reduce the cost of IoT application development. Hence it would be able to reduce overhead cost, and at the same time productivity would also increase. Android Things is very much secure and reliable than other software. It is also very much efficient in providing a hardcore defense against those malicious bugs. Another significant advantage is that any product built with Android Things would get 3 years of system updates automatically over the air, and it is free. Apart from all of these, Google has also provided a communication protocol named 'Weave'. With the help of this protocol, IoT appliances can share information without Wi-Fi.

7 Disadvantages of Using Android Things

Though it has many advantages, there persist some disadvantages also. One of the major disadvantages is that we all know this OS is basically developed for IoT applications, but recently, Google has announced that Android Things flourished in the field of smart speakers and displays [34, 35]. According to news articles, Google is not focusing on IoT with Android Things, the reason might be increasing competition in the IoT market as each, and everyone wants to invest in this sector. As a Google product, Android Things supports almost all kind of Google services, but it does not support those services which are applicable to the mobile world or requires authentication of users.

8 Applications of Using Android Things as an OS

The projects that are built using Android Things have a wide range of applications. However, its main application lies in the field of IoT, projects like automatic alarm systems, women safety devices, environment monitoring, and many things. Moreover, almost all kinds of IoT projects can be built with the help of Android Things. It has also collaborated with Byte flies to build wearable devices that would measure health data. Apart from these applications, Android Things can also be used in smart speakers like Google Home, Polka, and also in smart displays like Lenovo, LG displays, and many things [34]. Researchers also discovered that machine learning [36] could also be used along with Android Things to build projects and hence had opened the door to a new dimension in the field of IoT projects [37, 38].

9 Further Applications of Android Things

The interference of Google into the IoT world would have a hell lot of impact on society because the society would be now having a popular platform for programming merging with one of the largest and trustworthy companies in terms of everything. Google is still working on it for its betterment, and as per many researchers and analysts, it could spread its roots into many dimensions with ease and reliability. Some of the further applications of Android Things are given below:

- In the present scenario, the main focus of IoT is to building a smart motorway system which would reduce the traffic jams and congestion of vehicles on roads. Android Things would be a smarter approach as it is introduced to work with low powered devices like sensors and to operate in less time.

- Android Things software is open to all developers, and since we can use the old software to build Android Things apps so we can say that the Android kingdom would flourish because most of the apps used in smartphones are android supported; hence, this can be an excellent opportunity for Android Developers.
- Android Things can also be used in air quality monitoring systems. This system can measure parameters like humidity, temperature, carbon dioxide, carbon monoxide, and other harmful gases.

Apart from IoT applications, Google has confirmed that Android Things OS can also be used in smart displays like Lenovo, LG, and many things and smart speakers like Polk, iHome, and many things, the main applications of such type of devices are fit bands, smartwatches, home assistance, etc. [34].

10 Conclusions

Android Things was modeled to work with IoT on various gadgets. If anyone has such devices that are empowered by Android Things, communication among these devices would be a lot easier. Unlike Android, we won't be able to detect the existence of Android Things though it runs in the background. For low powered devices, Android Things is a more stable and enhanced operating system. As Google would provide direct updates, Android Things supported devices get a highly secured and fast up-gradation facility in every month. This dominant OS would reduce software development works up to 90% for the developers as the Android Things apps can be built with existing knowledge of Android. Although Google initially planned to use Android Things in IoT applications, now it is no longer focusing on it. As Android Things has achieved great success in smart speakers and displays powered by Google Assistant, it is refocusing this immensely powerful OS in the direction of OEM partners to build smart products in these categories. The reason can be a highly competitive market against custom-built solutions or more success in other directions. Hence, Android Things would only focus its working on smart displays and speakers rather than hardware-based IoT products according to the news.

Acknowledgements Under R&D Grant-in-Aid scheme, this work is financially supported by The Institution of Engineers (India) [UG2020026; 2019]. The authors also profoundly recognize the support of the Bengal Institute of Technology.

References

1. J.S. Banerjee, A. Chakraborty, A. Chattopadhyay, Fuzzy based relay selection for secondary transmission in cooperative cognitive radio networks, in *Proceedings of OPTRONIX* (Springer, 2017), pp. 279–287

2. J.S. Banerjee et al., Relay node selection using analytical hierarchy process (AHP) for secondary transmission in multi-user cooperative cognitive radio systems, in *Proceedings of ETAEERE* (Springer, 2018), pp. 745–754
3. I. Pandey, H.S. Dutta, J.S. Banerjee, WBAN: a smart approach to next generation e-healthcare system, in *Proceedings of ICCMC 2019* (IEEE, 2019), pp. 344–349
4. M. Chibuye, J. Phiri, A remote sensor network using android things and cloud computing for the food reserve agency in Zambia. Int. J. Adv. Comput. Sci. Appl. (IJACSA) **8**(11), 411–418 (2017)
5. T. Cho, H. Kim, J.H. Yi, Security assessment of code obfuscation based on dynamic monitoring in android things. IEEE Access **5**, 6361–6371 (2017)
6. http://www.gartner.com/newsroom/id/3165317l. Accessed May 2019
7. S. Paul et al., A fuzzy AHP-based relay node selection protocol for wireless body area networks (WBAN), in *Proceedings of OPTRONIX 2017* (IEEE, 2018), pp. 1–6
8. S. Paul et al., The extent analysis based fuzzy AHP approach for relay selection in WBAN, in *Proceedings of CISC 2018* (AISC-Springer, 2018), pp. 331–341
9. J.S. Banerjee, A. Chakraborty, D. Goswami, A Survey on agri-crisis in India based on engineering aspects. Int. J. Data Model. Knowl. Manage. **3**(1–2), 71–76 (2013)
10. F. Azzola, *Android Things Projects* (Packt Publishing Ltd. 2017)
11. J. Banerjee, S. Maiti, S. Chakraborty, S. Dutta, A. Chakraborty, J.S. Banerjee, Impact of machine learning in various network security applications, in *Proceedings of ICCMC 2019* (IEEE, 2019), pp. 276–281
12. A. Chakraborty, J.S. Banerjee, A. Chattopadhyay, Non-uniform quantized data fusion rule for data rate saving and reducing control channel overhead for cooperative spectrum sensing in cognitive radio networks. Wireless Pers. Commun. **104**(2), 837–851 (2019)
13. J.S. Banerjee, A. Chakraborty, A. Chattopadhyay, Reliable best-relay selection for secondary transmission in co-operation based cognitive radio systems: a multi-criteria approach. J. Mech. Contin. Math. Sci. **13**(2), 24–42 (2018)
14. A. Chakraborty, J.S. Banerjee, A. Chattopadhyay, Non-uniform quantized data fusion rule alleviating control channel overhead for cooperative spectrum sensing in cognitive radio networks, in *Proceedings of IACC2017* (IEEE, 2017), pp. 210–215
15. O. Saha, A. Chakraborty, J.S. Banerjee, A fuzzy AHP approach to IT-based stream selection for admission in technical institutions in India, in *Emerging Technologies in Data Mining and Information Security* (Springer, Singapore, 2019), pp. 847–858
16. O. Saha, A. Chakraborty, J. S. Banerjee, A decision framework of IT-based stream selection using analytical hierarchy process (AHP) for admission in technical institutions, in *Proceedings of Optronix2017* (IEEE, 2017), pp. 1–6
17. M. Pavithra, S. Ashikha, D. Sharmila, Women security using IOT and android things (2018)
18. A. Chakraborty, J.S. Banerjee, A. Chattopadhyay, Malicious node restricted quantized data fusion scheme for trustworthy spectrum sensing in cognitive radio networks. J. Mech. Contin. Math. Sci. **15**(1), 39–56 (2020)
19. http://developer.android.com/reference. Accessed May 2019
20. https://developer.android.com/things/hardware/index.html. Accessed May 2019
21. https://source.android.com/source/index.html. Accessed May 2019
22. J.S. Banerjee, A. Chakraborty, A. Chattopadhyay, A novel best relay selection protocol for cooperative cognitive radio systems using fuzzy AHP. J. Mech. Contin. Math. Sci. **13**(2), 72–87 (2018)
23. J. Chattopadhyay, S. Kundu, A. Chakraborty, J.S. Banerjee, Facial expression recognition for human computer interaction, in *Proceedings of ICCVBIC 2018* (Springer (press), 2018)
24. J.S. Banerjee, D. Goswami, S. Nandi, OPNET: a new paradigm for simulation of advanced communication systems, in *Proceedings of International Conference on Contemporary Challenges in Management, Technology & Social Sciences, SEMS* (Lucknow, India, 2014), pp. 319–328
25. J.S. Banerjee, K. Karmakar, A hierarchy of micro and network mobility protocols, in *Proceedings of IEMCON2011* (IEEE, 2011), pp. 323–328

26. K. Karmakar, J.S. Banerjee, Different network micro-mobility protocols and their performance analysis. Int. J. Comput. Sci. Inf. Technol. **2**(5), 2165–2175 (2011)
27. J.S. Banerjee, A. Chakraborty, Fundamentals of software defined radio and cooperative spectrum sensing: a step ahead of cognitive radio networks, in *Handbook of Research on Software-Defined and Cognitive Radio Technologies for Dynamic Spectrum Management*, ed. by N. Kaabouch, W. Hu (IGI Global, USA, 2015), pp. 499–543
28. J.S. Banerjee, A. Chakraborty, Modeling of software defined radio architecture & cognitive radio, the next generation dynamic and smart spectrum access technology, in *Cognitive Radio Sensor Networks: Applications, Architectures, and Challenges*, ed. by M.H. Rehmani, Y. Faheem (IGI Global, USA, 2014), pp. 127–158
29. J.S. Banerjee, A. Chakraborty, K. Karmakar, Architecture of cognitive radio networks, in *Cognitive Radio Technology Applications for Wireless and Mobile Ad Hoc Networks*, ed. by N. Meghanathan, Y.B. Reddy (IGI Global, USA, 2013), pp. 125–152
30. J.S. Banerjee, K. Karmakar, A comparative study on cognitive radio implementation issues. Int. J. Comput. Appl. **45**(15), 44–51 (2012)
31. https://dzone.com/articles/coap-protocol-step-by-step-guide. Accessed May 2018
32. https://android-developers.googleblog.com/2017/12/lowpan-on-android-things.html?m=1. Accessed 2017
33. https://www.linkedin.com/pulse/android-things-iot-os-battles-%C5%9Famil-beden. Accessed 2017
34. https://www.androidpolice.com/2019/02/12/google-gives-up-on-android-things-as-an-iotplatform-now-will-just-be-for-smart-speakers-and-displays/. Accessed May 2019
35. https://arstechnica.com/gadgets/2019/02/android-things-is-no-longer-for-things-focuses-onsmart-speakers-and-displays/. Accessed May 2019
36. A. Chakraborty, J.S Banerjee, An Advance Q Learning (AQL) approach for path planning and obstacle avoidance of a mobile robot. Int. J. Intell. Mechatron. Rob. **3**(1), 53–73
37. D. Das, I. Pandey, A. Chakraborty, J.S. Banerjee, Analysis of implementation factors of 3D printer: the key enabling technology for making prototypes of the engineering design and manufacturing. Int. J. Comput. Appl. 8–14 (2017)
38. D. Das, I. Pandey, J.S. Banerjee, An in-depth study of implementation issues of 3D printer, in *Proceedings of MICRO 2016 Conference on Microelectronics, Circuits and Systems* (2016), pp. 45–49

Energy-Aware VM Migration in Cloud Computing

Shashi Bhushan Singh Yadav and Mala Kalra

Abstract The continuous growth of cloud data centers is accompanied by enormous amounts of energy consumption leading to carbon dioxide (CO_2) emissions making the environment unfriendly. Dynamic VM consolidation is an effective policy that can reduce energy consumption in data centers. It reduces the number of servers used complying with Quality of Service (QoS) constraints. This paper presents two-phase approach: First, Power-Aware Placement of VMs provides the least increase in power consumption. Second, a function of CPU utilization and memory utilization with double-threshold policy is used to estimate host current utilization. It chooses VMs for migration based on their CPU and memory utilization, thus reducing the chances of SLA violation and number of migrations as minimum as possible. The energy consumption is optimized by controlling the resource utilization and shifting the idle servers to sleep state. Simulation results depict that our proposed approach significantly reduces energy consumption in dynamic workload scenarios when compared with other algorithms.

Keywords Cloud computing · VM consolidation · VM placement · Energy-aware · SLA violation

1 Introduction

Cloud computing has emerged as a new model that delivers platform, infrastructure, and software applications as a service, which are readily available to customers around the globe on the basis of the pay-as-you-go model. The popular IT firms, such as Amazon, Facebook, and Google comprises huge data centers all over the world to provide services of cloud while managing the growing data of diverse applications. These cloud data centers consume extreme quantity of energy which

S. B. S. Yadav (✉) · M. Kalra
Department of Computer Science & Engineering, NITTTR, Chandigarh, India
e-mail: shashiyadav38@gmail.com

M. Kalra
e-mail: malakalra2004@gmail.com

M. Dutta et al. (eds.), *Proceedings of International Conference on IoT Inclusive Life (ICIIL 2019), NITTTR Chandigarh, India*, Lecture Notes in Networks and Systems 116,
https://doi.org/10.1007/978-981-15-3020-3_32

results in high operating cost for cloud service providers. In 2014, the environmental protection agency reported 100 billion kWh energy consumption of U.S. data center while infrastructure energy cost was accounted to be 75% of total operating cost [1].

Hence, to improve the energy efficiency of data centers and to minimize the environmental degradation, there is a need to make efficient use of cloud resources with minimum SLA violation by reducing the active servers to a minimum. VM migration is one of the measures to improve the server's performance and maximize energy efficiency. It is a method of migrating VMs from over-utilized hosts to under-utilized hosts to balance hosts utilization [2]. The VM migration problem is divided into two parts: (i) VM placement and (ii) VM selection. The placement phase requires deciding which VM to be deployed to which host according to their utilization, and selection phase requires detecting overloaded and underloaded hosts to decide which VM is to be moved from an overloaded host and where to be placed.

However, VM migration can lead to performance degradation in many ways such as increased response times, failures, time-outs, etc. Thus, there is a need to devise an optimized solution so that QoS demands can be met.

The objective of the proposed study is to minimize the energy consumption of data center resources with reduced number of migrations and less SLA violations. For this, we have presented VM placement and VM selection approaches with double-threshold policy to identify the utilization levels of servers. The contribution of our work is as follows:

- Power-aware placement algorithm is designed for VM allocation to available servers in data centers.
- The energy consumption has been minimized through new VM selection policy that performs minimum migrations, thereby optimizing the server utilization and SLA violations.
- Simulation results depict the proposed algorithms that minimize energy consumption, number of migrations, and SLA violations, comparing it with other techniques.

The rest of the paper is organized as follows: Sect. 2 describes related work; Sect. 3 explains the problem formulation and proposed solution; Sect. 4 discusses the performance evaluation, experimental setup and results and analysis whereas Sect. 5 concludes the research with the future scope.

2 Background Details and Related Work

Beloglazov et al. [3] have presented the green cloud model for efficient energy management and proposed two well-known algorithms for VM migration. The MBFD algorithm allocates the VMs such that there is minimal increment in power consumption of host and further this allocation is optimized by monitoring host's current utilization and migrating some or all VMs from overloaded and underloaded host to provide the normal utilization of each host.

Zhou et al. [4] have presented the energy-saving algorithm that depends on three-threshold scheme for identifying the resource utilization according to its workload. The authors have proposed five VM selection policies for migrating VMs and among the five policies, MIMT is proved to be the best policy for efficiently migrating the VMs for decreasing energy consumption.

Wei et al. [5] have presented two problems in VM live migration such as to determine the best VM for migration and selection of best host to reallocate the VM after migration. The two models are presented for energy-aware migration and energy-efficient load dispatch model.

Ferreto et al. [6] have presented an algorithm for VM consolidation to reduce number of migrations. For this, the authors presented a linear programming based heuristic to reduce the migrations and prioritizing the VM that should not be migrated if it is running with a steady workload.

Shribman et al. [7] have presented a problem of live migration failure due to inefficient migration of memory intensive application. Several approaches have been proposed for pre-copy and post-copy live migration to control the memory-based failures during live migration.

Monil and Rahman [8] have proposed a Fuzzy logic based VM consolidation technique to meet QoS. The fuzzy model is used to choose VM from an overloaded server incorporating a fuzzy migration control method and for optimization of the VM selection.

Xiao et al. [9] have proposed VM consolidation to allocate resources dynamically in cloud data center. The concept of skewness is presented to combine various types of workloads to minimum number of servers and improve overall server utilization to prevent overloading and a large amount of energy consumption.

3 Problem Formulation and Proposed Solution

3.1 Power Consumption Model

The consumption of power by hosts in cloud data center comprises the CPU, memory, storage, disk, and network resources utilization. Out of these, CPU and memory utilization are major factors of power consumption. When a host is running in its idle state, i.e., zero workload condition it consumes 70% of the maximum power at busy state [10]. The power consumption during an idle period is termed as static energy consumption while power consumption at a busy period is termed as dynamic energy consumption. Hence, the linear relationship of the host with CPU and memory utilization is shown in Eq. 1.

$$P(u) = P_{idle} + (P_{max} - P_{idle}) \times (u) \tag{1}$$

where

$P(u)$ Total power consumption by host at utilization u
P_{idle} Power consumed at idle state
P_{max} Power consumed during busy state (maximum utilization).

Due to variable workload, the utilization of host can vary with respect of time, and host utilization can be seen as a function of time.

Hence, the consumption of energy of a host is computed as the integration of total power consumed during host total utilization with respect to change in time.

$$EC = \int_{t1}^{t2} P(u(t))\Delta t \quad (2)$$

3.2 Problem Description

VM placement and migration have always been a challenging task for the last couple of years. If appropriate VM is not selected it may result in SLA violation or may increase the number of migrations which may further negatively affect the energy consumption. Therefore, VM placement problem is to find an optimal allocation of $\text{VMs} = \{\text{vm}_1, \text{vm}_2, \text{vm}_3 \ldots \text{vm}_m\}$ to the $\text{Hosts} = \{\text{h}_1, \text{h}_2, \text{h}_3 \ldots \text{h}_n\}$ with distinct resources such as CPU, memory, disk, and storage in cloud data center. Thus, optimizing the total energy consumption of allocation is performed while keeping the VM migration and SLA violation as minimum as possible. Mathematically it can be formulated as

$$\begin{aligned}&\text{Given: (VMs, Hosts)}\\&\text{Minimize: } A = (\text{EC}, \text{Mig}, \text{SLA}_\text{v})\end{aligned}$$

where A is an allocation with minimum energy consumption (EC) such that number of migrations (Mig), and SLA violation (SLA_v) are minimum.

3.3 Proposed Approach

The proposed algorithm efficiently utilizes the VM consolidation problem which comprises VM placement, host workload detection, and VM migration in the available computing resources in the cloud data center and thus minimizes the energy consumption. When VMs are allocated to host machines, the host machines become overloaded or underloaded. In such a case, VM selection and migration algorithms are required which can migrate the virtual machines to the other hosts with specific

VM placement strategy without significant loss of performance degradation and minimizing the energy consumption while meeting SLA parameter agreed with the user. By discovering the workload and utilization of currently running hosts by several VMs of different processing capacities, the number of active hosts can be reduced after migrating some or all of the VMs from heavily loaded and lightly loaded hosts, respectively. VM allocation can be categorized into two phases:

VM Placement: This is the phase when the user sends its requests for computing its application that requires VM provisioning and allocation of VMs on heterogeneous hosts available in the data center. To solve this problem, the proposed approach applies the enhancement to the existing algorithm MBFD [3]. In our approach, we sort all VMs according to their CPU and memory utilization requirements and find a host that leads to a minimal increase in power consumption with this allocation as shown in Algorithm 1.

Algorithm 1: Power-Aware Placement (PAP)

```
Input: Host_List, VM_List
Output: VM_placement
Sort VM_List in decreasing order of CPU utilization and Memory utilization
for each VM in VM_List do
VM_cpu ← CPU required by VM
VM_mm ← memory required by VM
min_power ← maximum power
Allocated_host ← NULL
for each Host in Host_List do
if VM_cpu <= Host available CPU && VM_mm <= Host available Memory then
    place the selected VM on Host
    utilized_power ← estimated power (Host, VM)
    if utilized_power < min_power then
      min_power ← utilized_power
      Allocated_host ← Host
    end if
 end if
end for
if Allocated_host ≠ NULL then
    place VM to Allocated_host
end if
end for
return VM_placement
```

VM Selection: This is the phase where optimization is being performed on the current VM placement for the effective utilization of hosts. The first part can be seen as a significant problem of finding the minimum number of hosts for VMs with various CPU, memory, and power utilization factors (Algorithm 1). To further optimize this allocation, we need to identify hosts whose utilizations are not efficient and this utilization is computed depending on the total utilization by the VMs allocated to a host. The total utilization of a host can be divided into three classes: overloaded host, average-loaded host, and underloaded host.

Algorithm 2: Minimum Sum of CPU Utilization and Memory Size (MSCM))

```
Input: Host_List
Output: Migration_List
for each Host in Host_List do
Host_util ← Host utilization
VM_List ← allocated VM on Host
sort VM_List in increasing order of CPU utilization and Memory utilization
if Host_util > UP_TH then
  D ← Host_util - UP_TH
  for each VM in VM_List do
    VM_util ← VM utilization
    if VM_util ≥ D then
      Host_util ← Host_util - VM_util
      add VM to Migration_List
      VM_List = VM_List - VM
    end if
  end for
end if
if Host_util ≤ LOW_TH then
  add VM_List to Migration_List
  VM_List ← NULL
  switch host to sleep mode
end if
if Host_util > LOW_TH && Host_util < UP_TH
  Return Migration_List ← NULL
end if
end for
return Migration_List
```

Double-Threshold Policy: To determine the current utilization of a host, the upper, lower, and average utilization thresholds are set for a host. The basic idea is to maintain the overall utilization of a host between these thresholds. These three thresholds can be explained as follows:

(i) Upper-Threshold: If the host current utilization is above the maximum value of upper utilization threshold then few VMs are migrated to the less utilization host to avoid performance degradation. It can be defined as

$$Host_{Util} > UP_{TH} \tag{3}$$

where $Host_{Util}$ is the host current utilization and UP_{TH} is the upper threshold.

(ii) Average-Threshold: If the host current utilization is equal to average threshold, then the migration of VM remains unchanged as the host is running at its moderate utilization. It can be defined as

$$LOW_{TH} < Host_{Util} \leq UP_{TH} \tag{4}$$

where LOW_{TH} is the lower threshold.

(iii) Lower-Threshold: If the host current utilization is under the minimum value of lower utilization threshold then all the VMs are migrated to the average utilization host and the current host is turned to sleep mode to save energy consumption. It can be defined as

$$Host_{Util} < LOW_{TH} \tag{5}$$

VM Selection Policy

Algorithm 2 describes the VM selection policy to transfer VM from the overloaded and underloaded hosts. The host utilization and total number of VMs assigned to that host are computed (Lines 1–3). Then these VMs are being arranged in increasing sequence of their CPU and memory utilization (Line 4). After sorting VMs the host upper utilization violation is computed, so that from its allocated VMs such VMs can be selected for migration which can bring down its utilization below the upper threshold (Lines 5–6).

In terms of CPU utilization, the VM with low CPU utilization is selected in comparison to high CPU utilization to minimize migration time. During live migration transferring CPU state and memory state of a VM are the two most important factors that need to consider [11].

For example, if a host has two VMs, one with 4 GB memory and another with 2 GB memory, then a VM having a memory size of 2 GB will be selected for transfer to minimize the migration time [12].

Thus, in Lines 8–9 both CPU utilization and memory utilization are considered for the suitable VM selection on the host. As memory state is more significant than CPU state, MSCM policy assigns weightage (w_i) of 40% to CPU utilization and 60% to memory utilization of VM during its selection, as shown in Eq. 6.

$$VM[w_1(CPU_{Util}\,Size) + w_2(Mem_{Util}\,Size)] \tag{6}$$

In the case of overloaded host, a VM with low CPU utilization and low memory usage will be selected for migration in comparison to other VMs allocated to that host. The selection of such VM satisfies the following condition as shown in Eq. 7.

$$VM[w_1(CPU_{Util}\,Size) + w_2(Mem_{Util}\,Size)] \geq Host_{Util} - UP_{TH},\ \forall VM_i \in Host_j \tag{7}$$

where VM_i is the group of all allocated VMs on $Host_j$.

Let VM_i be a group of all VMs currently running on $Host_j$, MSCM policy finds a set |Mig| that consists of the number of VM to be migrated at different threshold violations as defined in Eq. 8.

$$|Mig| = \begin{cases} min(VM_i),\ Host_{Util} > UP_{TH} \\ \forall(VM_i),\ Host_{Util} \leq LOW_{TH} \\ Null,\ Otherwise \end{cases} \tag{8}$$

Lines 10–12 describes updating of host utilization after VM migration and return a migration list comprising of VMs that are required for migration. If a host is underloaded then all the VMs are appended to the migration list and the respective host is turned into sleep mode for saving idle energy consumption (Lines 16–20). Lastly, if a host is operating at a normal utilization then there will be no VM migration and a host is considered at its best-fit utilization (Lines 21–25).

4 Performance Evaluation

To validate the proposed algorithm, we present the comparative analysis of MSCM algorithm with RC and HPG [3]. For the simulation of the proposed approach, we choose MATLAB as a simulation platform. MATLAB is a widely used development environment to simulate large-scale cloud applications and services. For evaluating the feasibility of the proposed algorithm, three comparison metrics are chosen as described below.

4.1 Performance Metrics

- **Number of Migrations**: It is defined as the number of migrations performed from overloaded or underloaded host to best-fit utilization host.
- **Energy Consumption**: It is defined as the total energy consumed by host due to its utilization by allocated VMs as shown in Eq. 9.

$$Energy\ Consumption = \sum_{j=1}^{n} EC_j \tag{9}$$

  where n is the number of hosts and EC signifies the total energy consumption of jth host.
- **SLA Violation**: The failure of providing Quality of Service is called SLA violation. There are two causes of SLA violation [13]:

  1. SLA violation due to overloading (SLA_{OL}): It is defined as the ratio of time the host experienced the state of overloading to the time host remained active as shown in Eq. 10.

$$SLA_{OL} = \frac{1}{n}\sum_{i=1}^{n}\frac{TOL_i}{TA_i} \tag{10}$$

     where TOL is the time of overloading, TA is the host active time and n shows the number of hosts.

2. Performance Degradation due to Migration (PDM): It is the ratio of performance degradation in utilization by VM during migration to the whole capacity requested by VM as shown in Eq. 11.

$$PDM = \frac{1}{m}\sum_{j=1}^{m}\frac{0.1 \times u_j(t)dt}{CR_j} \tag{11}$$

where $u(t)$ is the VM utilization at time (t) with estimation of 10% performance degradation during migration, *CR* is the requested capacity of VM and m is the number of VMs. Thus the combined metric for SLA violation can be computed as defined in Eq. 12.

$$SLA_V = SLA_{OL} \times PDM \tag{12}$$

4.2 Experimental Setup

It is assumed that a user submits requests for VM provisioning that can accomplish their applications' different workloads. Initially, the VMs are allocated to hosts as per their utilization demands. The maximum power consumption of host is considered to be 250 W at maximum utilization and when it is idle, the power consumption is 175 W which is significant idle energy consumption to be minimized.

In proposed study, we have assumed a data center which comprises 200 hosts and 615 VMs. The CPU capacity of hosts is varied from 1000 to 3000 MIPS and memory capacity is 8 GB. The bandwidth of each host is 1000 Mbps. The CPU capacity of each VMs is varied from 250 to 1000 MIPS with memory requirement of 128 MB.

4.3 Results and Analysis

The comparative evaluation of MSCM algorithm is presented with RC and HPG [3] by varying the number of VMs and hosts. The experiment is executed 10 times and the results obtained are the average of these 10 executions.

The overall comparative results from our evaluation study are summarized in Table 1 for different algorithms with respect to different number of hosts and VMs.

With the variation of host and VMs, the minimum numbers of migrations by MSCM are 16.32% while maximum numbers of migrations are 42.85% than RC. In comparison to HPG the minimum and maximum numbers of migrations by MSCM are 24.77% and 55.55%, respectively.

Similarly, the minimum energy consumption for MSCM is 9.78% and maximum energy consumption is 38.36% than RC while in comparison to HPG the minimum

Table 1 Simulation results with variation in hosts and VMs

Algorithm	Total hosts	Total VMs	VM migrations	Energy consumption (kWh)	SLA violation (%)
MSCM	20	30	4	0.723	0.3
	60	160	32	3.723	1.8
	100	290	47	5.481	3.3
	140	420	55	6.829	5.2
	180	550	76	9.932	6.7
	200	615	82	10.541	7.3
RC	20	30	7	1.173	0.8
	60	160	38	4.543	2.4
	100	290	57	6.272	4.4
	140	420	67	7.978	6.8
	180	550	87	10.709	8.2
	200	615	98	11.684	8.9
HPG	20	30	9	1.682	1.1
	60	160	44	4.974	2.9
	100	290	68	7.877	4.8
	140	420	79	9.862	7.2
	180	550	98	12.245	9.1
	200	615	109	13.041	9.8

energy consumed by MSCM is 19.17% and maximum energy consumed by MSCM is 57.01%.

When number of host and VMs are varied, MSCM contributes to least percent of SLA violation. The minimum SLA violation by MSCM is 17.97% and maximum SLA violation is 62.5% than RC while in comparison to HPG the minimum SLA violation by MSCM is 25.51% and maximum SLA violation by MSCM is 72.72%.

Figure 1 depicts the number of VM migrations performed. It is observed that MSCM reduces the number of migrations by 16.38% than RC and 27.34% than HPG with the variation of hosts and VMs. Figure 2 represents the energy consumption using three VM selection algorithms. MSCM saves 12.18% energy than RC and 25.12% than HPG with the variation of hosts and VMs. Figure 3 represents the reduction in SLA violation with the variation of hosts and VMs. MSCM shows an improvement in minimizing SLA violation by 21.90% and 29.55% when compared to RC and HPG, respectively.

From the obtained results, we can conclude that MSCM outperforms the other two algorithms with respect to number of VM migrations, energy consumption, and SLA violation.

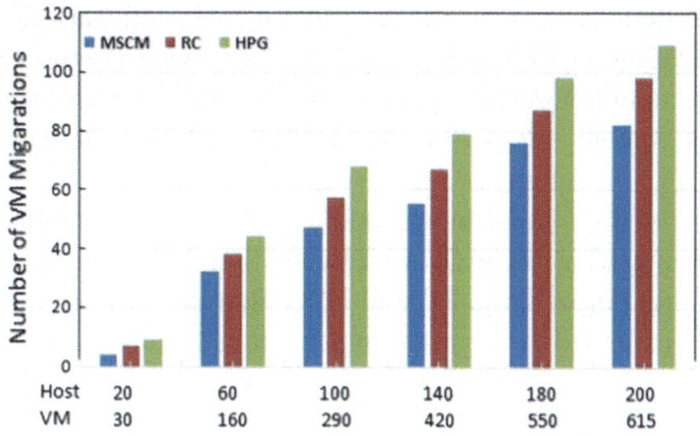

Fig. 1 Analysis of VM migrations with variation in hosts and VMs

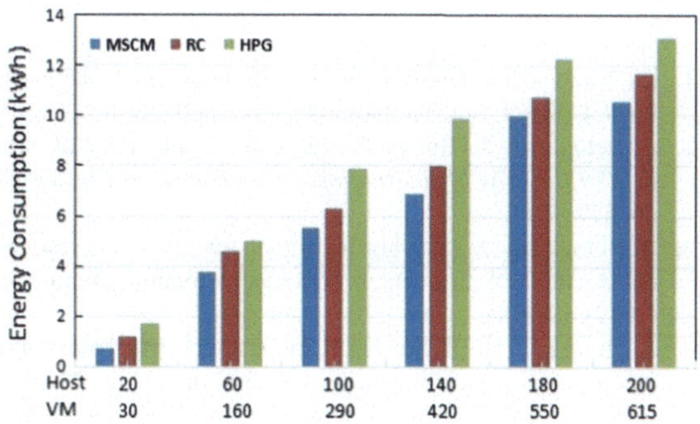

Fig. 2 Analysis of energy consumption with variation in hosts and VMs

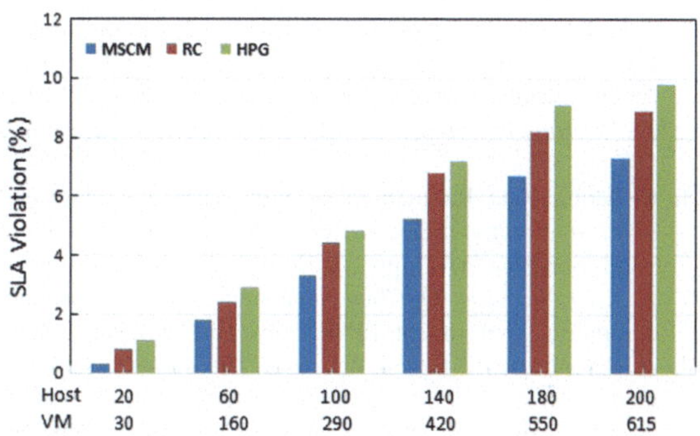

Fig. 3 Analysis of SLA violation with variation in hosts and VMs

5 Conclusion and Future Work

Energy consumption in cloud data centers is one of the most challenging issues. Therefore, we have proposed the novel approach of VM placement and VM migration for efficient handling of resource utilization which results in low energy consumption by the cloud resources. PAP algorithm addresses the significant problem of finding the appropriate hosts for VMs with varied CPU, memory, and power utilization requirements while MSCM is used to optimize the host utilization followed by double-threshold policy to determine if the host is in overloading, underloading, or average utilization state. MSCM sort VMs in increasing order with respect to their CPU utilization and memory utilization from overloaded hosts to minimize the number of migrations thus reducing the energy consumption with minimum performance degradation due to SLA violation. Simulation results demonstrate that MSCM algorithm outperforms RC and HPG algorithms. In future, the proposed algorithm can be extended by incorporating machine learning techniques to improve its efficiency.

References

1. I.S. Dhanoa, S.S. Khurmi, Energy-efficient virtual machine live migration in cloud data centers. Int. J. Comput. Sci. Technol **5**(1), 43–47 (2014)
2. A. Beloglazov, R. Buyya, Energy efficient allocation of virtual machines in cloud data centers, in *10th IEEE/ACM International Conference on Cluster, Cloud and Grid Computing* (2010), pp. 577–578
3. A. Beloglazov, J. Abawajy, R. Buyya, Energy-aware resource allocation heuristics for efficient management of data centers for cloud computing. Futur. Gener. Comput. Syst. **28**(5), 755–68 (2012). Elsevier
4. Z. Zhou, Z.G. Hu, T. Song, J.Y. Yu, A novel virtual machine deployment algorithm with energy efficiency in cloud computing. J. Cent. South Univ. **22**(3), 974–983 (2015)
5. B. Wei, C. Lin, X. Kong, Energy optimized modeling for live migration in virtual data center, in *Proceedings of International Conference on Computer Science and Network Technology (ICCSNT)*, vol. 4 (IEEE, 2011), pp. 2311–2315
6. T. Ferreto, M.A.S. Netto, R.N. Calheiros, C.A.F. De Rose, Server consolidation with migration control for virtualized data centers. Futur. Gener. Comput. Syst. **27**, 1027–1034 (2011)
7. A. Shribman, B. Hudzia, Pre-copy and post-copy VM live migration for memory intensive applications, in *European Conference on Parallel Processing* (Springer, Berlin, 2013), pp. 539–547
8. M. Monil, R.M. Rahman, Fuzzy logic based energy aware VM consolidation, in *Proceedings of 8th International Conference on Internet and Distributed Computing Systems (IDCS)*, vol. 9258 (Springer, 2015), pp. 31–38
9. Z. Xiao, W. Song, Q. Chen, Dynamic resource allocation using virtual machines for cloud computing environment. IEEE Trans. Parallel Distrib. Syst. **24**(6), 1107–1117 (2013)
10. A. Beloglazov, R. Buyya, Y.C. Lee, A. Zomaya, A taxonomy and survey of energy-efficient data centers and cloud computing systems. Adv. Comput. **82**(11), 47–111 (2011). Elsevier
11. A. Al-Dulaimy, W. Itani, R. Zantout, A. Zekri, Type-aware virtual machine management for energy efficient cloud data centers. Sustain. Comput.: Inform. Syst. **19**, 185–203 (2018). Elsevier
12. Z. Zhou, Z. Hu, K. Li, Virtual machine placement algorithm for both energy-awareness and SLA violation reduction in cloud data centers. J. Sci. Program. **2016**(1), 1–11 (2016). Hindawi, New York
13. A. Beloglazov, R. Buyya, Optimal online deterministic algorithms and adaptive heuristics for energy and performance efficient dynamic consolidation of virtual machines in cloud data centers. Concurr. Comput.: Pract. Exp. **24**(13), 1397–1420 (2012). Wiley

Deadline Constrained Energy-Efficient Workflow Scheduling Heuristic for Cloud

Shalu Saharawat and Mala Kalra

Abstract Cloud computing facilitates execution of large-scale scientific workflow applications by providing heterogeneous virtualized resources that can be provisioned dynamically. The proliferation of cloud data centers has introduced a serious challenge of enormous energy consumption. Hence, the key concern for executing performance-driven workflows is to devise a scheduling heuristic which can improve utilization of cloud resources and thus helps to reduce power dissipation while adhering to the user's Quality of Service (QoS) demands. In this paper, we propose a scheduling approach aiming for optimization of makespan, resource utilization, and energy consumption under given deadlines. The proposed heuristic D-DEWS exploits the novel policy of list-based scheduling incorporated with dynamic voltage and frequency scaling (DVFS). It scales the discrete operating frequencies of resources to minimum possible levels, such that a given workflow gets completed within a user-defined deadline with minimum energy consumption. Simulation conducted with synthetic workflows demonstrates the proficiency of proposed heuristic in achieving a significant trade-off between energy savings and performance with deadline compliance. The results obtained confirm that the proposed heuristic outperforms other state-of-the-art algorithms.

Keywords Cloud computing · Energy efficient · Workflow scheduling · Heterogeneous · DVFS · Heuristic algorithm

1 Introduction

The last few decades have shown tremendous growth in computing demands which impel the entrepreneurs to shift their enterprise workloads to large-scale cloud data centers as it offers on-demand provisioning of the heterogeneous resource pool,

S. Saharawat (✉) · M. Kalra
Department of Computer Science & Engineering, NITTTR, Chandigarh, India
e-mail: shalu.nitttrchd@gmail.com

M. Kalra
e-mail: malakalra2004@gmail.com

M. Dutta et al. (eds.), *Proceedings of International Conference on IoT Inclusive Life (ICIIL 2019), NITTTR Chandigarh, India*, Lecture Notes in Networks and Systems 116,
https://doi.org/10.1007/978-981-15-3020-3_33

elasticity along with the low economical cost. However, cloud data centers consume a lot of energy leading to footprints of carbon dioxide (CO_2) emissions making the environment unfriendly. Thus to manage this energy consumption, data centers require additional expenses in the deployment of cooling systems, PDU, UPS, etc., which results in the high total cost of ownership (TCO) for cloud providers. A recent study estimates that the information technology (IT) sector is consuming 7% of the global electricity, and if the current IT usage trends continue, then surely the energy consumption is expected to be raised by 13% in 2030 [1]. Therefore, Green cloud computing is now a topic of concern for cloud service providers (CSPs) to minimize large investment in data centers and moving toward sustainable development. One of the appropriate measures to minimize energy dissipation in large-scale cloud data centers is the virtualization technology. Virtualization is playing a significant role in physical servers-consolidation to minimum as it enables the installation of several virtual machines (VMs) onto a single server which allows efficient management of data centers with a minimum number of resources and maximum workload handling.

Further, techniques like DVFS have been proven to be very beneficial with its capability of lowering supply voltage and clock frequency of virtualized resources for energy savings [2, 3].

Also, some researchers observed that scaling down the resources operating frequencies while tasks are executing may result in degradation of performance (increased execution time) [4, 5]. The motivation for this research is based on the above-mentioned observation to investigate the various frequency transitions effects on tasks' execution time while performance constraint is imposed on workflow and minimizing virtual machines energy consumption along with maximizing resource utilization.

This paper presents a novel energy-aware scheduling heuristic named D-DEWS which aims to minimize the energy consumption of cloud resources with maximum resource utilization while still meeting the QoS demands. The proposed approach can be categorized in two phases: (1) the task-mapping phase: prioritizing the execution of tasks and allocating the tasks to the available virtual machines, and (2) the scaling phase: scaling down the maximum operating frequency of resources to a best-lower optimal frequency aiming for reduction in energy consumption while complying with deadlines.

The significant contributions of this research are summarized below.

1. Scheduling the workflow on the cloud with optimal execution time.
2. Incorporating the DVFS mechanism to further minimize the energy consumption of prior generated schedule by identifying and utilizing slacks between various tasks without exceeding the pre-imposed workflow deadline.
3. To validate the proposed heuristic, comparing it with other state-of-the-art algorithms.

The rest of the paper is organized as follows. Section 2 describes related work, Sect. 3 presents models used and problem formulation, Sect. 4 explains proposed approach, Sect. 5 discusses experimental settings and result analysis, whereas Sect. 6 concludes the research with the future directions.

2 Related Work

2.1 Scheduling in Heterogeneous Computing Environment

Based on existing researches, workflow scheduling approaches can be categorized into several classes, such as duplication-based scheduling, cluster-based scheduling, and list scheduling. Among those, the list-based scheduling algorithms are most popular, and performance efficient with minimum time complexity. Heterogeneous-Earliest-Finish-Time (HEFT) algorithm [6] is an effective example of list scheduling among existing work with time complexity, $O(E \times R)$ for E edges and R resources. Some noticeable achievements in list scheduling algorithms addressed several resource and performance constraints in workflow scheduling [7–10]. The major difference between the existing researches and the proposed study is that these heuristics only emphasized minimizing the execution time of tasks and did not pay much attention to imbalanced resource utilization and increased energy consumption.

2.2 Energy Optimization

Power management policies in a heterogeneous distributed computing environment and especially energy-aware workflow scheduling schemes using DVFS technique have been the key focus in many existing studies [11–13].

Kim et al. [2] worked on DVFS-enabled cluster for scheduling bag-of-tasks to reduce energy consumption considering time and space shared policies within a user-specified deadline constraints.

Kimura et al. [14] provided a scaling algorithm to find out the appropriate voltage-frequency levels called as gear to uniformly extend only those tasks which are having a slack, so that overall execution time is not increased.

Lee et al. [11] proposed two energy-aware policies to schedule workflow application on multiprocessor systems with the use of DVS. The author presented a new approach for minimizing tasks' execution energy while conserving the execution time for overall less energy consumption of a schedule.

Huang et al. [12] proposed an approach of slacking of noncritical tasks of workflow applications and allocating globally at uniform frequencies to meet the deadline and achieve higher savings on energy.

Rountree et al. [13] highlight the problem of increasing the current makespan with the use of DVFS. The author states that increasing the execution time of high-performance parallel tasks with the use of DVS is not desirable and presented a runtime strategy of predicting slacks and utilizing these slacks in such a way which will lead to a minimal delay in overall execution time.

Wang et al. [15] reduced energy consumption by scheduling tasks to get an energy-efficient schedule first and second scaling the frequency of critical and noncritical tasks so that schedule length does not exceed deadline constraint defined in SLA negotiation with a user for the green cloud.

Aupy et al. [16] studied the problem of changing various speed levels of a processor and states DVFS as a classical technique which leads to NP-hard solutions. Therefore, the author extends the DVFS model with new VDD hopping-model that can update the execution time of tasks in polynomial time.

Qiu et al. [17] identified the problem of huge power consumption for chip multiprocessors and proposed a three-stage DVFS mechanism for balancing the performance and energy minimization objective simultaneously through various graph concatenation to create an updated large graph. The frequency scaling is applied to the updated task graph to enhance the power efficiency of the multiprocessor system. The execution time stretching of a task to fill their slacks is performed by observing the length of the critical path and the status of power consumption of resources.

In contrast to these studies, the presented study achieves a significant trade-off between makespan and energy consumption by scaling the frequencies of each task on each resource so that idle period length of resources is minimized; depending upon data dependencies among tasks and limited voltage-frequency states of distinct resources, to reduce overall energy consumption.

3 Models and Problem Formulation

3.1 System Model

In this study, we consider a single cloud data center comprising a set of heterogeneous virtual machines: $VM = \{VM_j\}$, such that every VM depicts distinct performance in terms of processing speed (MIPS). Also, each VM is DVFS-enabled which indicates that it can run at discrete voltage-frequency $levels(s)$. For each VM, the supply voltage level is depicted as $V = \{v_s\}$, and operating frequency as $F = \{f_s\}$. Each VM resource stays at its lowest frequency with minimum voltage ($V_{low,s} > 0$) when there is no task running on it or if there is an idle slot in between tasks execution but it still consumes a significant amount of energy. The time spent during inter-change of frequency states in VM resources can be ignored because it is a negligible amount of time if compared with the tasks execution times (e.g., *10–150* μs [13]).

3.2 Cloud Application Model

Cloud workflow application is the series of parallel task execution which can be depicted as a directed acyclic graph (DAG), $W = (T, E)$, where $T = \{T_1, T_2 \ldots\ldots T_k\}$ is a group of tasks and $E = \{E_1, E_2 \ldots\ldots E_k\}$ defines a set of connecting edges that defines the control and data dependencies among the tasks as shown in Fig. 1. These edges consist of data transfer cost (*DTC*) between multiple tasks. The relationship ($T_i < T_k$) explains the precedence constraint among two different tasks of a workflow which suggest that task T_k cannot be scheduled before its predecessor task T_i. The task execution time on each available resource is expressed using expected computation cost matrix, $ECC(T_i,\ VM_j)$ as shown in Table 1.

To execute a workflow application DAG, few assumptions are made

(a) The level-wise execution of tasks is performed. In every level, each task is executed according to precedence order computed by a novel priority policy.
(b) Communication cost between (T_i and T_k) is considered zero if these dependent tasks are scheduled on the identical resource.

Next, some of the important definitions used in the proposed study are explained as follows:

1. $ET(T_i, VM_j)$ defines the *execution time* of a task on a resource, which can be estimated as shown in Eq. 1.

$$ET(T_i, VM_j) = \frac{Len(T_{i,})}{(PC(VM_j) * (1 - PD(VM_j)))} \tag{1}$$

where $Len(T_{i,})$ is the length of task measured in *Million Instructions (MI)* and $PC(VM_j)$ is the processing capacity of virtual machine measured in *Million Instructions Per Second (MIPS)*. Performance deviation of VMs, $PD(VM_j)$ is modeled as given in [18]. The performance of VM varies by almost 24% due to virtualization and heterogeneity of virtual machines, respectively [19].

Fig. 1 An example workflow

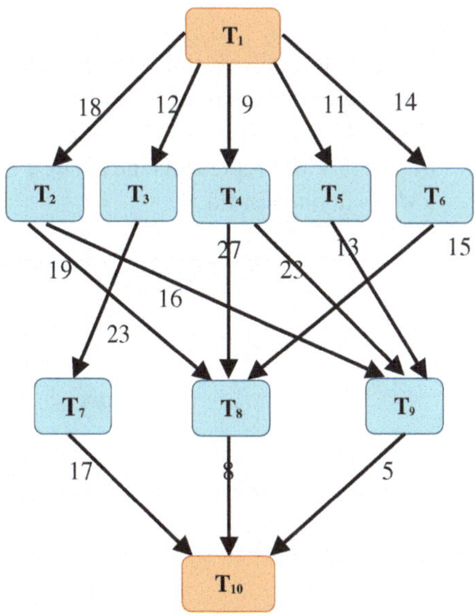

Table 1 Expected Computation Cost (ECC) matrix

Task	VM_1	VM_2	VM_3
T_1	14	16	9
T_2	13	19	18
T_3	11	13	19
T_4	13	8	17
T_5	12	13	10
T_6	13	16	9
T_7	7	15	11
T_8	6	11	14
T_9	18	12	20
T_{10}	21	7	16

2. $DTC(T_i, T_k)$ defines *data transfer cost* or communication time among two precedence constraint tasks. It can be computed as shown in Eq. 2.

$$DTC\left(T_{i,}, T_k\right) = \frac{O\_File\ size(T_i)}{BW} \tag{2}$$

where *O_Filesize* defines the size of the output file for transfer from a predecessor task to a successor task and *BW* defines the available bandwidth.
3. $EST(T_k, VM_j)$ represents the *earliest starting time* of a workflow task T_k on VM_j as defined in Eq. 3.

$$EST\left(T_k, VM_j\right) = max\left\{avail\left[VM_j\right], \max_{T_i \in pred(T_k)} \left(ACT(T_i) + DTC(T_i, T_k)\right)\right\} \tag{3}$$

where T_k is a current task to be scheduled on Virtual machine VM_j, $avail[VM_j]$ is the time of availability of VM_j. $ACT(T_i)$ is the actual completion time of task T_i on VM_j and $DTC(T_i, T_k)$ is the data transfer time from T_i to T_k.
4. $ECT(T_k, VM_j)$ represents the *earliest completion time* of a workflow task T_k on VM_j as defined in Eq. 4.

$$ECT\left(T_k, VM_j\right) = EST\left(T_k, VM_j\right) + ET\left(T_k, VM_j\right) \tag{4}$$

5. $M(W)$ defines the makespan of a workflow (W) as shown in Eq. 5.

$$M(W) = max(ACT(T_{Exit})) \tag{5}$$

where $ACT_{T_{Exit}}$ signifies the *actual completion time* of an exit task of the workflow.

3.3 Energy Model

Dynamic voltage and frequency scaling (DVFS) is recognized as a practical approach to manage the power consumption of cloud resources in large data centers [12].

By bringing down resources supply voltage and clock frequency during some periods, such as intra-communication or idle periods, significant reductions in power consumption of resources can be achieved.

Power dissipation while a resource is busy in executing tasks, accounts significant energy consumption, and the DVFS technique can efficiently decrease this power dissipation [14].

The dynamic power consumption is defined by the relationship of operating voltage and frequency of a resource as shown in Eq. 6.

$$P_{(R_{BUSY})} = K \cdot v^2 \cdot f \tag{6}$$

where K is a device-related constant, i.e., total capacitance.

Hence, energy consumption when a resource is busy can be computed using Eq. 7.

$$E_{(R_{BUSY})} = P_{(R_{BUSY})} \cdot \Delta T_B \tag{7}$$

where ΔT_B is the time when a resource is executing a task.

Similarly, when a resource is in idle state, the resource cannot be operating at zero voltage and frequency; instead, the resource is being scaled down to its minimum voltage and frequency level to save its energy.

Hence, Energy consumption while a resource is idle can be computed using Eq. 8.

$$E_{(R_{IDLE})} = P_{(R_{IDLE})} \cdot \Delta T_I \tag{8}$$

where ΔT_I is the idle period.

Therefore, the total power consumption in both the operating phases of a resource can be described as shown in Eq. 9.

$$E_{(R_{TOTAL})} = E_{(R_{BUSY})} + E_{(R_{IDLE})} \tag{9}$$

3.4 Problem Definition

The presented study focuses on two research objectives:

- The task to resource mapping

Each workflow task is mapped to best-suitable cloud resource so that the makespan of a workflow can be minimized.

- Energy-performance trade-off dynamic scheduling

The research study is based on the SLA negotiation between users and cloud service providers for the green cloud. The user provides the maximum acceptable deadline for completing its workflow application. Therefore, the goal is to lessen the energy consumption of cloud resources while keeping the execution time of workflow within an allowed deadline.

Based on the above definitions, the research problem can be defined as

"Given a workflow W and a cloud with available resources R, find a feasible schedule S (M, EC, RU, Map)," which

- *minimizes the makespan (M) of workflow W*
- *minimizes energy consumption (EC) with efficient resource utilization (RU).*

Mathematically, it can be formulated as

$$\begin{aligned}&\text{Minimize: } \mathrm{S} = (\mathrm{M}, \mathrm{EC})\\&\text{Maximize: } \mathrm{S} = (\mathrm{RU})\\&\text{Subject to: } \mathrm{M} \leq \mathrm{D}\end{aligned}$$

where D is the user-defined deadline which is determined by green cloud SLA negotiation.

4 Proposed Approach

In the proposed model, a user submits a workflow application specifying its deadline. An initial assignment of workflow tasks on heterogeneous resources is performed that takes into account the tasks expected computation cost on heterogeneous resources and data transfer cost among tasks. After acquiring initial task mapping and assuming the resources are operating at their maximum frequency, the execution of a task can be slowed down by scaling their assigned frequencies to the best-possible lower frequencies. It utilizes the idle time slots on each resource such that execution gets completed by the user given deadline and returns an energy-efficient tasks-mapping operating with enhanced resource utilization.

4.1 Algorithm Design

This subsection discusses the proposed algorithm DVFS-enabled Deadline constrained Energy-efficient Workflow Scheduling (D-DEWS) as shown in *Algorithm 1*. It aims to lessen the makespan of a workflow and energy consumption of resources complying with the specified deadline while increasing the utilization of cloud resources. D-DEWS has two major phases: (1) task mapping, and (2) Energy-aware frequency scaling.

Algorithm1: ***DVFS-enabled Deadline constrained Energy-efficient Workflow Scheduling (D-DEWS)***

Input:

A Workflow: $W\ (T, E)$ *with the user-defined deadline* (D)
Set of DVFS Enabled Processors: VM

Output:

Schedule S

1. Compute $\boldsymbol{S\_Priority}()$ of each task by traversing Workflow W upward from the exit task
2. Sort each task in Precedence Queue (Q) by decreasing order of $S\_Priority$ in each level
3. while there are unscheduled tasks in the Q do
4. Select the first task T_i from the Q
5. for each virtual machine VM_j do
6. Compute $EST\ (T_i, VM_j)$ and $ECT\ (T_i, VM_j)$ using Eqs. 3 and 4
7. Assign task T_i to VM_j that minimizes the $ECT\ (T_i, VM_j)$
8. end for
9. end while
10. Get Curr Plan = Schedule S
11. Compute Makespan (M_{old}) of schedule S using Eq. 5
12. Compute Energy Consumption (EC_{old}) of schedule S using Eq. 8
13. Compute task-slacking phase $\boldsymbol{Frequency\_Scaling}()$ using DVFS technique
14. Compute Energy Consumption (EC_{new}) of Energy optimized schedule S
15. if $EC_{new} < EC_{old}$
16. Get Curr Plan = Energy optimized schedule S
17. end if
18. Compute (M_{new}) and RU
19. Return schedule S $((M_{new}), (EC_{new}), RU, Map)$

Phase 1: Task Mapping

This phase is implemented in (Lines 1–12). It begins with prioritization of each task which is computed using $S\_Priority$ (*Algorithm* 2). Each task priority is computed based on its score. The score is the sum of the *average execution time (AET)* of a task (T_i) with the average of *data transmission cost* of a task (T_i) with its predecessor and successor to the total number of predecessor and successor of a task (T_i) in a workflow.

The score of each task is computed as shown in Eq. 10.

$$Score\ \ (T_i) = AET(T_i) + \frac{\left(\sum DTC(Pred(T_i)) + \sum DTC(Succ(T_i))\right)}{Count\ \ (Pred(T_i) + Succ(T_i))} \tag{10}$$

where $Pred(T_i)$ are the predecessors and $Succ(T_i)$ are the successors of task (T_i). $AET(T_i)$ is the average execution time of each task nodes on the j number of virtual machines (VMs). It is calculated as shown in Eq. 11.

$$AET(T_i) = \sum_{1}^{j} \frac{ET(T_{i,j})}{VM_j} \tag{11}$$

Further task to resource mapping is done in (Lines 4–10). Tasks are picked from *Precedence Queue* and allocated to a VM which minimizes its completion time. Lastly, based on the feasible schedule (S) obtained and user agreed extension factors, the deadline (D) can be computed according to Eq. 12 [15].

$$D = M_F \times (1 + \beta) \tag{12}$$

where M_F defines the minimum makespan of the feasible schedule S, and β signifies the user allowed extension or deadline factor and $\beta \geq 0$.

Algorithm 2: *S_Priority()*

```
1. Compute Average Computation Cost AET (T_i) of each task using Eq. 11
2. Arrange a list of tasks in reverse topological order
3. Compute Score (T_i) of each task using Eq. 10
4. for each task (T_i) in reverse topological order list
5.    P_Max = 0
6.    for each successor T_j of T_i do
7.       if (Score (T_i) + S_Priority (T_j)) > P_Max then
8.          P_Max = Score (T_i) + S_Priority (T_j)
9.       end if
10.   end for
11. S_Priority (T_i) = P_Max
12. end for
```

Phase 2: Energy-Efficient Frequency Scaling

(Line 13, *Algorithm 1*) calls *frequency_scaling(), Algorithm 3* to scale the pre-assigned frequencies of tasks on VMs by distributing the slack time to the tasks at best-possible minimum frequencies in an energy-efficient manner.

This phase aims to utilize the idle slots (*Slacks*) in the VMs and reassigning them using DVFS technique. For this, the overall allowed deadline D is distributed as a sub-deadline among each task. The sub-deadline is the latest allowable finish time (*LFT*) for each task on each VM and it is calculated using Eq. 13 (*Line 1, Algorithm 3*).

$$LFT\ allowed(T_i) = \begin{cases} D, \ if\ T_i = T_{exit} \\ \min_{(T_j) \in succ(T_i)} \begin{pmatrix} LFT\ allowed(T_j) - ET(T_j) \\ -DTC(T_i, T_j) \end{pmatrix}, & otherwise \end{cases} \tag{13}$$

With the use of *LFT allowed*, the *Slack time* is computed *(Line 2)* using Eq. 14.

$$Slack(T_i) = LFT\ allowed(T_i) - EST(T_i) - ET(T_i) \tag{14}$$

where $Slack(T_i)$ denotes the maximum time by which a task can be slowed down by adding it to its execution time, without exceeding the workflow deadline.

The next step follows the optimization of pre-assigned frequencies of each task according to their *LFT* and slack times. To scale the frequencies of tasks a reverse counter on *voltage-frequency levels(s)* of VM_j is set to pick the best-optimal lower frequency (*Line 7*). At each iteration, the updated execution time $ET(f_{low})$ of task (T_i) is calculated using Eq. 15 (*Line 8*).

$$ET_i(f_{low}) = \frac{ET_i(f_{max}) \times f_{max}(VM_j)}{f_{low}(VM_j)} \tag{15}$$

where $ET_i(f_{low})$ is the updated execution time of task (T_i) at low frequency (f_{low}). Next, the latest allowed start time $LST(f_{low})$ of task (T_i) is computed, as shown in Eq. 16 (*Line 9*).

Algorithm 3: *Frequency_Scaling()*

Input:

Schedule S (operating at $f_{max}(VM_j)$) of Algorithm 1 with Makespan (M_{old}),
Deadline (D)
Set of DVFS Enabled Processors: VM

Output:

Energy optimized Schedule S (Assignment of tasks to best-optimal voltage-frequency levels (s) of VM_j)

1. Compute *LFT* (T_i) of each task using Eq. 13
2. Compute *Slack* (T_i) of each task using Eq. 14
3. for each task (T_i) on VM_j in schedule *S*
4.   if $task\ (T_i) == T_{Entry}$
5. $LST\ (f_{low}) = 0$
6.   end if
7.   for k = *s* ; k ≥ 0 ; k -- do /* Pick best optimal frequency in volt-freq levels of VM_j */
8.   Compute $ET\ (f_{low})$ of task (T_i) using Eq. 15
9.   Compute $LST\ (f_{low})$ of task (T_i) using Eq. 16
10.   Compute $LFT\ (f_{low})$ of task (T_i) using Eq. 17
11.   if $LFT\ (f_{low}) \le LFT\ allowed\ (T_i)$ and $Energy\ Gain > \theta$
12.   Best-optimal frequency $f_{low}\ (T_i, VM_j) = f(VM_{j,s})$
13.   Update execution time $ET\ (f_{max})$ to $ET\ (f_{low})$
14.   Update $EST\ (f_{max})\ to\ LST\ (f_{low})$
15.   Update $LFT\ allowed\ (T_i)\ to\ LFT\ (f_{low})$
16.   Return $Slack\ covered$ as $[ET\ (f_{low}) - ET\ (f_{max})]$
17.   break
18.   end if
19.   Assign task (T_i) to optimized-list
20. $LST\ (T_i) = T_{(i-1)}[(LFT\ (f_{low})]$
21. $LFT\ (T_i) = [LST\ (T_i) + ET\ (f_{max})]$
22.   Return $Slack\ covered = NULL$
23.   end for
24. end for

$$(LST(f_{low})T_i) = max_{(T_j)\in pred(T_i)}\left(\left(LST(f_{low})T_j\right) + \left(ET(f_{low})T_j\right) + DTC\left(T_i, T_j\right)\right) \tag{16}$$

If a task is an entry task, latest allowed start time (LST) is set equal to zero (*Line 4–6*).

Finally, *Line 10* computes the $LFT(f_{low})$ of task (T_i) as shown in Eq. 17 and compared with its *LFT allowed* (Line 11).

$$(LFT(f_{low})T_i) = (LST(f_{low})T_i) + ET_i(f_{low}) \tag{17}$$

If $LFT(f_{low})$ is less than or equal to its allowed sub-deadline, and energy gain is obtained then immediately the new execution time slot of task (T_i) is updated as $\left[(LST(f_{low})T_i),\ (LFT(f_{low})T_i)\right]$ and the corresponding low frequency is saved as a best-optimal frequency for the updated time slot of the task. In each iteration, the energy gain of each task is calculated as shown in Eq. 18, to determine how much energy saving is achieved after scaling a task to the lower frequency. A threshold value (θ) of 0.01% is chosen in line 11, so that for frequency scaling only those tasks can be selected if scaling gives an energy gain (*Lines 11–18*). And, *Lines (3–7)* are repeated to update the time slot of each task in *schedule S* until there is no task left for scaling. If energy saving is obtained the energy-optimized schedule is accepted. If there is no such frequency available that can extend task execution time within its sub-deadline then the task will not be scaled and it is concluded that lowering frequency levels of tasks in *schedule S* is not energy efficient and the current task continues with the previously assigned frequency in *schedule S*.

$$Energy\ \ Gain(T_i) = \frac{E_{i_{fmax}} - E_{i_{flow}}}{E_{i_{fmax}}} \tag{18}$$

where $E_{f_{max}}$ is the energy consumption of task (T_i) operating at high-frequency in initial *schedule S* and $E_{f_{low}}$ is the energy consumption of task (T_i) at low frequency while performing frequency scaling.

5 Performance Evaluation

To validate the performance of the proposed approach, comparative analysis of D-DEWS is presented with two baseline algorithms HEFT [6] and EES [12]. MATLAB [20] is used as a simulation platform. MATLAB is a widely used development environment to simulate cloud applications and services.

The comparative evaluation is based on the following performance metrics:

1. ***Makespan*** is defined as the schedule length of a scheduled workflow as explained above in Eq. 5. The scheduling algorithm that gives the lowest makespan is considered more efficient to performance.

2. ***Energy Consumption*** is defined as the total power consumption by the cloud resources during workflow execution. The energy consumption of workflow (W) is computed as shown in Eq. 19.

$$EC(W) = \sum_{j=1}^{n} E_{TOTAL}(T_i) \tag{19}$$

where $E_{TOTAL}(T_i)$ is the sum of busy and idle power consumption of VMs in processing n number of workflow tasks.
3. ***Average Resource Utilization*** is the fraction of resource utilization percentage to the total number of resources in the cloud.

   Resource utilization is the fraction of resource total busy time to the time resource remained switched on as shown in Eq. 20 [8].

$$RU = \frac{VM_{BUSY_{TIME}}}{VM_{AVAILABLE_{TIME}}} \times 100 \tag{20}$$

   Hence, the average resource utilization can be calculated as shown in Eq. 21 [21].

$$ARU = \frac{\sum_{j=1}^{n} RU_j}{n} \tag{21}$$

where n represents the total number of active VMs to schedule a workflow.

5.1 Experimental Setup

In the presented study, we have chosen randomly generated workflows of various sizes. Table 2 presents the parameter settings for conducting experiments. Three groups of heterogeneous DVFS-enabled virtual machines are simulated for the proposed approach with associated voltage-frequency pairs as presented in Table 3.

Table 2 Simulation parameters

Parameter	Value
Datacenter	1
VM types	3
VMs processing capacity (MIPS)	800, 1000, 1200
VMs bandwidth (Mbps)	1000
Workflow size	10, 50, 100, 150, 300, 400
Task size (MI)	3900–10800
Output file size (Mb)	5000–27000

Table 3 The voltage/frequency pairs [22]

	Pair 1		Pair 2		Pair 3	
Level (s)	Supply voltage (V)	Frequency states (GHz)	Supply voltage (V)	Frequency states (GHz)	Supply voltage (V)	Frequency states (GHz)
0	1.5	2.0	1.2	1.8	1.484	1.4
1	1.4	1.8	1.15	1.6	1.463	1.2
2	1.3	1.6	1.1	1.4	1.308	1.0
3	1.2	1.4	1.05	1.2	1.18	0.8
4	1.1	1.2	1.0	1.0	0.956	0.6
5	1.0	1.0	0.9	0.8		
6	0.9	0.8				

5.2 *Results and Analysis*

The comparative results after assessment of the presented study are summarized in Figs. 2 and 3, followed by the evaluation under different scenarios.

Scenario 1: (*Varying the workflow size, Deadline Factor is fixed,* $\beta = 0$)
Figure 2a depicts the makespan of scheduled workflow with various sizes. The maximum improvement in average makespan is 3.20 and 2% in comparison to HEFT and EES. The makespan of D-DEWS is lower than HEFT due to the incorporation of improved priority policy. Also, when compared with EES the makespan of D-DEWS is less due to a distinctive scaling approach with minimum increase in overall execution time.

The analysis of energy consumption by workflows with different sizes is shown in Fig. 2b. For workflow size 100, D-DEWS performs 23.22% better than HEFT and 15.62% than EES. If a large workflow of size 400 is considered, D-DEWS performs 17.65% better than HEFT and 14.84% than EES.

Figure 2c depicts the results of average resource utilization of workflows varying number of tasks. HEFT heuristic shows the minimum utilization for different workflows while EES and D-DEWS increase resource utilization effectively. In the case of D-DEWS, it is observed that it outperforms EES and achieves maximum utilization of 78% for the large workflow consisting of 400 tasks. For workflow size 100, D-DEWS performs 26.16% better than HEFT and 5.71% than EES. If a large workflow of size 400 is considered, D-DEWS performs 27.71% better than HEFT and 7.63% than EES.

Scenario 2: (*Varying the deadline factor* (β)*, workflow size is fixed (400 tasks)*)
Figure 3a depicts the makespan of randomly generated workflow with 400 tasks considering various deadlines. The HEFT schedule is not affected by deadlines as it is not bound with the deadline constraints and does not aim frequency scaling. Scaling has a significant effect on the other two algorithms. EES extends schedule length to

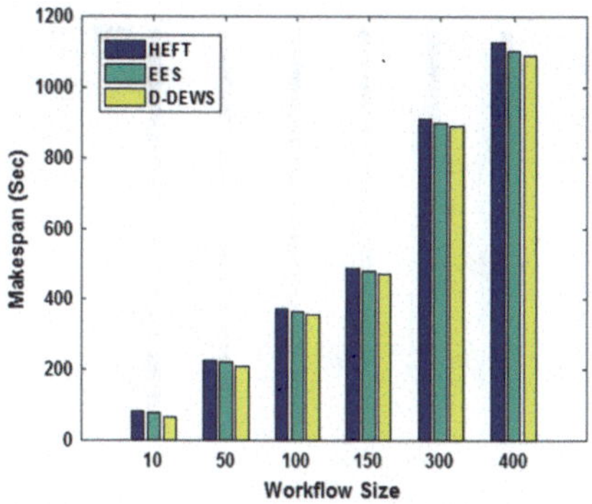

(a) Makespan analysis

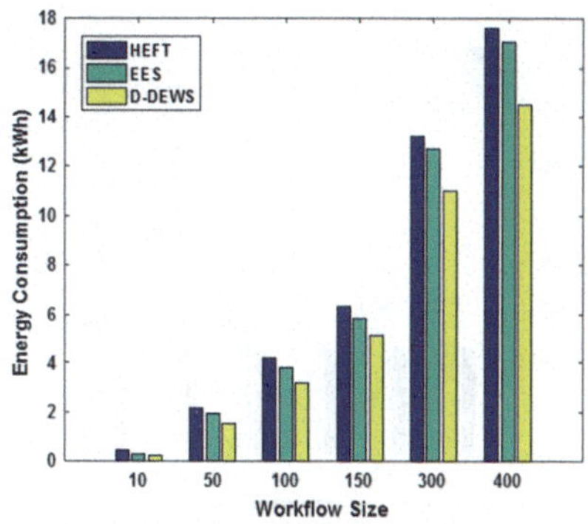

(b) Energy consumption analysis

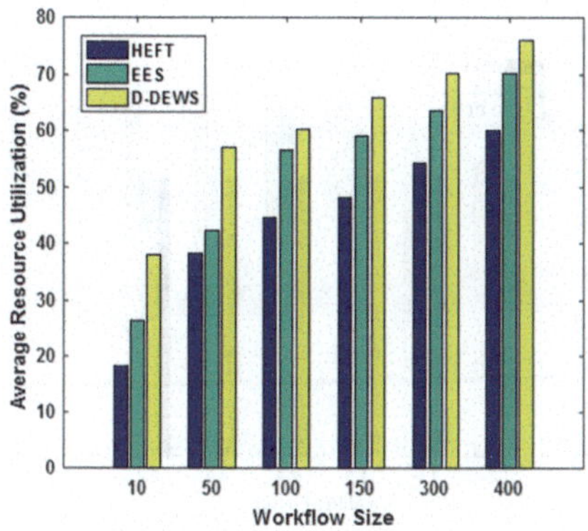

(c) Average resource utilization analysis

Fig. 2 Evaluation of random workflows with different sizes

conserve energy complying with the deadlines. D-DEWS leads to less increment of overall execution time and results in shorter schedules along with energy savings.

Figure 3b, depicts the results obtained for the optimization of energy consumption at various deadlines. The energy-saving ratio of D-DEWS is 38.13% better than HEFT and 17.67% than EES. This indicates D-DEWS presents better energy-performance trade-off than the other two algorithms.

Figure 3c shows an improvement in average resource utilization. The results reveal that D-DEWS increases resource utilization by 18.17% than HEFT and 10% than EES. It is observed that after a certain value of β the energy-saving ratio and resource utilization will tend to decrease due to the limited availability of slack times in large workflows.

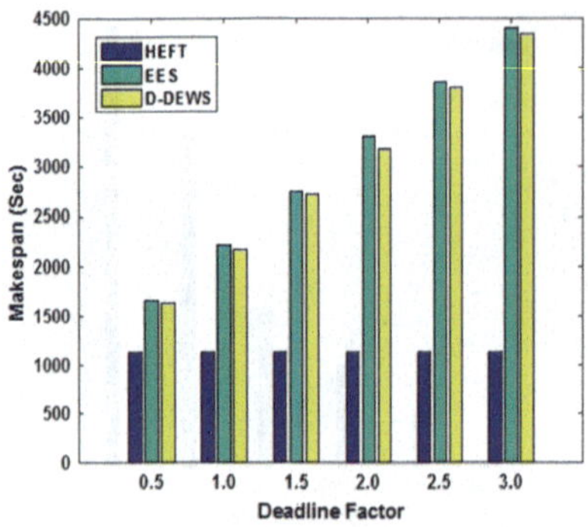

(a) Makespan analysis

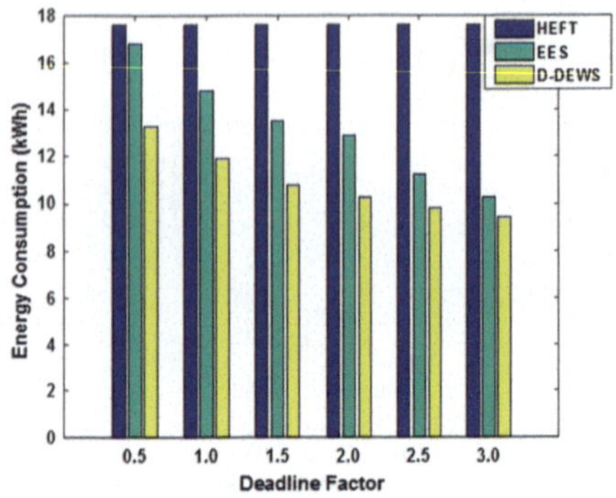

(b) Energy consumption analysis

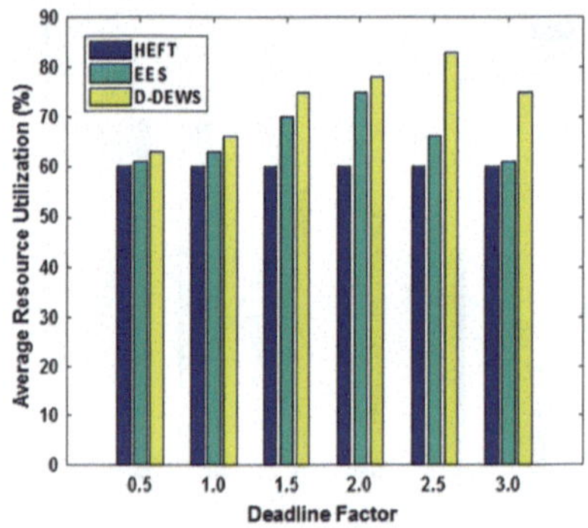

(c) Average resource utilization analysis

Fig. 3 Evaluation of random workflows with different deadlines

6 Conclusion and Future Scope

The proposed study has explored the critical issue of energy escalation in cloud data centers. Designing an efficient scheduling method that can increase energy efficiency of cloud resources while ensuring the performance within a certain time-frame, has become increasingly attractive. The proposed heuristic adopted a novel scheduling policy combined with DVFS technique for scheduling deadline constrained workflows.

It investigates a challenge that how to allocate tasks to the fastest resources and set a suitable frequency for each resource to optimize the total energy consumption. Simulation results prove D-DEWS as a promising approach having the potential to minimize a considerable amount of energy without a loss in performance and achieving enhanced resource utilization.

In future, we plan to extend the work taking into account the other power-consuming components such as memory, considering the overhead of data transfer cost. The proposed algorithm can be enhanced including other QoS parameters such as cost under budget constraints. Lastly, we intend to implement the approach in a multi-cloud environment.

References

1. M. Avgerinou, P. Bertoldi, L. Castellazzi, Trends in data centre energy consumption under the european code of conduct for data centre energy efficiency. MDPI AG (2017). http://dx.doi.org/10.3390/en10101470
2. K. Kim, R. Buyya, J. Kim, Power-aware scheduling of bag-of-tasks applications with deadline constraints on DVS-enabled clusters, in *Proceedings of the 7th IEEE International Symposium on Cluster Computing and the Grid* (IEEE Computer Society Washington, DC, USA, 2007), pp. 541–548
3. G. Stavrinides, H. Karatza, An energy-efficient, QoS-aware and cost-effective scheduling approach for real-time workflow applications in cloud computing systems utilizing DVFS and approximate computations. Futur. Gener. Comput. Syst. **96**, 216–226 (2019)
4. S. Wang, Z. Qian, J. Yuan, I. You, A DVFS based energy-efficient tasks scheduling in a data center. IEEE Access **5**, 13090–13102 (2017)
5. M. Etinski, J. Corbalan, J. Labarta, M. Valero, Understanding the future of energy-performance trade-off via DVFS in HPC environments. JPDC **72**, 579–590 (2012)
6. H. Topcuouglu, S. Hariri, M.Y. Wu, Performance effective and low-complexity task scheduling for heterogeneous computing. IEEE Trans. Parallel Distrib. Syst. **13**(3), 260–274 (2002)
7. M.A. Rodriguez, R. Buyya, A taxonomy and survey on scheduling algorithms for scientific workflows in IaaS cloud computing environments. Concurr. Comput.: Pract. Exp. **29**(8) (2017)
8. A. Rehman, S.S. Hussain, Z. Rehman, S. Zia, S. Shamshirband, Multi-objective approach of energy efficient workflow scheduling in cloud environments. Concurr. Comput.: Pract. Exp. **31**(8) (2018)
9. L. Liu, Q. Fan, R. Buyya, A deadline-constrained multi-objective task scheduling algorithm in mobile cloud environments. IEEE Access **6**, 52982–52996 (2018)
10. S. Abrishami, M. Naghibzadeh, D.H.J. Epema, Deadline-constrained workflow scheduling algorithms for infrastructure as a service clouds. Futur. Gener. Comput. Syst. **29**,158–169 (2013)
11. Y.C. Lee, A. Zomaya, Energy conscious scheduling for distributed computing systems under different operating conditions. Parallel Distrib. Syst. IEEE Trans. **22**(8), 1374–1381 (2011)
12. Q. Huang, S. Su, J. Li, P. Xu, K. Shuang, X. Huang, Enhanced energy-efficient scheduling for parallel applications in cloud, in *12th IEEE/ACM International Symposium Cluster, Cloud and Grid Computing (CCGrid)* (2012), pp. 781–786
13. B. Rountree, D.K. Lownenthal, B.R. de Supinski, M. Schulz, V.W. Freeh, T. Bletsch, Adagio: making DVS practical for complex HPC applications, in *Proceedings of the 23rd ICS* (ACM, 2009), pp. 460–469
14. H. Kimura, M. Sato, Y. Hotta, T. Boku, D. Takahashi, Empirical study on reducing the energy of parallel programs using slack reclamation by DVFS in a power-scalable high performance cluster, in *IEEE International Conference on Cluster Computing* (Barcelona, 2006), pp. 1–10
15. L. Wang, G. Von Laszewski, J. Dayal, F. Wang, Towards energy-aware scheduling for precedence constrained parallel tasks in a cluster with DVFS, in *Proceedings of the 10th IEEE/ACM CCGrid* (IEEE, 2010), pp. 368–377
16. G. Aupy, A. Benoit, F. Dufossé, Y. Robert, Reclaiming the energy of a schedule: models and algorithms. Concurr. Comput.: Pract. Exp. **25**(11), 1505–1523 (2013)
17. M. Qiu, Z. Ming, J. Li, S. Liu, B. Wang, Z. Lu, Three-phase time-aware energy minimization with DVFS and unrolling for chip multiprocessors. JSA **58**(10), 439–445 (2012)
18. M.A. Rodriguez, R. Buyya, Deadline based resource provisioning and scheduling algorithm for scientific workflows on clouds. IEEE Trans. Cloud Comput. **2**(2), 222–235 (2014)

19. J. Schad, J. Dittrich, J.-A. Quiané-Ruiz, Runtime measurements in the cloud: observing, analyzing, and reducing variance. Proc. VLDB Endow. **3**(1–2), 460–471 (2010)
20. https://in.mathworks.com/academia/research.html
21. Y. Fang, F. Wang, J. Ge, A task scheduling algorithm based on load balancing in cloud computing, in *International Conference on Web Information Systems and Mining WISM LNCS*, vol. 6318 (Springer, Berlin, 2010), pp. 271–277
22. Z. Tang, L. Qi, Z. Cheng, K. Li, S.U. Khan, K. Li, An energy-efficient task-scheduling algorithm in DVFS-enabled cloud environment. J. Grid Comput. **14**(1), 55–74 (2016). Springer, Netherlands

Homodyne Detection in WDM FSO System—A Better Solution to Mitigate Scintillation Effects

Neha Rani, Preeti Singh and Pardeep Kaur

Abstract Free space optical communication is a cost-efficient method for accessing high-bandwidth applications which has gained more attention with latest commercialization achievements. Fading induced by turbulence is a serious issue in FSO link which harshly deteriorates link performance. In order to reduce the turbulence effect caused by scintillation, various researches have been carried out in the domain of FSO network. This research work has employed one such technique in order to reduce the scintillation effect, which involves the use of homodyne detection in conjunction with the multibeam technique in WDM FSO system. Performance comparison of the single beam and multibeam WDM FSO system using homodyne detection in context of BER, Q factor and eye height have been explored for the scintillating effect in this paper.

Keywords FSO · Single beam · Multibeam · WDM · Homodyne

1 Introduction

Free Space Optics means the transmission of information by optical radiation over the free space. Intensity/phase or frequency of an optical carrier can be modulated for the information signal. Line of sight (LOS) is the essential demand for FSO communication to take place between transmitter and receiver. High optical bandwidth available in FSO link allows high data rate. No digging cost and unlicensed spectrum are the other advantages of FSO innovation [1]. The FSO units are compact, handy, and are easy to redeploy. The contracted beamwidth of the laser beam offers the advantage of high security in FSO system as compared to prevailing RF and microwave communication [2]. FSO communication finds its application in many areas such as storage area network (SAN), Last mile access, Military applications, etc. [3]. With so many advantages, FSO system performance is limited by various

N. Rani (✉) · P. Singh · P. Kaur
Department of Electronics and Communication Engineering, University Institute of Engineering and Technology, Panjab University, Chandigarh, India
e-mail: nehaluthra.ec@gmail.com

M. Dutta et al. (eds.), *Proceedings of International Conference on IoT Inclusive Life (ICIIL 2019), NITTTR Chandigarh, India*, Lecture Notes in Networks and Systems 116,
https://doi.org/10.1007/978-981-15-3020-3_34

environment factors such as snow, haze, rain, etc., which makes the FSO system a random phenomenon [4]. The system is mainly disadvantaged by the scintillation which can affect the FSO performance even over the relatively shorter distance.

1.1 Scintillation Effect

Free space optical (FSO) communication is a cost-efficient method with access to high bandwidth that has gained more attention with latest commercialization achievements. A serious issue in FSO links is the fading induced by turbulence which harshly deteriorates the FSO link performance. The optical turbulence in the received signal is called scintillation (Fig. 1).

Atmospheric changes associated with heating, temperature, and pressure results in the formation of cells of variable sizes, refractive indices, and temperature which leads to the fluctuation in amplitude, phase, and direction of the signal. These cells of variable sizes ranging from 10 cm to 1 km called as eddies act like lens from which when the signal passes undergo scattering and multipath fading. In addition, there can be distortion of beam wave front that results in defocusing of the beam. The interaction between thermal turbulence in the propagation medium leads to beam dancing at the receiver, hence it needs to be alleviated for the successful transmission.

2 Related Study

Various techniques to diminish this effect of atmospheric turbulence have been proposed. Among them, large aperture receivers and diversity in transmission and/or reception are the two major approaches commonly used in viable FSO systems. In literature, and in real-world systems, aperture averaging approach has been widely taken into account [5]. The receiver's aperture dimension more than the correlation length of fading can be viewed as a straightforward type of spatial diversity. When the size of aperture size is considerably larger than length of correlation, i.e., $0 \gg 0$, then aperture averaging leads to alleviation of detrimental effect of turbulence-induced fading. Intensity variation can also be minimalized by making the receiver observation time larger than the correlation time (time averaging) [6]. Though, it is not always feasible to lean on these techniques because of receiver size and bit rate constraints. Temporal and Spatial domain detection techniques can be utilized to shrink the atmospheric turbulence. In Spatial domain where multiple receivers can be

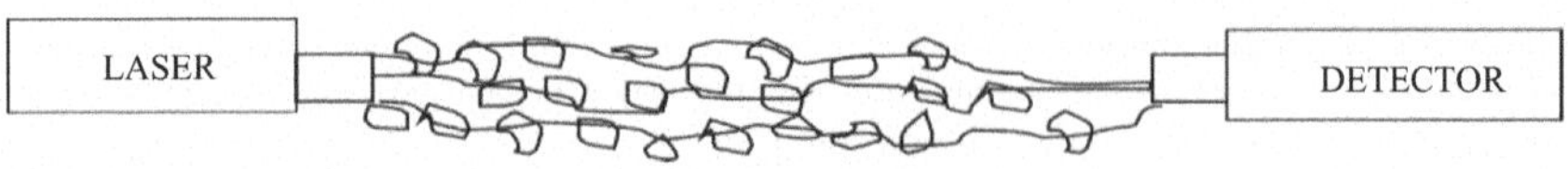

Fig. 1 Scintillation due to cells smaller than beam size

employed, an optimal maximum likelihood (ML) detection technique making use of Spatial correlation performs much better than the Equal-gain combining (ECG). But these techniques are limited in the systems having small receiver aperture relative to the fading correlation length [7]. Diversity combining techniques manifests to be a high-quality driving force to lessen the effect of turbulence and to design speedy and bandwidth-efficient long haul FSO transmission system. Diversity technique involves identical transmission from sources to receivers. If these different routes are uncorrelated, each transmission route finds an autonomous fading, enhancing general system efficiency as the probability of all routes experiencing a deep fading at the same time is very small. The harmful impact of atmospheric turbulence and beam wander can be reduced by using MIMO system. The alleviation of scintillation in MIMO system is achieved with the aid of multiple apertures and multiple input lasers, in that way creating a MIMO channel [8]. Multibeam WDM FSO system making use of spatially diverse transmitters can be used to enhance the link distance in presence of atmospheric attenuation and turbulence [9, 10], also the transfer rate of physical optical link could be increased by means of multichannel technology WDM [11]. Coherent detection is another solution demonstrated in [12, 13] to suppress scintillation effectively. Coherent on–off key (OOK) FSO Communication System is simple to operate and requires no extra DSP algorithm for the detection of the received signal to retrieve the phase information. The enhanced quality of signal makes the DSP receiver effective in reducing the fading effect [14]. Different Diversity schemes in coherent and direct detection system can be used to enhance the system performance under atmospheric turbulence. In the system using diversity direct detection, diversity cannot be increased beyond an optimum value as the effect of background noise come into existence, outage probability is also increased. But if diversity increases, background noise and the signal of interference continue to be the same in the coherent diversity scheme. The ability to reduce background noise has demonstrated that diversity coherent detection delivers considerably better efficiency [15].

3 Simulation Setup

In the present work, coherent-based homodyne detection is employed in receiver section of pre-existing single beam (SB) as well as multibeam WDM FSO system [10] with an aim to reduce the scintillation effect. Both systems are designed in Optisystem environment. Performance comparison of both the system is done in context of BER and Quality factor.

3.1 System Model 1

The layout design of single beam WDM FSO system using homodyne detection in Optisystem environment is displayed in Fig. 2. The portion of the transmitter comprises CW laser. The fork element duplicates the signal produced from the laser supply so that a multiplexer can be provided to separate the signal into wavelength carriers. PRBS is employed for the generation of codes that respond to the data signal. The NRZ pulse generator provides electric pulses of the signal produced by the PRBS with the format of NRZ pulse generation. The modulation is carried out by the Mach–Zehnder modulator (MZM). Signal is transmitted under highly turbulent conditions. For high turbulence, value of refractive index structure parameter C_n^2 is taken as 10^{-13}. Transmitted signals are demultiplexed and extracted using homodyne detection. Homodyne detection is used at the receiver due to its superiority over the direct detection.

The received signal is combined with a reference wave from local oscillator having identical phase and frequency as that of received signal as shown in Fig. 3. It is a relatively easy way of amplifying the photo current by rising the local oscillator power. This detection delivers better SNR by increasing the local oscillator power.

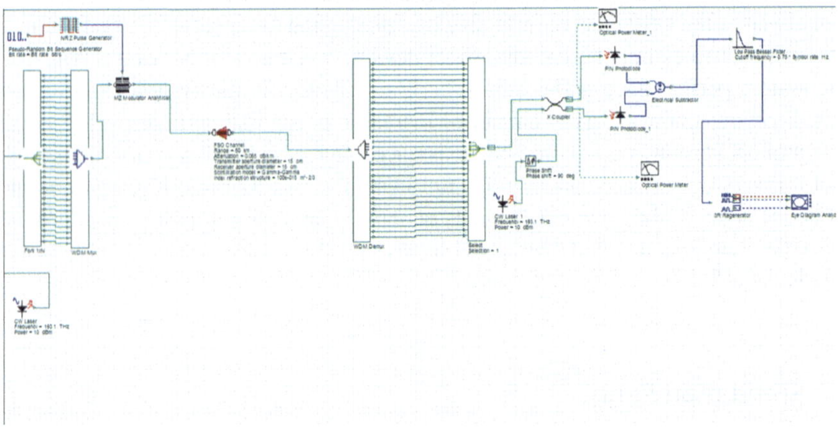

Fig. 2 SB-WDM FSO system using homodyne detection designed in optisystem

Fig. 3 Homodyne detection [16]

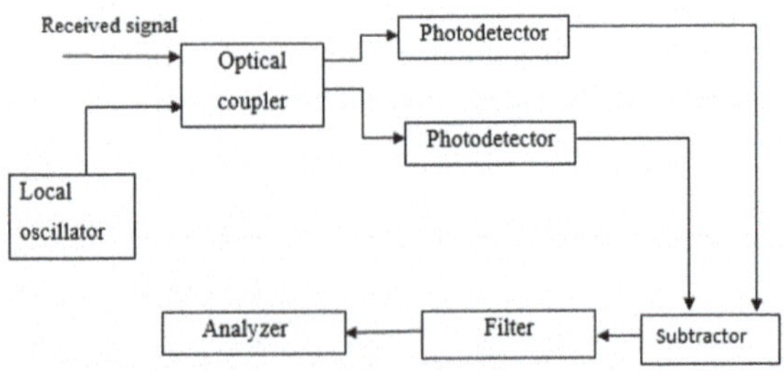

3.2 System Model 2

Multibeam System varies only in the manner of transmission after modulation from single beam. Since it is a multibeam scheme, the signal is transmitted with various spatial transmitting apertures. As Fig. 4 shows, after MZM modulator, fork simulates four distinct transmission apertures and one receiver aperture that creates it into a 4×1 WDM FSO scheme. Signal is transmitted over the four channels having different structure parameter of refractive index as 10^{-13}, 10^{-14}, 10^{-15} and 10^{-16} m$^{-2/3}$, respectively. The CW laser power of 10 dBm works at frequency of 1550 nm with information rate of 10 Gb/s. Diameter of receiver and transmitter aperture is taken as 15 cm. In the analysis, geometric loss was also taken into account and divergence of beam is considered as 2 mrad.

Different models are accessible to the model the FSO link impacted by turbulence. These models offer the probability density function (PDF) for received signal after the turbulent atmospheric circumstances are passed through. Log-normal model is used for FSO channel, if it is influenced by weak turbulence. K-turbulence model and negative exponential model are utilized in the event of powerful turbulence in the channel [13]. Gamma–gamma model is used in this research work by considering the turbulence from moderate to high. Gamma–Gamma distribution probability density function (pdf) is given by

$$f(I) = \frac{2(\alpha\beta)^{\alpha+\frac{\beta}{2}}}{\Gamma(\alpha)\,\Gamma(\beta)} I^{\frac{\alpha+\beta}{2}-1} K_{\alpha-\beta}(2\sqrt{\alpha\beta I}) \tag{1}$$

where K_x (.) is x ordered, second kind of modified Bessel function. The parameters α and β are the amount of minor and major scale eddies calculated from the following equations

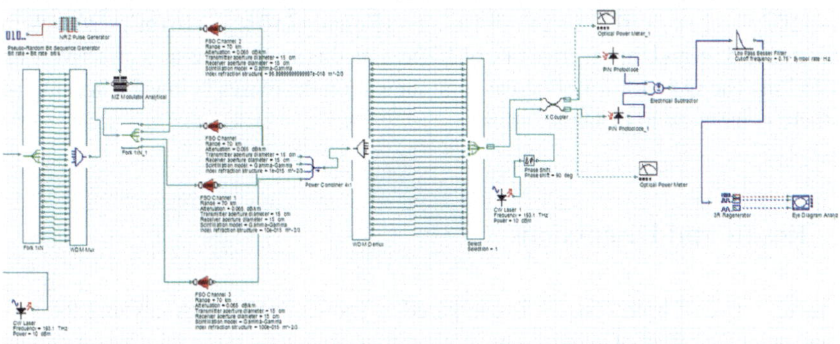

Fig. 4 MB-WDM FSO system using homodyne detection designed in optisystem

$$\alpha = \exp\left[\frac{0.49\sigma_r^3}{\left(1 + 1.11\sigma_r^{12/5}\right)^{\frac{5}{6}}}\right] - 1 \tag{2}$$

$$\beta = \exp\left[\frac{0.51\sigma_r^3}{\left(1 + 0.69\sigma_r^{12/5}\right)^{\frac{5}{6}}}\right] - 1 \tag{3}$$

σ_r^2 is Rytov variance given by equation

$$\sigma_r^2 = 1.23C_n^2 k^{\frac{7}{6}} z^{11/6} \tag{4}$$

4 System Analysis

The scintillation effect relies on the structure parameter of refractive index C_n^2, which is provided to the FSO channel as a parameter and intensity of signal is reduced based on C_n^2 value. For highly turbulent conditions with refractive index structure parameter of 〖10〗^(−13), maximum link length attained by the single beam WDM FSO system (SB WDM FSO) with a satisfactory quality factor is 10 km. Achieved quality factor and BER for this range are 6.13 and 4.01e−10, respectively. Attenuation value is kept as 0.065 dB/km assuming the clear weather conditions.

Multibeam WDM FSO system utilizes four system beams separately and therefore, suffer from distinct scintillation levels depending on the index structure refractive parameter. Due to their different propagation routes, beams are subjected to different scintillation eddies indicating C_n^2 values for the four beams as 〖10〗^(−13), 〖10〗^(−14), 〖10〗^(−15), and 〖10〗^(−16) m^(−2/3). With Q factor of 8.98 and the bit error rate of 1.26e−19, this system operates effectively up to 25 km. The Q factor for effective communication decreases below the acceptable if there is increase in distance further.

5 Result and Discussions

Q factor value of both systems is compared as shown in Fig. 5. Analyzing the systems under the effect of scintillation, both graphs obviously favor multibeam system performance. The two systems were equated with the help of eye diagrams as well. The eye opening in Fig. 6 indicates that Q factor of multibeam WDM FSO system at 10 km is much greater (the blue curve in the diagram), also eye height for multibeam system is 144 μm whereas the eye height for single beam systems is 110 μm. Performance of both the systems is summarized in Table 1.

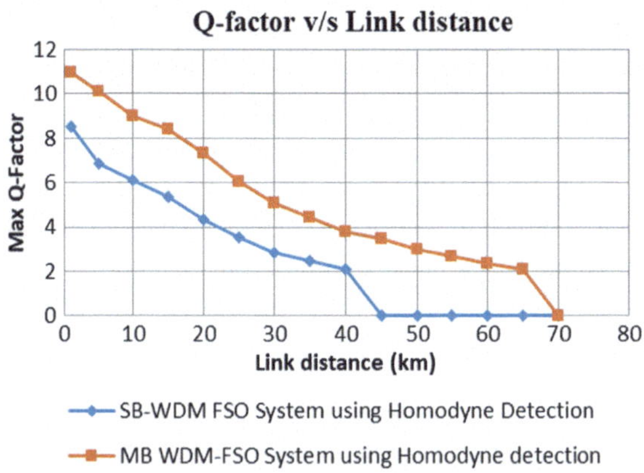

Fig. 5 Comparison of single beam and multibeam WDM FSO system using homodyne detection in terms of Q factor

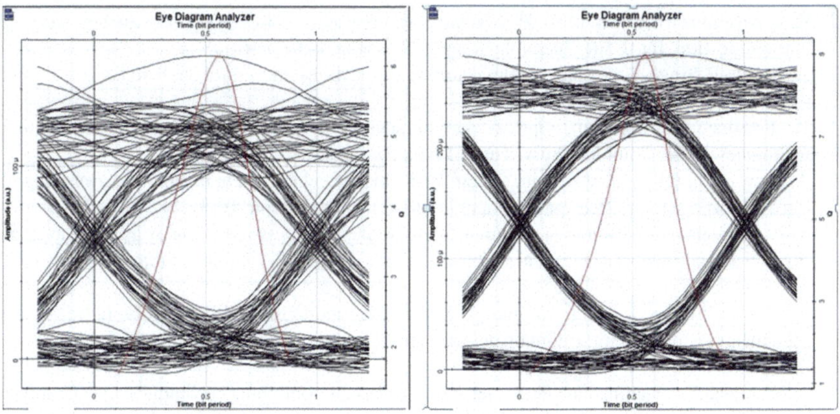

Fig. 6 Eye diagram of single beam WDM FSO system (left) and multibeam WDM FSO system (right) using homodyne detection

Table 1 Comparison of single beam and multibeam WDM FSO system using homodyne detection

System	C_n^(2) (m^(−2/3))	Max distance (km)	Min BER	Q factor
SB WDM FSO system using Homodyne detection	10^{-13}	10	4.01e−10	6.13
MB WDM FSO system using Homodyne detection	10^{-13}, 10^{-14}, 10^{-15}, 10^{-16}	25	7.27e−10	6.04

6 Conclusion

Analysis demonstrates that a multibeam WDM FSO system using homodyne detection transmits up to 25 km when simulated under the effect of scintillation. It is much larger than the single beam WDM FSO system employing homodyne detection, for which link length is confined to only 10 km in the identical conditions. Considering the effect of scintillation, performance of multibeam system is far better than a single beam system. Therefore, it has a prospect for use in the FSO applications requiring high reliability.

References

1. H. Willebrand, B.S. Ghuman, *Free Space Optics: Enabling Optical Connectivity in Today's Networks* (SAMS publishing, 2002)
2. R.A. Alsemmeari, S.T. Bakhsh, H. Alsemmeari, Free space optics vs radio frequency wireless communication. Int. J. Inf. Technol. Comput. Sci. **8**(9), 1–8 (2016)
3. A. Malik, P. Singh, Free space optics: current applications and future challenges. Int. J. Opt. (2015)
4. H. Kaushal, G. Kaddoum, Optical communication in space: challenges and mitigation techniques. IEEE Commun. Surv. Tutor. **19**(1), 57–96 (2016)
5. A. Viswanath, P. Gopal, V.K. Jain, S. Kar, Performance enhancement by aperture averaging in terrestrial and satellite free space optical links. IET Optoelectron. **10**(3), 111–117 (2016)
6. M.A. Khalighi, N. Schwartz, N. Aitamer, S. Bourennane, Fading reduction by aperture averaging and spatial diversity in optical wireless systems. IEEE/OSA J. Opt. Commun. Netw. **1**(6), 580–593 (2009)
7. X. Zhu, J.M. Kahn, J. Wang, Mitigation of turbulence-induced scintillation noise in free-space optical links using temporal-domain detection techniques. IEEE Photonics Technol. Lett. **15**(4), 623–625 (2003)
8. A.A. Johnsi, V. Saminadan, Performance of diversity combining techniques for fso-mimo system, in *2013 International Conference on Communication and Signal Processing*, Apr 2013 (IEEE, 2013), pp. 479–483
9. M. Grover, P. Singh, P. Kaur, C. Madhu, Multibeam WDM-FSO system: an optimum solution for clear and hazy weather conditions. Wireless Pers. Commun. **97**(4), 5783–5795 (2017)
10. M. Grover, P. Singh, P. Kaur, Mitigation of scintillation effects in WDM FSO system using multibeam technique. J. Telecommun. Inf. Technol.
11. N. Dayal, P. Singh, P. Kaur, Long range cost-effective WDM-FSO system using hybrid optical amplifiers. Wireless Pers. Commun. **97**(4), 6055–6067 (2017)
12. N. Cvijetic, D. Qian, J. Yu, Y.K. Huang, T. Wang, Polarization-multiplexed optical wireless transmission with coherent detection. J. Lightwave Technol. **28**(8), 1218–1227 (2010)
13. B. Zheng, S. Tong, Performance simulation of heterodyne synchronous receiving system in coherent optical communication, in *Selected Papers from Conferences of the Photoelectronic Technology Committee of the Chinese Society of Astronautics 2014, Part I*, vol. 9521, Mar 2015 (International Society for Optics and Photonics, 2015), p. 952106
14. Z. Wang, W.D. Zhong, C. Yu, Performance improvement of OOK free-space optical communication systems by coherent detection and dynamic decision threshold in atmospheric turbulence conditions. IEEE Photonics Technol. Lett. **24**(22), 2035–2037 (2012)
15. E.J. Lee, V.W. Chan, Power gain of homodyne detection over direct detection receivers for free space optical communication in the presence of interference, in *2008 Conference on Lasers*

and Electro-Optics and 2008 Conference on Quantum Electronics and Laser Science, May 2008 (IEEE, 2008), pp. 1–2
16. J.K. Sahota, D. Dhawan, Reducing the effect of scintillation in FSO system using coherent based homodyne detection. Optik **171**, 20–26 (2018)

SSCCJ: System for Source to Source Conversion from C++ to Java for Efficient Computing in IoT Era

Preeti Bhatt, Harmunish Taneja and Kavita Taneja

Abstract Automatic language conversion can easily find its use in the industry as from time to time projects get migrated from one language to another for better suitability to new requirements. This requires manually translating the code from one language to another, which is tedious and time consuming. In this paper, we propose a system for converting C++ source code to Java code such that readability is preserved. Past researches are not able to translate features like pointers, multiple inheritance, friend function, etc. Since there are no direct equivalents present in Java, this paper attempts to substitute the features including Multiple Inheritance, Preprocessor Directives, and Friend Functions. These converted codes will provide ease to computer programmers thereby reducing their programming efforts. This paper represents the conversion of existing C++ high-level language code to Java language source code to provide the ease to the software engineer working on diverse platforms in the present era of Internet of Things.

Keywords Intermediate language · Transpiler · Source to source conversion · Programming language converter · Parser generator · Code migration · Internet of things (IoT)

1 Introduction

The compilers are most commonly used to translate from higher level source code to the machine-level code. Higher-level languages have better program readability and hence maintenance of code becomes easy. Over time, various programming

P. Bhatt · K. Taneja
DCSA, Panjab University, Chandigarh, India
e-mail: preeti.gcg@gmail.com

K. Taneja
e-mail: kavitatane@gmail.com

H. Taneja (✉)
DCSA DAV College, Chandigarh, India
e-mail: harmunish.taneja@gmail.com

M. Dutta et al. (eds.), *Proceedings of International Conference on IoT Inclusive Life (ICIIL 2019), NITTTR Chandigarh, India*, Lecture Notes in Networks and Systems 116,
https://doi.org/10.1007/978-981-15-3020-3_35

languages emerge starting from the procedural programming paradigm to an object-oriented programming paradigm. The problem occurs when code that is already existed in a language needs re-implementation or migration using different programming languages. With time, new languages are designed and it is the need of the hour to migrate projects from one programming language to another. The problem arises in writing and testing old standards again using different languages [1]. Hence this demands some tools to perform these operations during conversion. IoT is a system of interrelated computing devices/machines/objects or persons that are provided with unique identifiers. IoT works on the principle of transferring data over communication devices without requiring human to human or human to computer interaction. In today's fast-changing era of technological advances, the new technologies and software are coming to provide effective any time anywhere communication. The traditional software packages/programs need to be aligned with new technologies or present computer languages. Source code conversion from one language to another language helps the computer programmers to reduce this job especially for the projects which are already working on the platform which are not in sync with present technological advances. The energy efficient computing in the era of IoT is the need of the hour and various architectures and protocols are exploited for computing using caching technique to achieve energy efficiency in diverse applications related to ad hoc networks with restricted mobility [2] thereby saving the programming efforts of the engineers. This approach can be used with multi-channel communication in IoT applications which provides increased network performance by efficiently utilizing the available bandwidth [3]. Java is a language of network programming and is used for secure communication over networks. Migration from one language to another is often required by Industry. This is generally tedious and expensive. Automating such a task can simplify the migrating process.

2 Related Work

For a better conversion process, the system should have a better syntax replacement system as well as maintain the structure. For this purpose, [1] has proposed two compilers that need to be designed, the First compiler converts the given source code into its matching Intermediate language and the second compiler converts Intermediate code in that language into another high-level language source code. Ribic and Salihbegovic [4] proposed the idea of implementation of single-code programming language that eliminates the need for source code at all by keeping native machine code. C++ to Java Conversion tool [5] converts any C+ program containing pointers into its equivalent code. SED (Stream Editor) Scripts [6] introduced to Create Grammar Rule are supported by many C++ Libraries. A converted segment from C++ Structure to Java exploits public variables only. Similarly, unions can be implemented in Java. In C++, Container Classes need to be defined to implement the Template concept. The approach may be used for translating a function template by creating one class that handles all integrated objects. While solving pointer Arithmetic, a

Table 1 Existing open source conversion tools and their features

Tools	Features
Ephedra [20]	• C to Java conversion • Casting analysis tools and Java API used
C2J++ [20]	• C++ to Java conversion tool • Data type and data flow conversion
C2J [20, 21]	• C to Java conversion tool • C2J comes with large C runtime library
C++2Java [5]	• C++ to java conversion tool • Creates some classes automatically
CAPPUCCINO [22]	• C++ to Java conversion tool • Translate operator overloading, templates

special event dependent routine is called from special input [7]. C++ header files need generic, i.e., ANSI C standards, K&R C [8] and use macro named __cplusplus. CA Plex [9], a language migration tool was introduced which supports both pattern libraries and C++ Action Debugger. Source to Source Data Transformation [10] uses high-level grammar for re-writing rules, optimizing compilers and language generator. Trudel et al. [11] proposed a translator which can translate C features like pointers into Java, and support Abstract Syntax Tree (AST). Martin and Muller [12] proposed the integration of C program library into Java program code. Cordy et al. and Cordy [13, 14] proposed the compiler that helps in language transformation and grammar generation. Wang et al. [15] proposed the idea of using cluster technology, component modeling, and partition allocation which includes hash algorithm and requests classification methods for converting the C++ standalone program into J2EE (Java 2 Express Edition). Liu and Liu [16] proposed the solution of a tree to string parsing problem by translating the parse tree using a tree-based translator into a target string. As recommended by the new C/C++ standard, remove commands that are unsupported in the target code and should be removed from legacy code [17]. Yamakami [18] proposed the idea of the migrating system using APIs and platform architecture without changing existing structure and function. Table 1 shows various conversion tools and their features.

3 Gaps in Related Work

The commercial converters that are available are not able to convert all programming language features. Most of them are not able to substitute unsupported features, e.g., tangiblesoftwaresolution.com [5]. The work that has been done in this area uses compilers to convert, although sacrificing readability in the process. Those who can convert keeping readability intact, i.e., source to source translators are not able to substitute unsupported features. Previous studies on automatic translation can cause semantic changes. As a solution, some tools suggested to label the generated code while implementing Pointes. Table 2 shows the limitations of various existing conversion tools.

Most of the exiting tools failed to do the complete automatic conversion on C++ features that are unsupported in Java which includes Multiple Inheritance, Friend Function, Operator Overloading, Function Template, Pointers, etc.

4 Challenges

C++ has many features that are unsupported by Java, and each feature presents its own challenges. In order to translate these unsupported features, the code has to be transformed in such a way that each unsupported feature will be removed while the code being effectively the same. The base system proposed here relies on the fact that a substitute is available in the target language. It has the following challenges:

- Identification of grammar tokens
- Preservation of structure
- Preservation of User defined information.

5 SSCCJ: System for Code Conversion

Code conversion between languages requires not only inter-conversion of Syntax but also require maintenance of structure while performing transformations and user defined information such as Variable names as far as possible. For this purpose, we need to establish a translation system. Figure 1 shows the prototype of the proposed System.

The System will convert all the tokens from source language to its equivalent into the target language. It will also preserve structure and user defined information as much as possible. This system will require formal language specifications (known as Grammar) of source and target languages, as well as a Dictionary mapping of source to target tokens. This system primarily will contain 2 sub-systems:

(A) **Syntax Replacement System**

This sub-system will be responsible for identifying tokens as well as replacing them with perfect substitutions. Identification can be done via parsers. Substitutes can be mapped in an appropriate data structure for syntax replacement. However, adequate source code transformation will be required in case a perfect substitute is unavailable. Figure 2 shows the structure of Syntax Replacement System.

Table 2 Limitations of various existing conversion tools

Tools	Pointers	Readability	Operator overloading	External libraries	goto	Multiple inheritances	Replace function template	Union	Replace typedefs
Ephedra	×	+	×	×	×	×	×	×	×
C2J++	×	+	×	×	×	×	×	×	×
C2J	×	–	×	×	×	×	×	×	×
C++2Java	×	+	√	×	×	×	×	×	×
Cappuccino	×	+	√	×	×	×	×	×	√

Tools	Class template	Replace enumerations	Insert required casts	Replace class template	#define to constant variable	Friend function	Replace unsigned data types	Insert access controls
Ephedra	×	×	×	×	×	×	√	√
C2J++	×	×	×	×	×	×	√	√
C2J	×	×	×	×	×	×	√	√
C++2Java	×	×	×	×	×	×	√	√
Cappuccino	√	√	√	√	√	×	√	√

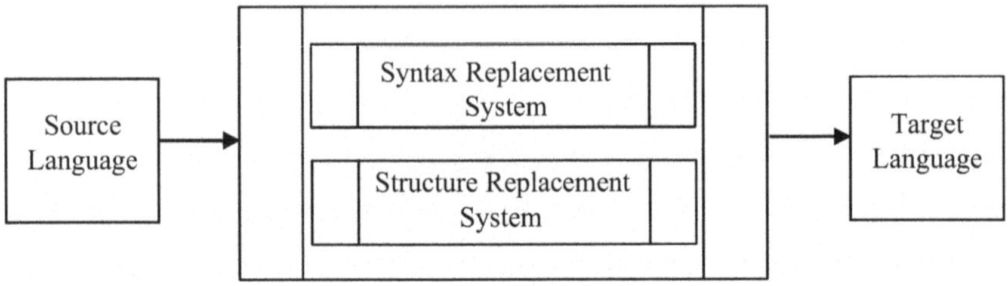

Fig. 1 Translation system

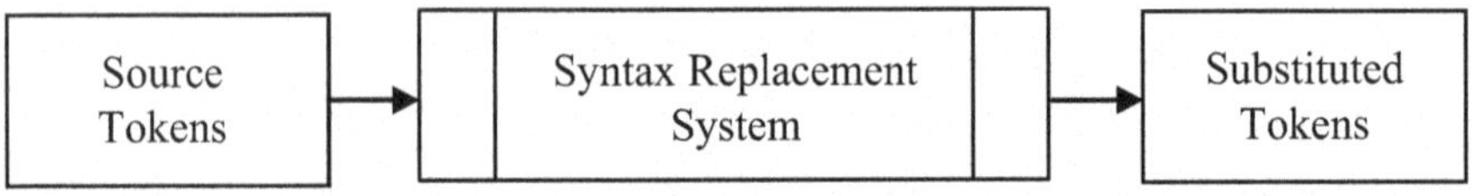

Fig. 2 Syntax replacement system

(B) **Structure Replacement System**

This sub-system will be responsible for identifying structures, scopes, and indentations in the source program. The structure of a program can be determined via parsers. The replacement can be determined by analyzing source and target language grammars. In case of conversion between C++ and java, the basic structure and scope will be almost identical, though structural features like Multiple Inheritance will pose a challenge to this sub-system. Figure 3 shows the structure of Structure Replacement System.

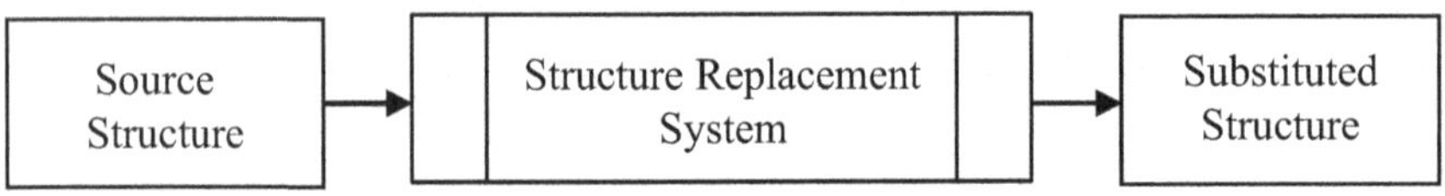

Fig. 3 Structure replacement system

Table 3 System abstract shows parameter as objectives, primary tools, requirements, sub-systems, applications

Objectives	Intra paradigm conversion
Primary tools	ANTLR 4, LLVM
Requirements	Source to target token dictionary
	Source and target grammar files
Sub-systems	Syntax replacement system
	Structure replacement system
Applications	Migration from one language/paradigm to another

The proposed system will be working on the above system abstract such as primary tools used, requirements, sub-systems, and applications. This proposed system is based on Intra Paradigm Conversion limited to object-oriented languages only, currently working on C++ to Java source code conversion. ANTLR 4 and LLVM or any other language parser may be used by this translation system. Translation System has sub-system called Syntax Replacement System and Structure Replacement System which may be interleaved.

5.1 Tools for Analysis of Code Conversion

Another Tool for Language Recognition (ANTLR) is a parser generator used to design compilers, language interpreters, and other converters [19]. The LLVM (Low Level Virtual Machine) is a set of compiler and tools which is designed around high-level intermediate language representation and supports various transformations that can be applied for optimization.

6 Conclusion

In this paper, we are attempting to find the solution of converting one programming language code into another such that readability is preserved and the unsupported features are appropriately substituted. We can create a substitution system which will substitute source tokens to target tokens and will apply appropriate algorithms to create a workaround for unsupported features. In the present era of Internet of Things, this conversion will help programmers to convert the programming code from C++ to Java as per their requirements and is used in computer for diverse application(s).

References

1. D. George, P. Girase, M. Gupta, P. Gupta, A. Sharma, Programming language inter-conversion. Int. J. Comput. Appl. **975**, 8887 (2010)
2. K. Taneja, H. Taneja, R. Bhullar, EEGRP: grid based and caching equipped energy efficient routing protocol for mobile ad-hoc networks with restricted mobility. Far East J. Electron. Commun. **3**, 185–199 (2016)
3. K. Taneja, H. Taneja, R. Kumar, Multi-channel medium access control protocols: review and comparison. J. Inf. Optim. Sci. **39**(1), 239–247 (2018)
4. S. Ribic, A. Salihbegovic, High level language translator with machine code as representation of the source code, in *29th International Conference on Information Technology Interfaces*, 25 June 2007 (IEEE, 2007), pp. 777–782
5. Tangible software solutions inc., C++ to Java converter, commercially available project, release v18.12.17 (2018). https://www.tangiblesoftwaresolutions.com/

6. S. Malabarba, P. Devanbu, A. Stearns, MoHCA-Java: a Tool for C++ to Java conversion support, in *Proceedings of 1999 International Conference on Software Engineering* (IEEE Cat. No. 99CB37002), 22 May 1999 (IEEE, 1999), pp. 650–653
7. I. Saini, M. Sharma, A. Singal, An analytical study of C++ to Java migration strategy using conversion tool, in *2014 International Journal of Computer Application and Technology* (2014), pp. 75–78
8. Forte Developer 6 Update 2, Sun Workshop 6 update 2. C++ Migration Guidem, Sun Microsystems Inc., July 2001
9. L. Alder, CA PLEX-Migrating from C++ to C# WPF client (2008)
10. K. Olmos, E. Visser, Composing source-to-source data-flow transformations with rewriting strategies and dependent dynamic rewrite rules, in *International Conference on Compiler Construction*, 4 Apr 2005 (Springer, Berlin, Heidelberg), pp. 204–220
11. M. Trudel, C.A. Furia, M. Nordio, Automatic C to OO translation with C2Eiffel, in *2012 19th Working Conference on Reverse Engineering*, 15 Oct 2012 (IEEE, 2012), pp. 501–502
12. J. Martin, H.A. Muller, Strategies for migration from C to Java, in *Proceedings of Fifth European Conference on Software Maintenance and Reengineering*, 14 March 2001 (IEEE, 2001), pp. 200–209
13. J.R. Cordy, T.R. Dean, A.J. Malton, K.A. Schneider, Software engineering by source transformation-experience with TXL, in *Proceedings of First IEEE International Workshop on Source Code Analysis and Manipulation*, 10 Nov 2001 (IEEE, 2001), pp. 168–178
14. J.R. Cordy, TXL source transformation in practice, in *2015 IEEE 22nd International Conference on Software Analysis, Evolution, and Reengineering (SANER)*, 2 Mar 2015 (IEEE, 2015), pp. 590–591
15. X. Wang, J. Sun, X. Yang, C. Huang, Z. He, S.R. Maddineni, Reengineering standalone C++ legacy systems into the J2EE partition distributed environment, in *Proceedings of the 28th International Conference on Software Engineering*, 28 May 2006 (ACM, 2006), pp. 525–533
16. Y. Liu, Q. Liu, Joint parsing and translation, in *Proceedings of 23rd International Conference on Computational Linguistics*, 23 Aug 2010 (Association for Computational Linguistics, 2010), pp. 707–715
17. S.M. Alnaeli, M. Sarnowski, M.S. Aman, A. Abdelgawad, K. Yelamarthi, Vulnerable C/C++ code usage in IoT software systems, in *2016 IEEE 3rd World Forum on Internet of Things (WF-IoT)*, 12 Dec 2016 (IEEE, 2016), pp. 348–352
18. T. Yamakami, Migration-oriented partial adaptation architecture for IoT-empowered city platform as a service, in *Twelfth International Conference on Digital Information Management (ICDIM)*, 12 Sept 2017 (IEEE, 2017), pp. 81–85
19. T. Parr, *The Definitive ANTLR Reference: Building Domain-Specific Languages*. The Pragmatic Bookshelf
20. J. Martin, Ephedra: a C to Java migration environment. Doctoral dissertation
21. NovoSoft's C2J Converter, Software development Services, tech.novosoft-us.com, release date 2001. http://tech.novosoft-us.com/product_c2j.jsp
22. F. Buddrus, J. Schödel, Cappuccino—a C++to Java translator, in *Proceedings of the 1998 ACM symposium on Applied Computing*, 27 Feb 1998 (ACM, 1998), pp. 660–665

Efficient VM Migration Policy in Cloud Computing Environment

Annie Pathania, Kiranbir Kaur and Prabhsimran Singh

Abstract Cloud computing provides resources on shared basis but resources do get exhausted, as more and more resource-dependent tasks are being executed on the cloud. This eventually leads to distortion and one possible solution to overcome this problem is migration. In this paper, we perform VM migration in an energy-efficient manner for which we calculate the load factor on all the individual servers. If the load exceeds the assigned threshold value, then that server is considered as the overloaded host, after this the random selection of the VMs is done from the under-loaded hosts and then the machine with less migration time and more utilization will be migrated to the destination host. Further, we compare our purposed technique with already established techniques. The comparison results in the form of Migration Time, Utilization, and Energy Consumption shows that the proposed technique performs better than the existing one.

Keywords Migration · Load balancing · Utilization · Energy consumption · VM sorting

1 Introduction

Cloud computing is a research area which deals with on-demand availability of computer resources. Almost all IT industries need flexible cloud computing platform with millions of physical hosts and devices. It is also the computing over Internet with no requirement of the infrastructure at the user's site. The main technology behind cloud computing is virtualization with the help of which the cloud providers offer the on-demand services to the users. These services are Pay-Per-Use services which

A. Pathania · K. Kaur · P. Singh (✉)
Department of Computer Engineering & Technology, Guru Nanak Dev University, Amritsar, India
e-mail: prabh_singh32@yahoo.com

A. Pathania
e-mail: apathania89@gmail.com

K. Kaur
e-mail: kiran.dcse@gndu.ac.in

M. Dutta et al. (eds.), *Proceedings of International Conference on IoT Inclusive Life (ICIIL 2019), NITTTR Chandigarh, India*, Lecture Notes in Networks and Systems 116,
https://doi.org/10.1007/978-981-15-3020-3_36

are flexible, reliable, and available. Instead of purchasing the high-cost infrastructure and maintaining it, organizations can make use of cloud platform. There exists pool of resources associated with cloud and these resources are labeled as local or global pool resources [1]. Local pool resources belong to same cluster and can be accessed by clients belonging to same cluster without authorization. While the global pool resources belong to distinct clusters, these resources are outside the premises of client. In order to access these resources proper authorization is required. The cost encountered while accessing such resources is high [2], and in order to tackle this issue, optimum allocation policies are worked upon where global pool resources are accessed only if local resources are unavailable.

However, reliability is a major concern in cloud computing since the resources and mass requirements may vary. The required resources for the desired period of time may not serve the purpose of client and hence reliability is an issue that is considered in this research. One of the possible mechanisms to enhance reliability is to make use of load balancing strategies [3]. Several mechanisms are prescribed for balancing the load on virtual machines (VMs) and though cloud is considered to be containing virtually infinite resources but still load asserted by various clients makes it fail in one way or the other [4].

This has led to the creation of large-scale data centers which contain large number of nodes which consumes enormous amount of electricity hence leading to environmental degradation. This inefficient usage of resources is the main reason for this high energy. The handling and managing of the over-provisioned resources result toward the high cost. Another problem is the servers, even the ideal servers consume up to 70–75% of the power. Hence keeping the servers underutilized is highly inefficient. High-energy consumption by the infrastructure leads to carbon dioxide emission which leads to greenhouse effects.

With the help of the virtualization, multiple virtual machines can be created on a single physical server which improves the utilization of the resources. This also leads to increased Return on Investments (ROI). Switching the ideal nodes to sleep or hibernate mode can help to achieve reduction in energy consumption. Live migration is one technique, using which the VMs can be consolidated to least number of nodes based upon their requirement. With application's increased demand that leads to rise in the resource utilization can cause performance degradation in VMs. If the application's resource requirements are not met, then the application faces increased time outs, failures, response times, etc. Hence the providers focus on the reduced energy consumption which is a part of Service Level Agreement (SLA).

This paper pays stress on determining accurate load on servers and then initiates migration depending upon the migration time and the utilization of the virtual machine. The load balancing strategies considered for evaluation are broker dependent. The enhancement to load balancing strategies leads to migration strategies. Rest of the paper is organized as follows: Sect. 2 discusses earlier work that focuses primarily on load balancing strategies. Section 3 provides in depth the research methodology followed. Section 4 provides performance analysis and results. Finally, we make concluding remarks in Sect. 5.

2 Review of Literature

This section is categorized into two subsections. In the first subsection, we discuss the various work done in load balancing strategies. While the second subsection discusses about migration.

(a) **Load Balancing Strategies**: In this subsection, we discuss the various contributions made in the field of load balancing in cloud environment.

Nathuji and Schwan [5] were the first to consider the issue of power management of virtualized data centers. Their proposed technique divided the resources into local and global level. The operating system's power management strategies were supported by local level. While the global manager gets the information from the local mangers regarding the current resource allocation and then applies its policies to decide if the VM placement needs to be considered.

Pinheiro et al. [6] worked on minimization of power consumption which serves multiple web applications. In their technique the idle nodes were turned off. The load of the nodes was monitored periodically to take the decision to turn off the idle nodes. This check helps to minimize the overall power consumption. However, their technique suffered from a major shortcoming as the algorithm used to run on master node, it created the problem of single point failure.

Chase et al. [7] worked on the problem of homogeneous nature of resources in Internet hosting centers. Since determining the resource demand of the application and allocating the resources is a major challenge. Therefore, they applied a technique where the services used to bid for quantity of resources. The system serves the request for each service by maintaining an active set of servers. By switching the ideal servers to power saving modes the energy consumption was reduced. The web workload was targeted by the system which includes noise in the data. This problem was addressed by applying "flip-flop" filters.

Kusic et al. [8] used heterogeneous environment to define the problem of the power management and addressed it using Limited Lookahead Control (LLC). To estimate the future requests and the future state, Kalam filters are applied to perform the reallocations. Due to model's complexity the execution time was more.

Zhu et al. [9] worked on the issue of automated resource allocation and capacity planning. They considered different time scale ranging from hours to seconds. This places workload onto the groups of servers, reallocates the VMs, allocates resources to the VMs, and applied an approach of setting up the fixed utilization threshold value, which was not that efficient.

Berral et al. [10] used the machine learning technique to optimize energy consumption and SLA. Their proposed approach was designed for the High-Performance Computing (HPC) where concept of deadline constraint occurs making it unsuitable for the mixed workload.

Keahey and Deshpande [11] considered the migration of the multiple virtual machines from multiple source hosts to multiple destination hosts taking network contention into consideration.

Beloglazov and Buyya [12] proposed an algorithm which is used to reduce energy consumption while maintaining high-level SLA. Their approach was based upon a novel adaptive heuristics for dynamic consolidation of VMs which is used to analyze the historical resource usage of the VM.

(b) **Migration**: Since, the literature discussed in the above subsection mainly focused on load balancing strategies that ultimately lead to migration strategies. Generally, load balancing strategies are categorized as the following [13–16]:

- Time Quantum-Based Load balancer
- Weighted Round Robin Load balancing
- Least Connection-based Load balancing
- Weighted Least Connection
- Agent-based Load balancing approach.

The migration is initiated if any of the above load balancing strategies fail. The migration strategies are commonly used for fault tolerance and increasing reliability. Migration is required if deteriorated virtual machines or servers are detected [17, 18]. The migration mechanism could be data oriented or virtual machine oriented [19, 20].

Data Migration [21, 22]: Data migration mechanism shifts the data from current server or virtual machine to next virtual machine. The data centers in the cloud act as resource provider and virtual machine mechanism divides the data centers into a set of virtual machines. Time-critical applications often require that virtual machine to complete the assigned task within prescribed deadline. This task execution requires data and if virtual machine is not executing task within deadline then data migration is also required. Moreover, data migration plays a critical role in reactive fault tolerance.

VM Migration [23, 24]: VM migration is required incase virtual machines become unusable. The VM migration mechanism could be live or offline. In offline migration, source and destination virtual machines must be switched off while performing transformation. This leads to increase in downtime and migration time. In live migration source and destination both are online in nature. This means that downtime and migration time shows great deviation as compared to offline deviation. Live VM migration, however, suffers from overhead problem of maintaining database and storage resources all the time. A possible solution to overcome this problem is to use hybrid migration that uses advantages of both offline and online migration.

Core Migration: The core level migration is within the local pool of virtual machines. The virtual machines are divided into cores as per requirements of the client. The dynamic requirements cause virtual machines to split into parts where each distinct part is known as core. Core migration mechanism initiates if one part of virtual machine fails then other core within same virtual machine is selected for migration. The core level migration is divided into static and dynamic parts. Static cores are fixed that means core division is fixed to specific quantity and in dynamic core division, core partitioning depends upon the requirements of the clients. The reliability in both the cases is high but overhead in dynamic cores is more as compared to static core migration.

So taking inspiration from above studies, migration mechanism in the proposed system is dynamic and online facility of virtual machines is used for decreasing downtime and migration time. Through this research we try to fulfill two main objectives: (a) To reduce the overall energy consumption in the cloud environment. (b) To increase the overall efficiency of the cloud system.

3 Research Methodology

Multiple hosts are considered during migration process. Physical machine selected as a host for migration does not perform any migration. In other words, this host is a static host that is optimal enough without deterioration affects. The simulation setup for the proposed mechanism is described in Table 1.

The simulation mechanism first of all initializes the Cloudsim. Once the Cloudsim is initialized, datacenter is created. The datacenters provide resources to the virtual machines. The virtual machines are given proportionate resources corresponding to the datacenter. The broker initialized for the simulation determines the optimal VM and assign cloudlets to the VM. The VM capacity is the critical parameter determining load over the host machine. As the load over the host increases, host may deteriorate. This requires a special host machine that is used for hosting the migrated task. This machine has high memory capacity for holding the task list that is received from deteriorated machines.

The entire implementation of the proposed work is done using Cloudsim 3.0. Host-based migration is considered in the proposed system for achieving optimality of result. With the decreasing migration time and the increasing utilization, Energy consumption is decreased to some extent and if the expected migration time of any VM is less as compared to the other VMs, it will be migrated to the destination host. The flowchart of the proposed work is given in Fig. 1.

The proposed mechanism follows the broker aware monitoring that allows the load to be balanced in case threshold load is exceeded. In case overloaded host is detected, host-level migration is initiated. The overloaded host list is maintained to prevent the load allocation on the overloaded host. The starvation problem and reliability are resolved using the proposed mechanism. This is demonstrated using the result and performance analysis.

Table 1 Simulation setup for the proposed mechanism

Nodes/software	Configuration
5 Hosts	I3 processor, with Varying virtual machines with each host
1 Physical Machine	Used for migration
Virtual Machines (VMs)	Varying distinct VMs
Netbeans with Cloudsim	Version 8.2 and 3.0

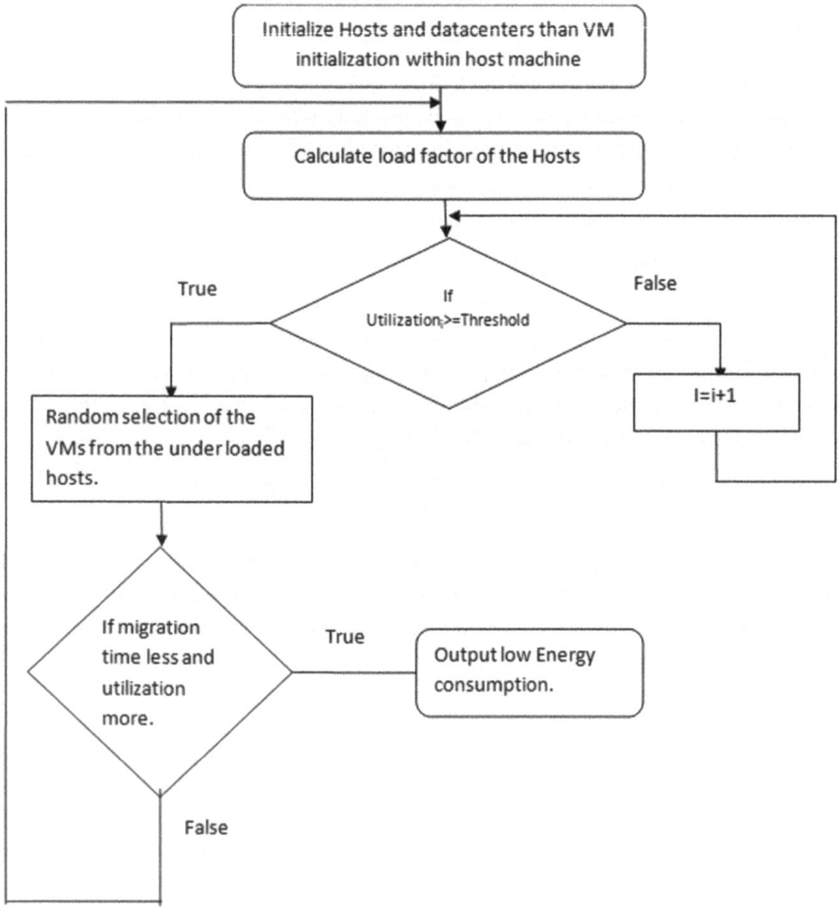

Fig. 1 Flowchart of the proposed work

4 Performance and Result

The performance and result analysis indicate the performance betterment in terms of downtime and migration time. The downtime indicates amount of time during which virtual machine does not perform any operation. Migration time indicates the amount of time that is required for transferring the memory and processing elements toward the destination host. Migration time and downtime are being optimized using the proposed mechanism.

To accomplish the task of comparing the efficiency of our algorithm, we have made use of several metrics. These metrics include Energy Consumption by the servers; another metric is VM migration time and the last metric is Utilization of the VMs which is considered high. With the combination of the decreased VM migration

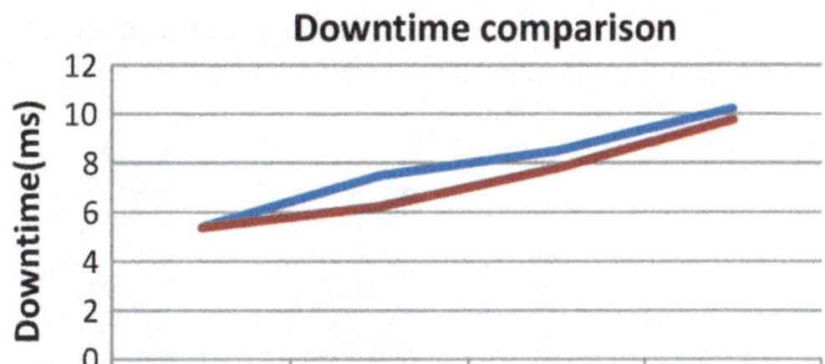

	With 100 Cloudlets	With 200 Cloudlet	With 300 Cloudlets	With 400 Cloudlets
Existing Work	5.4	7.5	8.5	10.25
Proposed With load balancing	5.4	6.25	7.8	9.8

Fig. 2 Downtime of existing and proposed mechanism

time and the increased utilization of the VM, Energy Consumption is decreased. The result of downtime comparison is shown in Fig. 2.

It is observed that at lower cloudlet count, the value of downtime does not show much variation but at high cloudlet count downtime shows significant deviation. The proposed mechanism reduces downtime as the load on the host increases. The migration time is compared against the existing migration time which is much greater at higher loads as compared to proposed mechanism. This is demonstrated through Fig. 3.

The migration time at significantly low load is similar for both existing and proposed mechanism. At higher load the values show deviation toward the proposed system. The Score factor is given in Eq. 1, where f1 is the migration time (factor 1) and f2 is the utilization (factor 2). α is the weight of the various VMs.

$$\mathrm{SF} = \alpha 1 f1 + (1 - \alpha 1) f2 \tag{1}$$

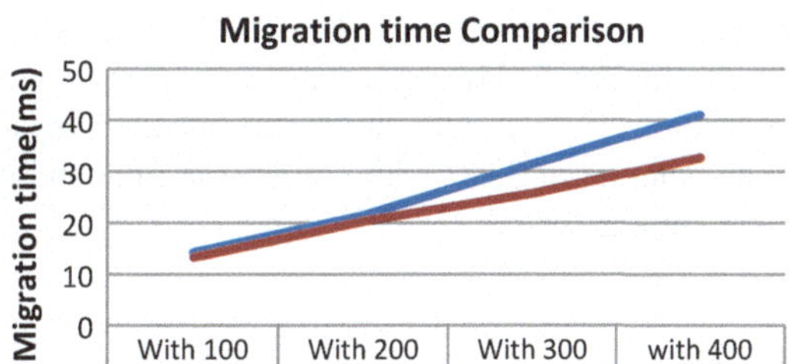

	With 100 Cloudlets	With 200 Cloudlets	With 300 cloudlets	with 400 cloudlets
Existing Approch	14.3	21.5	31.5	41.05
Proposed Approch	13.4	20.4	25.8	32.8

Fig. 3 Migration time comparison of existing and proposed system

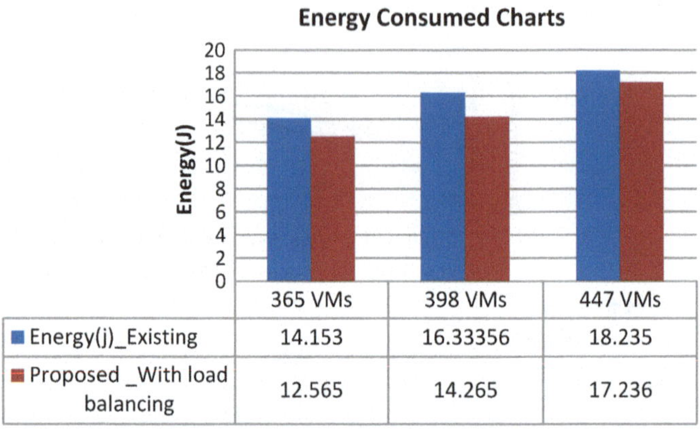

Fig. 4 Comparison of energy consumption

Table 2 Values of various parameters

Number of cloudlets	VM_Migration time (ms)	Utilization	Energy consumption (J)
100	10.235	53.247	2.3
200	15.365	56.784	6.75
300	19.525	52.486	6.24
400	23.365	54.172	7.51

The performance and result are observed in terms of energy consumed which are also optimized through the proposed system. For optimization, this factor must be reduced. By following our proposed approach energy consumption is significantly reduced. The plots in terms of energy consumed are given in Fig. 4. The results show that energy consumed is reduced significantly using our proposed approach.

Through this study, we also tried to established migration time, utilization and energy consumption. The results are shown in Table 2, while they are presented as a correlation plot in Fig. 5. Migration time and energy consumption share a strong positive association of 87%, utilization and energy consumption share a moderate positive association of 38%, while migration time and utilization share a weak negative association of 7%.

5 Conclusions

Nowadays cloud computing is providing a cheap alternative to organizations to carry out their operations. However, the problem of load balancing creates a major hindrance at performance level. The aim of this paper was to work on the principle of load balancing and migration. The load on the host is observed, and then threshold values

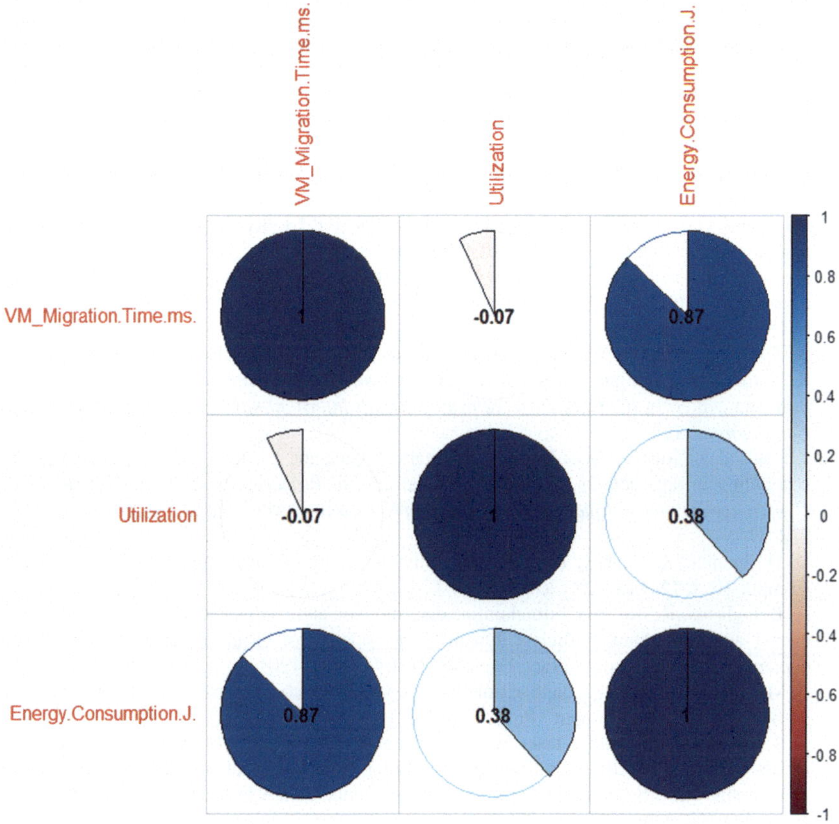

Fig. 5 Relationship plot

are maintained to check overloaded virtual machines. In addition, overloaded virtual machine host deterioration is checked. In case deterioration is detected, hosts which are not overloaded are considered and randomly the VMs are selected from those hosts. The randomly selected VMs migration time and the utilization are checked. The machine with the minimum migration time and the maximum utilization will be migrated to the destination host. The result obtained shows improvement at greater load and hence proves worth of study.

References

1. M. Kaur, S. Sharma, R. Kaur, Optimization of job scheduling in cloud computing environment. Int. J. Adv. Res. Comput. Sci. Softw. Eng. **4**(7) (2014)
2. B. Jennings, R. Stadler, Resource management in clouds: survey and research challenges. J. Netw. Syst. Manage. **23**(3), 567–619 (2015)

3. A. Zhou, S. Wang, Z. Zheng, C.H. Hsu, M.R. Lyu, F. Yang, On cloud service reliability enhancement with optimal resource usage. IEEE Trans. Cloud Comput. **4**(4), 452–466 (2014)
4. J. Li, S. Su, X. Cheng, M. Song, L. Ma, J. Wang, Cost-efficient coordinated scheduling for leasing cloud resources on hybrid workloads. Parallel Comput. **44**, 1–17 (2015)
5. R. Nathuji, K. Schwan, Virtualpower: coordinated power management in virtualized enterprise systems. ACM SIGOPS Oper. Syst. Rev. **41**(6), 265–278 (2007). ACM
6. E. Pinheiro, R. Bianchini, E.V. Carrera, T. Heath, Load balancing and unbalancing for power and performance in cluster-based systems (2001)
7. J.S. Chase, D.C. Anderson, P.N. Thakar, A.M. Vahdat, R.P. Doyle, Managing energy and server resources in hosting centers. ACM SIGOPS Oper. Syst. Rev. **35**(5), 103–116 (2001)
8. D. Kusic, J.O. Kephart, J.E. Hanson, N. Kandasamy, G. Jiang, Power and performance management of virtualized computing environments via lookahead control. Clust. Comput. **12**(1), 1–15 (2009)
9. X. Zhu, D. Young, B.J. Watson, Z. Wang, J. Rolia, S. Singhal, B. McKee, C. Hyser, D. Gmach, R. Gardner, T. Christian, Integrated capacity and workload management for the next generation data center, in *ICAC08: Proceedings of the 5th International Conference on Autonomic Computing* (2008)
10. J.L. Berral, Í. Goiri, R. Nou, F. Julià, J. Guitart, R. Gavaldà, J. Torres, Towards energy-aware scheduling in data centers using machine learning, in *Proceedings of the 1st International Conference on energy-Efficient Computing and Networking*, Apr 2010 (ACM, 2010), pp. 215–224
11. U. Deshpande, K. Keahey, Traffic-sensitive live migration of virtual machines. Futur. Gener. Comput. Syst. **72**, 118–128 (2017)
12. A. Beloglazov, R. Buyya, Optimal online deterministic algorithms and adaptive heuristics for energy and performance efficient dynamic consolidation of virtual machines in cloud data centers. Concurr. Comput.: Pract. Exp. **24**(13), 1397–1420 (2012)
13. B. Meroufel, G. Belalem, Adaptive time-based coordinated checkpointing for cloud computing workfl ows. Scalable Comput.: Pract. Exp. **15**(2), 153–168 (2014)
14. K. Li, Scheduling parallel tasks with energy and time constraints on multiple manycore processors in a cloud computing environment. Futur. Gener. Comput. Syst. **82**, 591–605 (2018)
15. A.V. Dastjerdi, R. Buyya, An autonomous time-dependent SLA negotiation strategy for cloud computing. Comput. J. **58**(11), 3202–3216 (2015)
16. J. Xu, S. Pears, A dynamic shadow approach to fault-tolerant mobile agents in an autonomic environment. R.-Time Syst. **32**(3), 235–252 (2006)
17. P.D. Patel, M. Karamta, M.D. Bhavsar, M.B. Potdar, Live virtual machine migration techniques in cloud computing: a survey. Int. J. Comput. Appl. 86(16) (2014)
18. D. Duolikun, S. Nakamura, R. Watanabe, T. Enokido, M. Takizawa, Energy-aware migration of virtual machines in a cluster, in *International Conference on Broadband and Wireless Computing, Communication and Applications*, Nov 2016 (Springer, Cham, 2016), pp. 21–32
19. B. Zhao, X. Chen, J. Zhu, Z. Zhu, Survivable control plane establishment with live control service backup and migration in SD-EONs. J. Opt. Commun. Netw. **8**(6), 371–381 (2016)
20. D. Duolikun, S. Nakamura, T. Enokido, M. Takizawa, Energy-efficient replication and migration of processes in a cluster, in *2015 Ninth International Conference on Complex, Intelligent, and Software Intensive Systems*, July 2015 (IEEE, 2015), pp. 118–125
21. J. Sekhar, G. Jeba, Energy efficient VM live migration in cloud data centers 1 (2013)
22. F. Curzi, M. Ryan, U.S. Patent No. 9,459,856. U.S. Patent and Trademark Office, Washington, DC (2016)
23. N.R. Katsipoulakis, K. Tsakalozos, A. Delis, Adaptive live VM migration in share-nothing IaaS-clouds with LiveFS, in *2013 IEEE 5th International Conference on Cloud Computing Technology and Science*, vol. 2, Dec 2013 (IEEE, 2013), pp. 293–298
24. Q. Wu, F. Ishikawa, Q. Zhu, Y. Xia Energy and migration cost-aware dynamic virtual machine consolidation in heterogeneous cloud datacenters. IEEE Trans. Serv. Comput. (2016)

Software-Defined Networks: Need of Emerging Networks and Technologies

Deepak Kumar and Jawahar Thakur

Abstract In recent years, the rapid growth of internet technologies has given rise to heterogeneous networking devices. SDN is a new networking architecture that can control, monitor, and configure devices on a large scale without manual intervention. In this paper, we started from the basics of SDN and current advancements that have been made till now. Here our main concern is to make data, control, and application plane efficient. Efficient packet handling and rule management techniques can improve the data plane performance. The choice of the controller and its placement are key factors for improving the performance of the control plane. SDN platform supports virtualization and because of this feature, it can create virtual links which makes it scalable. The scalability and security is the need of recent emerging technologies like BigData, cloud-based data centers, and IoT, which deal with enormous data generated by heterogeneous devices. Therefore it becomes necessary to offer machine learning-based SDN to effectively utilize the continuously generating data, which will help in making optimal decisions for resource management, routing optimization, placement of controller, identification of the elephant–mice flow, for tracing the malicious applications, and it can even detect or predict attacks. SDN has set up a benchmark that is currently difficult to achieve for any other networking paradigm.

Keywords OpenFlow · NFV · QoS · BigData · Cloud · IoT · Machine learning

1 Introduction

Software-Defined Networks is a new networking paradigm that has enabled the researchers to think beyond the limit. SDN [1] has not only solved the earlier unsolvable problems with ease but has also gained attention among researchers because of

D. Kumar (✉) · J. Thakur
Department of Computer Science, Himachal Pradesh University, Summer Hill, Shimla, India
e-mail: deepak.cs339@gmail.com

J. Thakur
e-mail: jawahar.hpu@gmail.com

M. Dutta et al. (eds.), *Proceedings of International Conference on IoT Inclusive Life (ICIIL 2019), NITTTR Chandigarh, India*, Lecture Notes in Networks and Systems 116,
https://doi.org/10.1007/978-981-15-3020-3_37

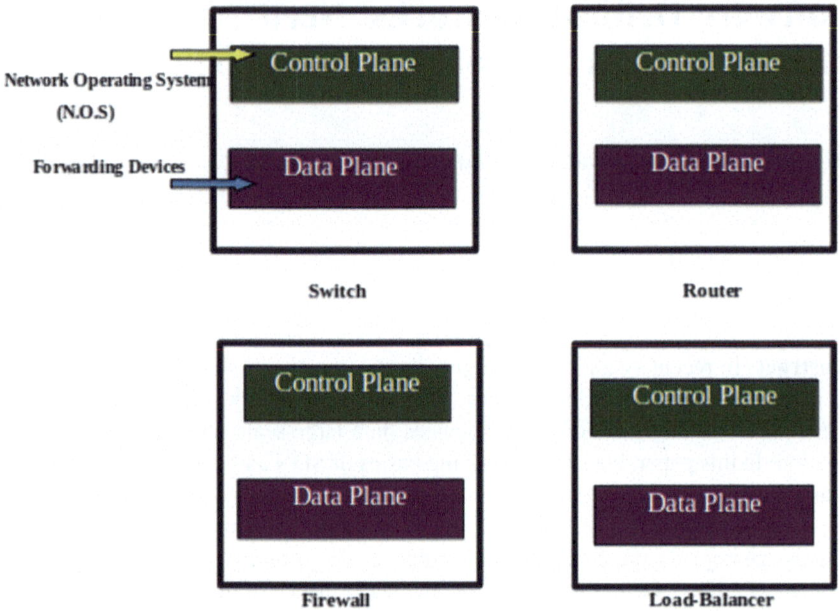

Fig. 1 Traditional networks coupled data plane and control plane

its programmatic feature, which makes it completely different and more advanced compared to traditional networks. SDN will affect the networking professionals because it requires different skill sets as compared to the skills required for conventional networks. If we try to look at the traditional network devices like switches, routers, firewalls, and load-balancer, etc., these devices consist of coupled data plane and control plane as shown in Fig. 1. Also, the software is an integrated part of the hardware and therefore vendor dependent interfaces are required. This tight coupling restricts to use the controller of own choice, another disadvantage of the traditional network is that each device offers limited functionality. The control plane is the costliest part and is the heart of the networking devices. The solution to these problems is SDN, and it is believed to have emerged from the 4D [2] project. It provides the centralized view of the entire network which makes the management task easier. The feature of SDN, i.e., separation of the control plane from data plane makes it different from the traditional networks as shown in Fig. 2. This decoupling is possible because of the OpenFlow [3] protocol and it allows us to choose the controller of our own choice. The shifted new control logic is known as SDN/OpenFlow controller. Having a data plane and SDN controller is not enough, because it is the applications which does actual management task, not the controller. SDN is changing the networking scenario with many benefits. If we talk about traditional networks as we know, each networking device has its functionality, whereas in SDN, each device can perform more than one functionalities, i.e., we can modify the features of devices just by changing the program code.

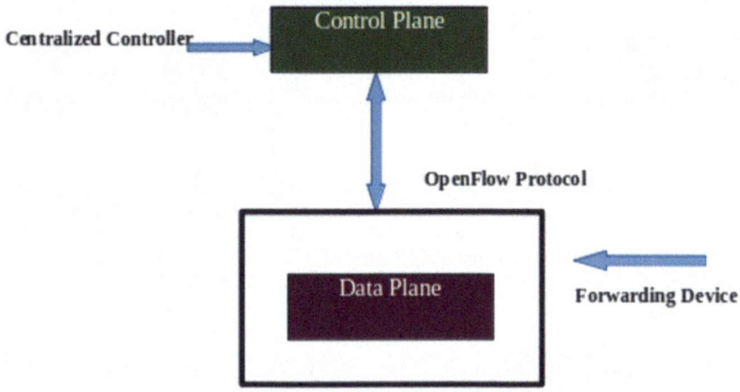

Fig. 2 Decoupled control plane and data plane in S.D.N switch

A single device can serve as different functionalities, by just running the different applications on the top of the application layer. So we can say that SDN has solved the problem of rigidity. By using a centralized controller, we can manage any number of the data plane, which simply means SDN supports scalability. Whenever we require new functionality, we can simply write the code, whereas in traditional networks to have the latest feature, the users have to wait for at least one year for a new version because of vendor-dependence. In simple words, we can say SDN supporting data plane devices will not be obsolete with the period, as a result, it will reduce future costs. In short, the SDN controller programmatically controls the behavior of data plane devices (switches, router, hub, firewall, load-balancer, NAT, etc.). The architecture of SDN can be seen in Fig. 3.

Data Plane: It is also known as the physical layer. It is the layer on which all physical networking devices work. In simple words, it is the part of the network through which packets are transmitted between devices. In the data plane, the devices used should support the OpenFlow functionality, which enables each device to behave or function differently in different circumstances. We can follow different topologies in the routing of packets from source to destination. The data plane devices are programmatically configured by the centralized controller.

Control plane: It is the plane that is just above the data plane and is separated. It is known as the heart of SDN architecture. It is the place where the controller is installed. Sometimes the controller is also called the centralized intelligent controller. It can configure the networking devices which works in data plane through its programmatic feature. This feature overcomes the problem of handling or configuring the physical devices manually. In short, it saves time and also reduces the complexity.

The control plane can contain one or more controllers depending upon the requirement. Currently, we have a variety of choices available to choose a controller. The choice of controller depends upon the platform you want to work on. Various controllers available are NOX [4], POX [5], RYU [6], FloodLight [7], and Maestro [8].

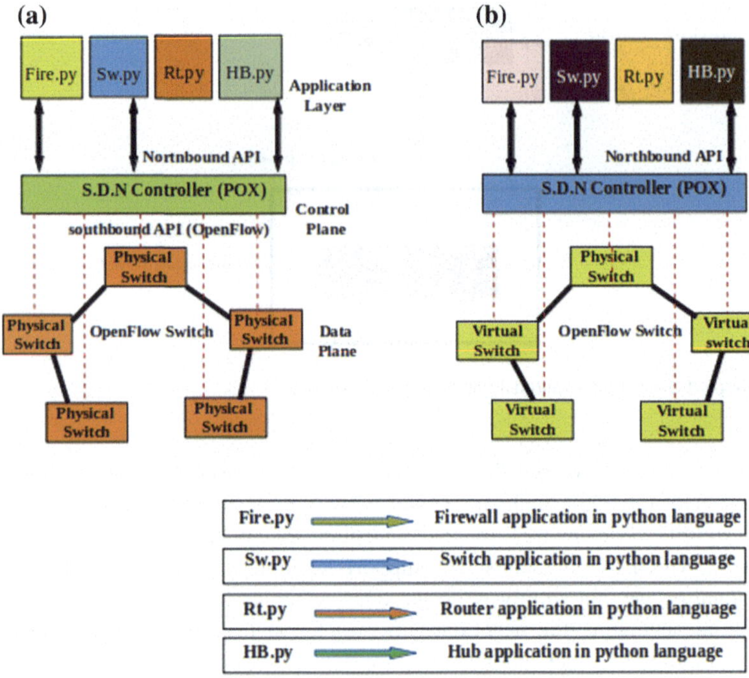

Fig. 3 S.D.N architecture. **a** Physical switch in data plane, **b** physical and virtual switch in data plane

Management Plane: Applications run on the top of the management plane. It is not the controller that does the management task, but it is done by the applications.

In SDN, the data plane devices can be a switch, router, firewall, hub, load-balancer, etc. In order to form a network, we can use any of the physical devices but in our example, we have taken a physical switch in Fig. 3a and we can also use the virtual switch as shown in Fig. 3b. In cloud computing, the 70–80% of the port or switches used are of a virtual type.

Each day networking or communicating devices are increasing which results in the growth of network; so difficulties of handling or managing such big networks are very much complex. At this point, traditional networks fail. SDN can adapt according to the network user's rapidly changing needs for network resources. SDN has made network management task easier through its programmatic feature. SDN concept has been successfully deployed in various projects like the B4 project [9] of Google, Ananta [10] Microsoft project, etc. SDN is the perfect solution for recent emerging networking technologies like cloud computing, IoT, NFV, 5G, etc.

The rest of this paper is organized as follows. The techniques for packet handling in SDN is presented in Sect. 2. The efficient rule management is presented in Sect. 3. The mechanism to make flow table efficient is briefly introduced in Sect. 4. The types of controller and their scalability are described in Sect. 5. In Sect. 6, we have presented the concept of SDN and parallel technologies. In Sect. 7, we have finally discussed the conclusion and scope.

2 Techniques for Packet handling in SDN

There are mainly three techniques that can be used for the handling of packet in network.

2.1 Reactive Technique

It is believed that the concept of the rules and policy management in SDN has originated from Ethane [11]. The reactive rule management technique consists of two steps. In **first** step, when the first packet came across the ingress physical switch, it simply forwards the packet to the controller in the control plane. In **second** step as the first packet is received by the controller, it then decides the routing path and informs the forwarding devices to update the new rules or policies. The number of control messages generated in a reactive approach is much more compared to the proactive approach [12]. The controller can change the topology or routing path at any time. Another disadvantage of the reactive approach is that cache rules across the forwarding switches can create problems like the need for large buffers, packet delay, and packet misses can lead to the switch's complexity. Flow entries placed across flow table using reactive approach are used in applications that provide best-effort services [13] such as online load balancing.

2.2 Proactive Technique

This technique reduces the load on the controller which minimizes the overhead and also the processing time. This affects the flexibility and also minimizes the decision making capabilities of the controller. However the aim is to keep the traffic in the data plane only. There is no need to ask the controller where to forward the packet. One of the scalable flow based architecture is difane [14]. The Fig. 4 illustrates how difane handles the flow of packet.

The controller finalizes the rules and distributes across the switches in the data plane including the authority switches. The purpose of the partitioning algorithm (which is the component of the controller) is to divide rules across authority switches and is run by the centralized controller. All packets in the data plane are handled by switches and forwarding of these packets takes place via authority switches. This makes SDN efficient and scalable by keeping all traffic in data plane. Wild card rules

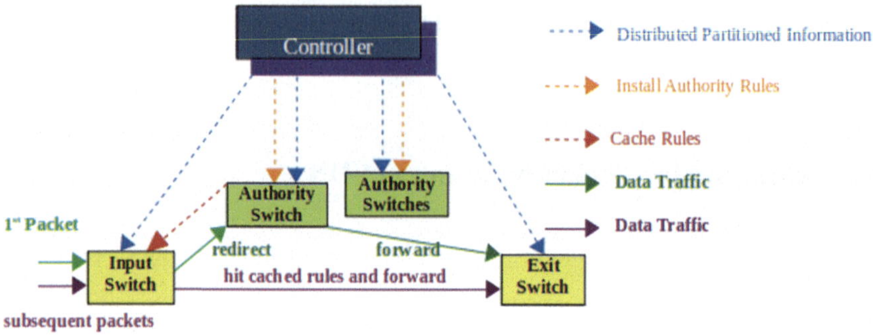

Fig. 4 Flow management architecture

are the key features of difane that performs necessary actions whenever matching of packet takes place.

The flow based switches either forward the packet or drop the packet, depending upon the rule i.e. each incoming packet matches against the packet header bits of each switch (If the packet is already present in the packet header field of the flow table then it will be discarded, otherwise will be accepted). It is not convenient to pre-installed all the rules in the advance in reactive approach, and this is not possible because of two main reasons:- first reason is rule changes over time, and the second reason is memory is limited. In order to make SDN efficient therefore switches themselves have to take intelligent decisions like where to forward the packet instead of asking the controller. By doing so we can overcome the problem of bottleneck across the controller and can keep the traffic in data plane only, and this is possible only if there is a permanent connectivity among forwarding devices or end hosts [12]. Using difane, we can handle the problems of packet misses very efficiently.

Proactive way of placing flow entries used in specific networking domain like data center networks or applications [13] which requires delay or bandwidth guarantees etc.

2.3 Hybrid Technique

In the hybrid technique [15], the controller can act as both the reactive as well as proactive way. Network administrator can follow the proactive way to install certain flow entries in the forwarding devices and controller can use reactive approach to update or delete new flow entries depending upon the incoming traffic.

3 Efficient Rule Management

Although the decoupling of data plane and control plane offers many functionalities, But this decoupling also leads to complexity in rule management. The reactive approach degrades the forwarding performance. The difane and cache flow helps to make the early proactive techniques even better by dividing the rule into several portions based upon rule-dependencies and memory capacity of OpenFlow switches. Even this proactive technique lacks to generate rules dynamically as the growth of new network states takes place. Therefore rule management scheme [16] must be intelligent enough to provide flexibility as well as it can improve the forwarding performance. This rule scheme consists of two types of rules:- (i) Rules for path. (ii) Rules for nodes.

3.1 Rules for Path

These rules usually consist of multiple rules which act together to implement particular routing policy in path. All possible paths are calculated in advance and pre-installed all path-based rules. Such rules can also ensure consistency.

3.2 Rules for Nodes

To provide the flexibility which is not supported by the proactive approach. For node-based rules, we need a reactive approach of SDN with additional features. The concept is that we need to divide the node-based rules into separate chunks and implement stratified matching to eliminate the duplicity of rules.

4 Mechanism to Make Flow Table Efficient

The general architecture of the OpenFlow switch [17] is as shown in Fig. 5.

OpenFlow enabled switch can contain more than one flow table. Apart from the flow table, the OpenFlow switch also contains the group table, meter table, and the OpenFlow logic (secure control channel). Here our prime focus is on the flow table. So we will discuss the flow table working in detail and will also discuss the mechanism to make data plane efficient.

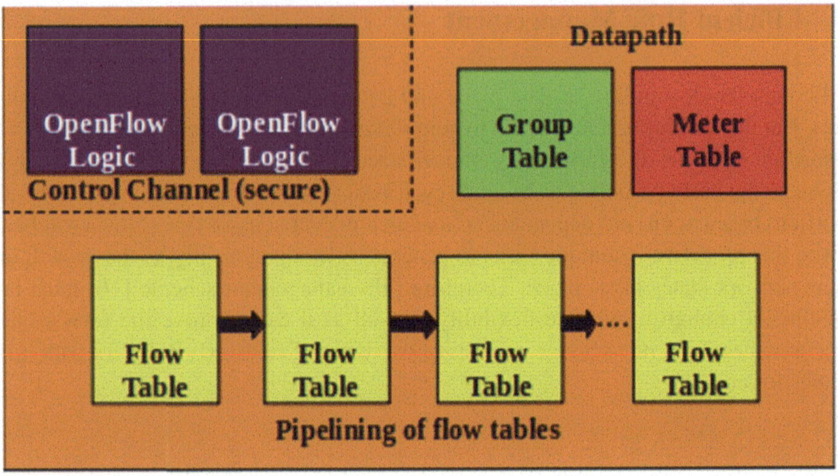

Fig. 5 OpenFlow switch

4.1 Flow Table

Each networking device maintains its flow table. The purpose of the flow table is to store information about each packet whether it is incoming or outgoing. (contain the traffic flow entries). Each flow table consist of the following fields (Fig. 6):-

- **Match Field**: The purpose of the match field is to match each incoming packet with the new updated information in the flow table. The match field further consists of the other fields like ingress port, ethernet source address, ethernet destination address, ethernet type, vlan_id, vlan_priority, ipv4 source address, IPv4 destination address, etc. as shown in Fig. 7. match field entries.
- **Priority Field**: It performs the precedence on each incoming packet based upon some prioity rule.
- **Counter Field**: As the name suggests it counts the each incoming packet. The value of the counter field increments by one whenever a new packet came. It also helps to calculate the life cycle of any packet.
- **Timeout Field**: It is maximum time limit pre-setted afterward no flow of packet is permitted.

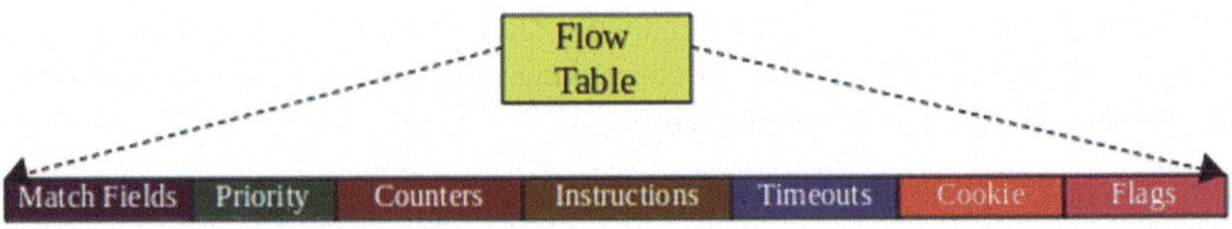

Fig. 6 Flow table fields

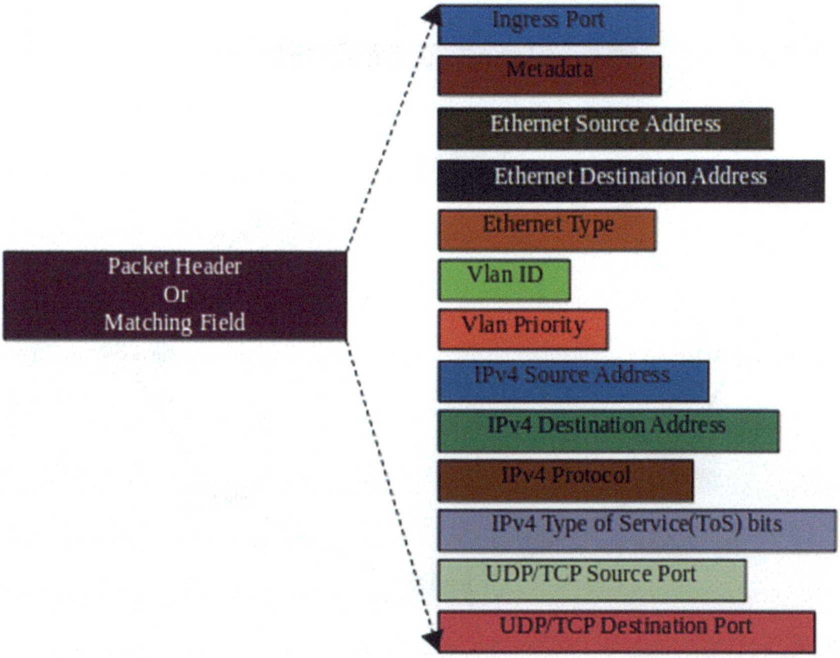

Fig. 7 Match field entries

- **Cookie**: It helps to identify traffic for modifcation or deletion operations. It is used by the controller to choose the opaque data and to filter the flow of packet traffic affected by updation, deletion, modification, and flow statistics requests.
- **Flags**: Flags helps us to manage the flow packet entries.

The flow table is the most important component of the OpenFlow switch. Each flow table consists of limited storage space. Flow entries are placed inside the flow table [18] either reactively or proactively as shown in Fig. 8. Because of the limited memory capacity of the flow table, it is impossible to deploy SDN in the data center network. We can make the flow table efficient by the following methods:

4.1.1 Reduce Flow Table Entries

One of the major roles of the flow table is to maintain the SDN services (including the QoS). Flow entries need to be minimized and the QoS need to be preserved at the same time, which is a difficult task. Earlier work on this also has failed to maintain the balance between the reduction of the flow entries [19–21] and to maintain the QoS well. The limited memory of the flow table in switches is due to the TCAM [18, 22] which is expensive enough. So because of this limited memory [23] of flow table, the SDN cannot be deployed in large data centers; because huge traffic flow

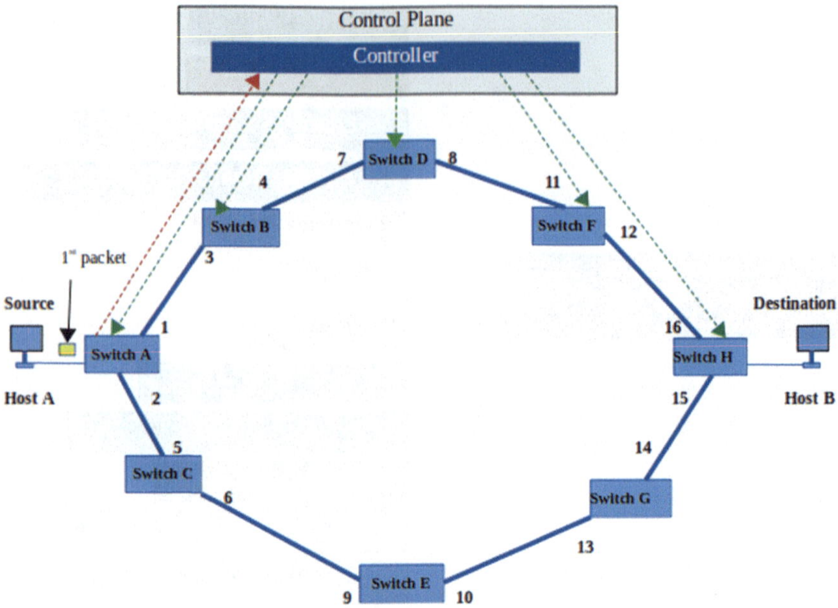

Fig. 8 Flow entries placed in the flow table inside the switch

takes place each second and flow table (with small memory) can't handle such large traffic.

When the flow table is full, then it is called as flow table overloaded problem, this may affect SDN services such as routing of the packet, flexible dynamic management, failure of packet transmission and even worse can happen like whole network breakdown. This problem can be minimized by restricting the new entries in the flow table. We can choose any combination of attributes and combine them into a new attribute; to minimize entries in the flow table. This happens due to fine-grained installation of rules across each switch (e.g. if two different packets are almost same and differ in its port number, both these packets will be considered as different flow entries.) Number of entries can be minimized by converting the fine-grained entries into the lesser number of coarse-grained through compression, and it makes difficult to collect statistical routing information by the controller. To address the problem of overloading across the flow table; we need a mechanism that must work on the following parameters.

- There must be **consistency** in routing of packet, before and after reduction of entries.
- All rules must be matched or executed according to their priority; to maintain the **absoluteness**.
- To maintain the **QoS** the entry of each flow must reside in at least one switch after the reduction.

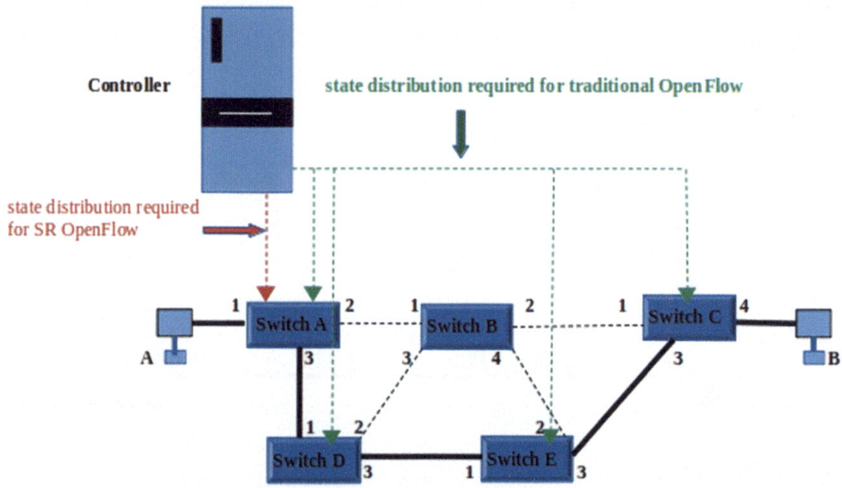

Fig. 9 Source routing

We need to **avoid extra routing**, which means there should not be matching taking place between the new traffic flow and the compressed entry after reduction.

4.1.2 Reducing Flow Table Use

Early per-hop based routing/forwarding was not efficient because the number of entries in the flow table was more. If we follow such **forwarding**, in which state information is updated across all the switches and this results in maintaining all flow tables of underlying switches. So to solve the problem of redundancy, we can use source routing [24].

Source Routing

It requires switches to be interconnected by following some topology as shown in Fig. 9 such as source routing. Each switch consists of more than one interface, e.g., switch A, C, D, and E consists of three interfaces (1, 2, and 3), whereas switch B consists of interfaces (1, 2, 3, and 4). The possible path between A and C switch follows like switch A–>switch D–>switch E–>switch C. Switches A, D, E, and C all use interface 3 to forward the packet. So path is alternating sequence switches and interfaces through which all the packets will be forwarded. Figure 9 shows how path can be conveyed as **3334**. By following the state routing approach, the new information is updated only across the ingress (switch A) switch. Switch A will form the packet header (packet header contains the path information) and will add it to the packet flow.

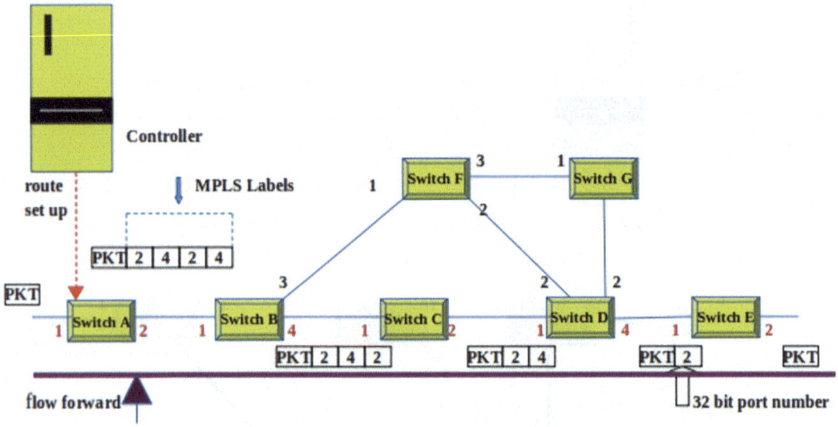

Fig. 10 MPLS based routing

The intermediate switches don't need to get the state information, they simply follow the packet header information for forwarding decisions. Source routing is better compared to traditional hop based forwarding because source-based routing is scalable. One of the advantages of the source routing is that it supports link failure recovery as well as reverse path traceback.

MPLS Label Based Routing

MPLS stands for multi-protocol label switching and was invented [25] in the year 1997 when ATM was the famous WAN (Wide Area Network) technology. MPLS was an efficient approach compared to the conventional IP routing [26]. We perform the forwarding of the packet including the packet header based on the MPLS label [27]. The size of the MPLS [24] packet header is 32 bits (because each MPLS label represents the port number and each port number takes 32 bits), and therefore total size depends upon the number of hops used. If we use 4 hops, the size will be 128 bits (32 bits for each port number, where is to forward the packet i.e. across each switch or hop there are multiple port number which can be used to forward the packet). The functioning of MPLS based routing can be seen in Fig. 10. When the packet came across Ingress switch, i.e., Switch A, controller sets up the route by adding the respective port number of the switches in the packet header field. MPLS label works as the last port number which is to be processed first.

The last MPLS label is 4, it means the switch B will forward the packet at port 4 and is received by switch C at port number 1. So as port number 4 is processed so we will delete it. Now the switch C will forward the packet at port 2 and it is received at port 1 of switch D. So now as the port 2 has processed so we will pop it out. Switch D will now pass the received packet at port 4 and switch E has received it at port 1. The port number of switch D has been processed so we will pop it out.

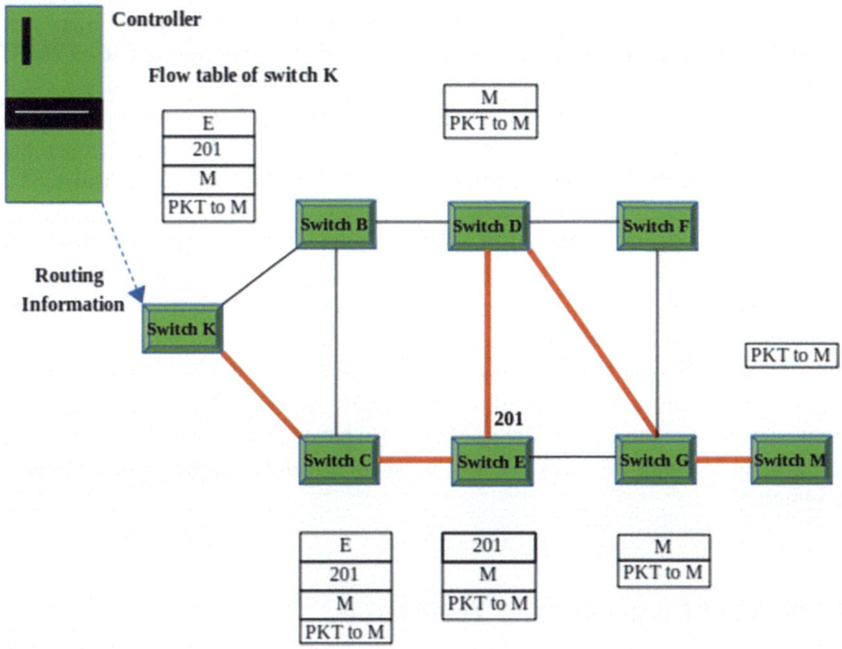

Fig. 11 Segment routing with SDN

The received packet Switch E will now forward at port 2 and the last port number is also processed so we will pop it. Finally, we remain with the packet that was sent and have received the same packet at the other end. The technique we have used here for packet forwarding to destination is MPLS based forwarding.

Segment Routing

Segment Routing (SR) is the ongoing research draft of IETF [28]. SR is the upgraded version of the source routing. In SR, the switch or node can take control of the packet through the list of instructions to be followed. The segment routing minimizes rules which results in minimizing the complexity of rule management, and it results in the reduction of the communication delay between the controller [27] and networking devices. SR supports any kind of instruction associated with the segment. Another advantage of SR is that it is control plane independent, i.e., it can support centralized, distributed, or hybrid. Traffic engineering is key factor for achieving the best effective utilization of the network [29]. The functioning of the SDN based SR can be understood from Fig. 11. Each switch or node is identified by its name, e.g., in this network, we have 8 switches (K, B, C, D, E, F, G, and M). These names represent the node segment. The number close to switch E, i.e., 201 is the port number (adjacency segment of switch E). The controller helps to calculate the routing path with the help of the routing module which is the component of the controller. If there is a demand for traffic flow through switch K to switch M, then the path (K-C-E-D-G-M) information is already encapsulated in the packet header in terms of the MPLS stack labels. Doing so eliminates the need to install the rules across the switches.

Switch K sees the topmost symbol in the stack and it is E which denotes the name of the switch E, so switch K forwards the packet to switch E through switch C. Switch C also sees the topmost symbol on the stack which is E, so switch C passes the packet to the Switch E. Now the packet is at switch E so the topmost symbol will pop out of the stack. Also, the switch E follows its adjacent port number 201, as now 201 port number is processed so we will pop it out. Now we have reached switch D and the topmost element in the stack is M, so we will forward it to the switch G (following the shortest path). At switch G, we again check the top of stack symbol which is M, again we will forward it to switch M. At switch M, we will check for topmost symbol in the stack which is M and also have reached switch M, so we will pop symbol M out of the stack. Finally, we remain with the packet which was sent to switch M.

5 Types of Controller and Their Issues

The idea of a centralized controller for SDN came from the 4D [2]. Ethane [11] enterprise-level networking architecture. It was the first architecture which includes a centralized single controller and the ethane switches, which fully depend upon the decision of the controller. Ethane can support approximately up to 10,000 networking devices [30]. One of the limitations of the ethane architecture was that it was not able to handle large traffic flow because of limited capacity which results in bottleneck across the controller. So it fails in providing scalability and performance which was required. Even if a single controller is able to control huge traffic still there are many reasons that show the networking architecture requires multiple controllers.

- Centralized controller as shown in Fig. 12 is the single point of failure because of either bottleneck problem or if it is attacked.
- Dividing the large network into multiple small subnetworks each of which can be controlled separately.
- If the network is geographically wide distributed.

To deal with the scalability of control plane we can follow the following methods:-

First: We can increase the performance of the control plane by providing more hardware resources or by performing the optimization method.

Second: We can reduce the load across the single controller and by transferring the functionality to other components.

Third: We can use multiple controllers that are logically centralized and physically distributed.

Therefore to improve the network performance of the centralized single controller, we can enhance its performance by adding more core to it so that it can handle large

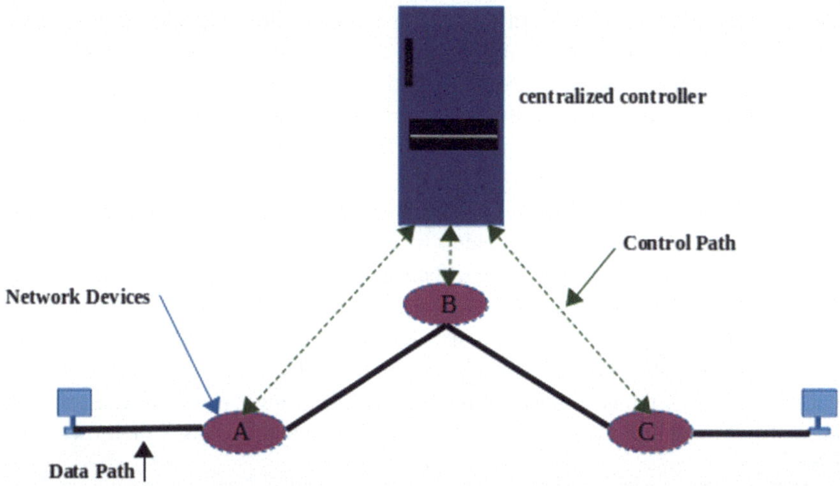

Fig. 12 Single centralized controller

flow request. The example of a multicore single controller is Maestro [8], Mcnettle [31], and Beacon [32]. Beacon performance increases when using the multicore feature with a single controller. When the number of core available is 12, then it can handle approx. 13,000,000 flow requests per sec. Mcnettle with 46 core and single controller can handle up to 4980 switches with maximum flow request of approx. 14,000,000 per sec. Maestro can handle approx. 600,000 flow requests per sec with 8 core single controller. Even though the multicore single controller performs better compared to the conventional single controller, still it consists of many problems like limited scalability and single point of failure. Karakus et al. [15, 33] NOX can handle 30,000 flow requests per sec with an average response time of approx. 10 ms. [30] The author Tootoonchian et al. [34] improved the performance of NOX by applying some optimization techniques, e.g., NOX single controller with 8 cores can handle up to approximately 1,600,000 flow requests with average response time of 2 ms [30]. Even after adding an extra core to the single controller, we achieve better performance but still, there is limited scalability. So to further improve the performance and scalability, the new networking architecture like Kandoo [35], Onix [36] were proposed. Which are of two types, i.e., architecture with one control plane and architecture with the multi-control plane. There occurs a contradiction between researchers, i.e., some researcher believes that a single controller is enough, they only need to address the performance issue, such control plane is known as a single control plane. The other type of control plane includes multiple controllers that coordinate with each other for the better management of the network. The example of the multiple controller networking architecture are Onix [36], DISCO [37], HyperFlow [38], OpenDayLight [39], and ONOS [40].

DISCO: An open-source control plane for SDN with extensible features. It uses a secure control channel to communicate with other controllers. The DISCO [38] controller architecture includes **intradomain** and **inter-domain**. The purpose of the intradomain is to collect the main features of the controller. The purpose of the inter-domain is to manage the information with other controllers.

Onix: It is an open-source control plane for distributed large scale networks. Onix important components are Onix API and NIB (Network Information Base). Onix API allows applications to control the state of the element with in-network though read or write function. Another function of this is to provide the consistencies among control application and network elements which might run on different instances [36] of Onix. To achieve scalability [15], it focuses on three concepts:- (1) Partitioning. (2) Aggregation. (3) Consistency and durability.

HyperFlow: HyperFlow is also a distributed OpenFlow supported control plane [38]. It provides logical central control over multiple distributed controllers and also scalability. One of the advantages of the HyperFlow minimizes the overall response time [30] of the control plane. This is because each controller takes their own local decision.

ONOS: It is physically distributed across multiple servers and logically centralized, which provideS the entire view of the network [41]. It is a upgraded version of Onix. ONOS provides scalability, high availability, and resiliency. It also provides software modularity [42]. The main of the ONOS is to provide the highest performance as much as possible for the scaled network operations.

5.1 Multiple Controller Architectures

We can use the multiple controllers either in:

(1) Hierarchical multi-controller architecture.
(2) Distributed multi-controller architecture.

5.1.1 Hierarchical Multi-controller Architecture

In hierarchical architecture [35, 43, 44] which consists of root and local controller as shown in Fig. 13. Kandoo [35] is also a hierarchical controller which includes the root and local controllers.

Root Controller

The root controller provides the centralized view of the entire network to manage the non-local applications. The root controller is more dynamic and powerful than local controllers. One of the major roles of the root controller is to configure the local controllers, switches, etc., and also collect network statistics information from them

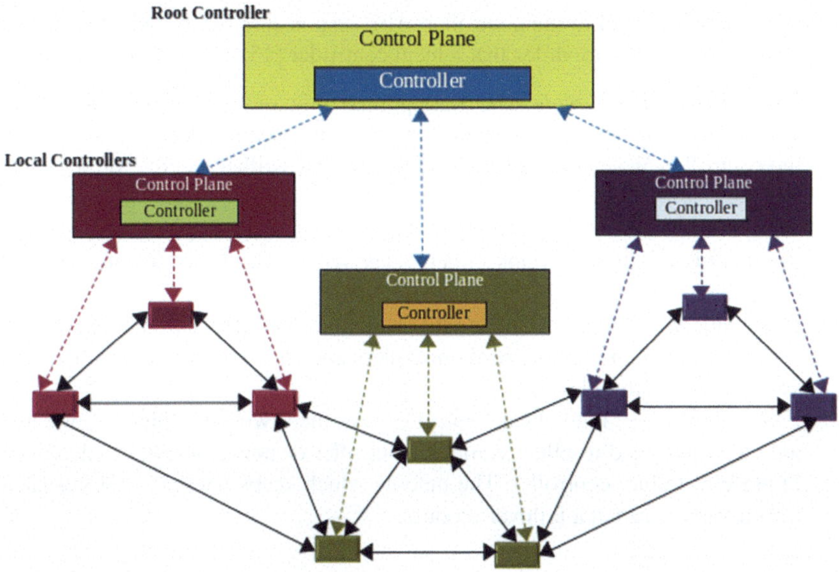

Fig. 13 Hierarchical controller architecture

[15]. It generates rules for network-wide policies. Following are the list of module that a root controller [45] need to maintain.

(a) *Monitoring Module*: One of the features of efficient networks is that they are consistent. So to achieve the desired level of consistency, there is a need to keep track of each device in real time. This helps us to find the accurate optimal point which reduces the corresponding network overhead.
(b) *Storage Module*: It stores the state information of each networking device.
(c) *Load Adaption Module*: Another important characteristic of efficient networks is the maximum utilization of the resources. The load adaption module follows the monitoring module to get real-time updates so that it can provide the dynamic provisioning of the network elements. It also helps in load balancing to efficiently utilize the local controllers.
(d) *Partition Module*: It helps to partition the load across the distributed controller.

Local Controller

The local controllers deal with the queries and requirements of local applications or control the behavior networking devices through NBI (North Bound Interface) protocol. The local controller can also be called as Zone controllers [45]. To share information, the local controller must coordinate with each other. So there is a need to establish a secure communication channel between the local or zone-based controllers. The local controller is responsible for the traffic flow management across the switches which are local to the controller. Also responsible for the selection of the route. Following are the modules that a local controller [15, 45] need to maintain:-

(a) *Path Computation Module*: The decoupling feature of the SDN has reduced the complexity of forwarding plane devices. So path selection decision is made by the controller. We can install rules across switches either reactive or proactive approaches.
(b) *Event Module*: Event is an activity that occurs frequently in a network. It keeps the records of all the events either generated by the hosts or any network component.
(c) *Communication Module*: This module maintains the information about peer to peer interaction among the local controllers and also with root controllers and switch at the bottom layer.
(d) *Fail Over Module*: If any local controller fails, then switches which got affected are either assigned to other available controller or new controller is deployed in place of failure controller. The module which keeps track of or stores such information is called a failover module.

5.1.2 Distributed Multi-controller Architecture

It is also known as flat controller architecture. In this, each controller manages the subnetwork of the entire network.

5.2 Dynamic Control Plane for SDN

There are a few reasons which show why there is a need of dynamic control plane:

- The controllers processing workload may vary from time to time.
- Traffic flow across control channels also varies from time to time.
- Need for high availability across the control plane.

To improve the scalability of SDN, it is necessary to provide the dynamic management of the control plane and this becomes more important when we talk about SD-IoT (software-defined Internet of Things), SD-WAN [46] (software-defined wide area network), and SD-IoV (software-defined internet of vehicles). The early research study was focused on the controller's performance in the small network which consists of the limited number of nodes.

Heller et al. [47] have mentioned the performance issues of the SDN controllers, it also includes the study of how many controllers are needed in an SDN network and how they should be placed to gain more performance. The most recent studies have been regarding deploying SDN control plane for cloud-based services. Cloud on its self is a huge network that continuously grows and each second requires dynamic management, traffic flow handling, load balancing, and live traffic monitoring. So

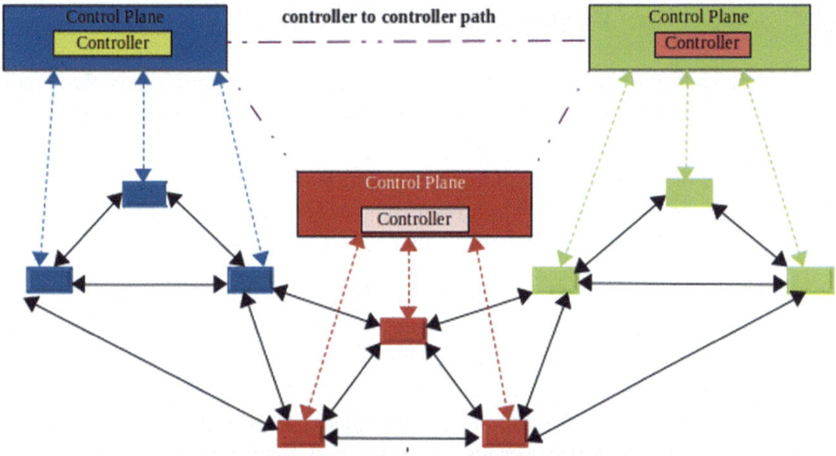

Fig. 14 Distributed controller architecture

SDN must support these services and scalability also which are required by the cloud network and it is possible only when SDN has a dynamic control plane. So this increases the demand for dynamic, robust, scalable, and flexible SDN control plane.

Kempf et al. [48] have mentioned how to monitor the controller utilization on the cloud. He has also mentioned the key point that the new controller can be issued or initialized on the cloud; it depends upon the workload on the processor.

Sharkh et al. [49] have studied how resource is allocated in the data centers networks running on the cloud. In [50], the author has discussed the meridian. Gorkemli et al. [46] have mentioned about cloud-based dynamic control plane architecture. The SDN supporting cloud-based dynamic control plane architecture is as shown in Fig. 14.

The components of dynamic control plane:

5.2.1 Controller Cloud Manager

It is said to be the brain of this dynamic control plane architecture. It consists of the following modules.

(a) *Controller process module*: It keeps track of the instances of the controller which are either activated or deactivated. It also allows various networking devices to discover and connect to the newly assigned controller.
(b) *Controller monitor module*: It monitors the performance and utilization, and location of each controller. Its main function is to distribute the traffic across the controllers.
(c) *Traffic monitor module*: Its main function is to determine how much traffic is across each controller and perform the load balancing for the effective utilization of the controllers.

5.2.2 Control Flow Monitoring

The amount of traffic travel across the controller and control doesn't have that much capacity to accommodate all the traffic so there is a need for flow management schemes like CCM (controller cloud manager). It can detect the bottleneck across the entire network with the help of the monitoring module. It offloads the controller suffering from bottleneck and distributes its load to another controller instances. The four different kinds of traffic flow offloading [46] are as follows:

(a) *Total control flow shift*: In this flow shifting technique either one switch or the multiple switches are disconnected from the earlier connected controller and reconnected them to either already existing controller or newly created controller.
(b) *Control flow split*: In this technique, particular type of traffic is routed to either already existing controller or just newly created controller.
(c) *Control flow reroute*: In this technique also, particular type of traffic flow is rerouted.
(d) *Service-based controller*: This technique focussed on introducing a new controller to provide specific service to control traffic. The rest of the traffic is served by the original controller.

5.3 Problem-Related to Placement of Controller

When we use multiple controllers in a network, we aim to have the best performance given by each controller. But what if, even after using multiple controllers we don't get the performance as was desired. So the important key point to remember is that, if we have multiple controllers then it does not always improve the network performance, the placement of the controller plays a major role to enhance the performance of the entire network. The wrong placement of controller results in problems such as **controller to switch** communication delay which results in flow rule set up delay, **controller to controller** communication delay, **fault tolerance**, and **overhead across** control plane. The controller placement problem was first observed in [47]. In recent studies [51, 52], author(s) also have discussed research issues related to the controller placement. In [53], author has proposed an algorithm for taking automated decisions for controller placement. The main goal of doing so is to make the SDN network robust. In [54, 55], author(s) have concluded that having multi controllers and their placement can affect the reliability of SDN networks. In [56], author has discussed how we can design a scalable control plane by following the characteristic of the controller placement problem. In [57, 58] author(s) presented the concept to optimally place the multi-controller.

6 SDN and Other Technologies

Because of the centralized control and programmatic features of SDN, traffic monitoring and dynamic update of flow rules has become easier. SDN applications have been deployed in other technologies like cloud computing, NFV(Network Function Virtualization), IoT (Internet of Things), and BigData. Apart from deploying SDN applications in other technology, SDN also supports deploying other technologies, e.g., Machine learning. Here we will discuss various technologies concerning SDN.

6.1 SDN and Machine Learning

Traditional networks were too complex, so it was difficult to use machine learning because of many heterogeneous devices. SDN has solved this problem by making management easier. Machine learning is the kind of learning in which the machine learns by itself without being programmed. It is mainly used in data mining. Using machine learning techniques (which perform data analysis, automated allocation of networking services, and network optimization), we can make the controller more intelligent for decision making. Learning capabilities embedded in the SDN controller enables it to learn automatically to make optimal decisions. Before understanding how machine learning works in SDN, we must overview the machine learning algorithms. Machine learning follows two steps; In the first step, we provide training and in the second step, there is a decision making as shown in Fig. 15.

Machine Learning techniques are as follows:-

6.1.1 Supervised Learning

In this, we trained the model using the labeled data set (predefined knowledge) to build an intelligent model (learns or finds a function which can convert a given input to appropriate output) [59] that can represent the relation between input and output (where input and output are already known). It is also known as the labeled learning technique. After the completion of training when the new input is given to the model, it is expected that trained model will give a desired output [60, 61]. Widely used supervised algorithm are KNN [61, 62], Decision Tree (DT) [63], Random Forest [64], Neural Networks (NN) [65] (Random Neural Networks (RNN) [66–68], Deep Neural Networks (DNN) [69–71], Convolutional Neural Networks (CNN) [72, 73], Recurrent Neural Networks [74–77]), Support Vector Machine(SVM) [78–80], Bayes Theory [81, 82], and Hidden Markov Model (HMM) [83, 84].

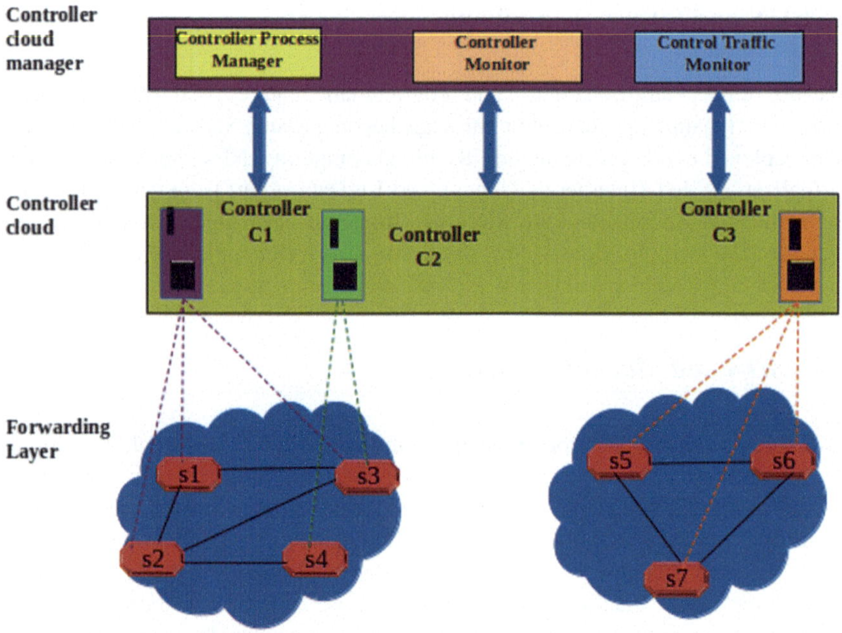

Fig. 15 Cloud-based dynamic control plane architecture

6.1.2 Unsupervised Learning

In the unsupervised learning algorithm, we trained the machine using the unlabeled data, it means only input is given there is no output. It is mainly used in aggregation of data and clustering [84, 85]. The unsupervised algorithms are like K-means [85, 86], Self Organizing Map (SOM) [87, 88], etc.

6.1.3 Semisupervised Learning

In this, we use both the labeled and unlabeled data [89, 90]. It offers many advantages like it is very difficult and expensive as well, to collect the labeled data whereas the collection of the unlabeled data is easier and cheap. Using the unlabeled data effectively in the training can enhance the performance of the trained model even better. Low-density separation and cluster-based assumptions are necessary for effective utilization of unlabeled data. The example of the semisupervised technique is pseudo labeling [91]. In this, we first trained the machine with the labeled data. Then we use the machine to predict the label for unlabeled data. The examples of semisupervised learning methods are transductive support vector machines (which is based upon the assumption of type low-density separation), expectation-maximization (based on cluster-based assumption), etc.

6.1.4 Reinforcement Learning

In reinforcement learning [92, 93], agent made interaction with its environment to receive information and in each agent–environment interaction, agent gets new updated information, and agent use this information in the best way to update its knowledge [94]. The advantage or disadvantage of an agent depends upon either it performs well or not [95]. The examples of reinforcement learning techniques are Deep Reinforcement Learning (DRL) [94, 96] and Reinforcement Learning-based Game Theory [97–100].

6.2 Machine Learning (ML) in SDN

ML-based SDN is more efficient and intelligent. In this section, we will discuss machine learning (ML) based algorithm for traffic flow identification or classification, ML-based routing optimization, ML-based resource management, ML-based security, ML-based resource management, etc.

6.2.1 ML-Based Identification of Traffic Flow

Early studies for traffic identification include a port-based approach and deep packet inspection(DPI) [101]. Port-based approach is based on the TCP, UDP port numbers, and are not effective for the current scenario and also most applications run on a dynamic port. DPI match the payload with the predefined pattern (where pattern are based on the regular expressions) to identify the traffic belongs to which application. Most of the DPI follows the digital signatures to distinguish different applications. One of the advantages of DPI is that it has higher classification accuracy. The disadvantages of DPI are like it can identify applications that follow the pattern. When the applications grow at a rapid rate, then it is difficult to keep the pattern updated. It also results in high computational cost. It cannot classify the traffic if it is in the encrypted form over the network [101]. If we apply machine learning-based techniques, then we can easily identify the encrypted traffic and is a cheaper alternative to the DPI approach. The first step in the ML-based traffic identification is the collection of traffic and the second step is to apply the ML techniques to extract the valuable information.

SDN controller can collect statistical data form underlying data plane devices to perform analysis. It is beneficial to implement ML approaches across the controller. Earlier studies have been made to identify the elephant–mice flow, application-aware flow, etc.

(a) *Identification of Elephant–Mice Flow*: Elephant flow occurs for a long time and consumes more bandwidth. The duration of mice flow is short. In [101], author has mentioned that in data centers 80% of the traffic flow is of type mice. To effectively manage the traffic; it is important to find out the location of elephant flow where it occurs. In [102], author has mentioned the issue regarding traffic flow across hybrid data centers. In [103], author has proposed the learning technique to identify the elephant flow. The ML-based techniques are used at the network edge levels to classify the traffic flow and this classification may help the centralized SDN controller to implement effective routing algorithms.

(b) *Identification of Application-Aware Flow*: In this, the aim is to identify those applications which are the main reason for traffic flows. In Amaral et al. [104], authors have studied application-aware traffic-aware identification at the enterprise-level network. In [105], the author has discussed how Deep Neural Networks (DNN) can be useful for recognizing various mobile based applications.

(c) *Identification of traffic based on QoS Requirements*: This aims for the identification of QoS classes for the packet flows. The applications are increasing day by day, so it is difficult to identify all applications. What we can do is we can classify applications according to QoS requirements. In Wang et al. [106] authors have proposed QoS based traffic classification.

6.2.2 ML-Based Routing Optimization

SDN controller is responsible for deciding and implementing routing decisions. If the controller will not handle routing decisions efficiently, then it may lead to transmission delays. Therefore routing decisions must be more efficient. Early algorithms (the Shortest Path First (SPF) and Heuristic Algorithms (e.g,. Ant Colony Optimization) for routing optimization were either complex in terms of computation [107, 108] or lacked in effective use of the available network resources [107]. Early studies suggest how we can overcome the problem of routing optimization using machine learning which includes supervised and reinforcement learning.

(a) *Supervised learning technique for routing optimization*: As we know the labeled data sets are used in supervised learning to make machine/system intelligent. It can help us to achieve optimal heuristic-based routing decisions in real-time. In [107], author has proposed supervised learning-based architecture for optimal routing as shown in Fig. 16. ML-based meta-layer solves the problem of dynamic routing, it can accept direct input as well as the output of the heuristic algorithm to get the accurate result. For having optimal routing, there must also have a feature of future prediction of traffic.

In [108], author has discussed about the NeuRoute [101] framework which uses LSTMRNN (Long Short Term Memory Recurrent Neural Networks) (Fig. 17).

(b) *Routing Optimization with Reinforcement Learning (RL)*: RL based technique is suitable for decision-based problems. Sendra et al. [109] have presented RL based distributed routing protocol. In [110], author has discussed routing optimization in SDN supporting data center network. A controller may use the

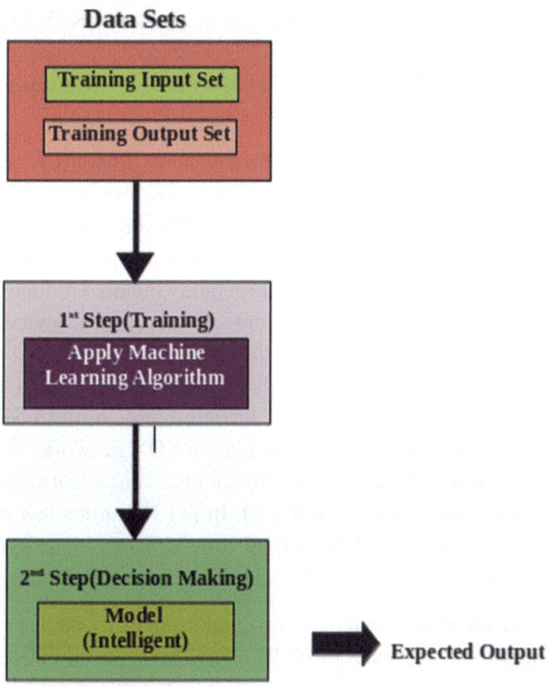

Fig. 16 Steps for machine learning

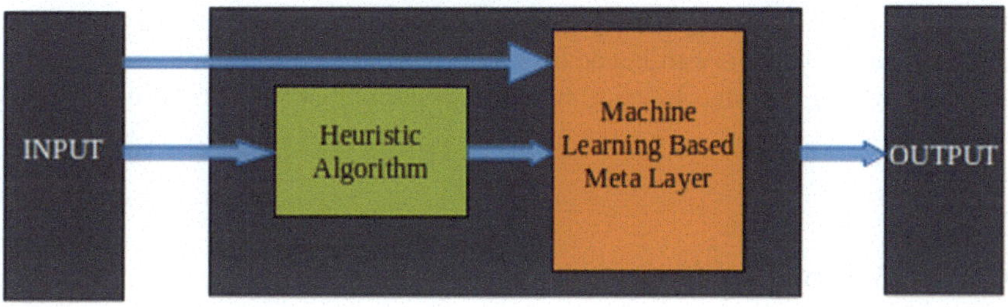

Fig. 17 Optimal routing architecture

ML-based techniques such as Random Neural Network (RNN) and RL (Reinforcement Learning) to provide optimal routing path between geographically distributed data centers [101]. Recent studies have been made for optimizing routing. In [110], author has presented a method called QoS based Adaptive Routing which uses the RL technique which performs packet forwarding efficiently. In [111], author has discussed about a model which can provide the best available path for all pair of source and destination.

6.2.3 ML-Based Resource Management (RM) for SDN

The SDN's global view and programmatic feature supports easier network management.

(a) *ML-Based Resource Management*: Resources available in data plane is like caching, computing, and network resources. So these resources need to be managed inefficient way to meet QoS standards. The main purpose of the caching resources is to maintain a copy of frequently accessed data. Doing so not only minimizes the transmission delay but also reduces the chances of duplicate data transmission. In the current scenario, user's device may get fail to handle all computational tasks due to the availability of limited resources and battery backup. Resource allocation in the data plane is mainly divided into two main categories: (a) Single-tenant based resource allocation in SDN network. (b) Multi-tenant based resource allocation in the SDN network. In He et al. [112], the author has discussed a framework to enhance the performance of SDN based vehicular ad hoc networks (VANET's). In [113], author has proposed a cloud-based architecture called RSU (Road-Side Unit) which provides various dynamic services (e.g., migrate and replicate services, etc.).

In **control plane,** the up-gradation in the network virtualization supports the further development of multi-tenant SDN based networks which focuses on sharing of available resources among multiple tenants in the data plane [101]. The widely used hypervisors are FlowVisor [114] and OpenVirteX [115], usually lies between data and control plane. Hypervisor can process and control traffic with limited computing resources between the forwarding devices and controller logic. Therefore it becomes necessary to use machine learning-based algorithms that need to be used in the hypervisor.

6.3 SDN Based NFV

The main key concept in both SDN and NFV is network abstraction. SDN is based on the separation of the controller from the data plane, to provide a centralized view of the network and to modify the behavior of networking devices through programmatic control. Whereas NFV is based on transferring the control from hardware devices (router, firewall, switch, hub, load-balancer) to software applications (router.py, firewall.py, switch.py, hub.py, and load-balancer.py) using virtualization. It reduces the capital and operating expenditure [116]. Its framework consists of a physical server (which includes CPU, storage memory, and RAM), hypervisor; It is a software that allows a computer to run and manage more virtual machines. It is also called as Virtual Machine Monitor (VMM). When we integrate SDN architecture with NFV, it is called Software-defined NFV architecture, which can manage network resources and functions [117]. When we use SDN with NFV, the aim is to maximize the performance, both SDN and NFV are not dependent on each other. SDN makes NFV more convincing and vice versa.

6.4 SDN Based IoT

It is a network of heterogeneous communicating computing devices equipped with wireless sensors, RFID, and actuators. The objective of IoT is to make each object communicate with other objects. In the future, almost everything will be the part of the internet which includes traditional communication devices (computers, laptops, smartphones, etc.), home appliances (refrigerator, television, doors, etc.). IoT can offer a wide variety of applications such as telecommunications, smart cities, vehicular networks, smart health services, and smart education. In [118], author has proposed IoT application as an intelligent transportation system. In [119], author has proposed IoT as smart health care. Currently billions of devices are part of the IoT network. Therefore, it becomes extremely important to make IoT networks secure and manageable. Therefore IoT networks require robust, secure, intelligent, efficient, flexible, and scalable architecture. SDN architecture holds all those features which are required by the IoT network. So currently SDN architecture is the best available choice for IoT. We can easily manage the network of heterogeneous devices via a centralized control feature of SDN. Also, IoT devices generate huge amounts of data, if we use machine learning modules in SDN which can effectively process the continuously generating data to make intelligent decisions.

7 Conclusion and Scopes

By overcoming the issues related to data, control, and application plane, we can make SDN's work efficiently. For a paradigm or architecture on which various technologies depend must not be limited to just being efficient, but it must also be robust. Therefore, it becomes necessarily important to use a machine learning algorithm that can keep track of each activity including the states of devices and flow packet through each device. Doing so will automatically improve the routing decisions, resource utilization, traffic filtering, etc. Because of its enormous features SDN is becoming very popular among researchers and networking industries because it includes all features required by current networking standards. Therefore, machine learning-based SDN architecture can directly be deployed in BigData, Cloud data centers, IoT, IoV (Internet of Vehicles), 5G, wireless mesh networks, and radio networks. Because such networks include lots of heterogeneous devices that require intelligent, secure, and scalable architecture for flexible and easier management, so SDN is the need of these emerging technologies and in all networking scenario's it fits in the best way. We can say SDN as the future of networks. SDN itself is a broad concept and its integration with other technologies makes it even more interesting for researchers to do research. So there is a wide scope of innovation as well as improvement.

Funding This research paper is not funded by any funding agency or organization.

Conflict of Interest There are no conflicts of interest among authors.

References

1. Software Defined Networks (SDN). https://www.opennetworking.org. Accessed 12 June 2019
2. A. Greenberg, G. Hjalmtysson, D.A. Maltz, A. Myers, J. Rexford, G. Xie, H. Yan, J. Zhanm, H. Zhang, A clean state 4D Approach to network control and management. ACM SIGCOMM Comput. Commun. Rev. **35**(5), 41–54 (2005)
3. N. McKeown, T. Anderson, H. Balakrishnan, G. Parulkar, L. Peterson, J. Rexford, S. Shenker, J. Turner, OpenFlow: enabling innovation in campus networks. SIGCOMMCCR **38**(2), 69–74 (2008)
4. N. Gude, T. Koponen, J. Pettit, B. Pfaff, M. Casado, N. McKeownm, S. Scott, NOX: towards an operating system for networks. ACM SIGCOMM Comput. Commun. Rev. **38**(3) (2008)
5. POX Controller. http://www.noxrepo.org/pox/about-pox. Accessed 14 June 2019
6. R. Kamath, R. Kondoker, K. Bayarou, F. Weber, Security analysis of OpenDaylight, ONOS, Rosemary and Ryu SDN controllers, in *2016 17th International Telecommunications Network Strategy and Planning Symposium (Networks)* (2016)
7. Floodlight. http://floodlight.openflowhub.org
8. Z. Cai, A.L. Cox, T.S. Eugene Ng, Maestro: a system for scalable OpenFlow control. Rice University Technical Report TR10-08 (2010)
9. S. Jain, A. Kumar, S. Mandal, J. Ong, L. Poutievski, A. Singh, S. Venkata, J. Wanderer, J. Zhou, M. Zhu, J. Zolla et al., B4: experience with a globally-deployed software defined wan, in *Proceedings of the ACM SIGCOMM 2013 Conference*, ACM, New York, USA, 2013, pp. 3–14
10. P. Patel, D. Bansal, L. Yuan, A. Murthy, A. Greenberg, D.A. Maltz, R. Kern, H. Kumar, M. Zikos, H. Wu, C. Kim, N. Karri, Ananta: cloud scale load balancing, in *Proceedings of the ACM Conference on Data Communication SIGCOMM'13*
11. M. Casado, M.J. Freedman, J. Pettit, J. Luo, N. McKeown, S. Shenker, Ethane: taking control of the enterprise, in *SIGCOMM'07* (2007), pp. 27–31
12. J. Li, J.H. Yoo, J.W.K. Hong, Dynamic control plane management for software-defined networks. Int. J. Netw. Manag. **26**(2), 111–130. https://doi.org/10.1002/nem.1924
13. Z. Guo, Y. Xu, M. Cello, J. Zhang, Z. Wang, M. Liu, H.J. Chao, JumpFlow: reducing flow table usage in software defined networks. Comput. Netw. **92**, 300–315 (2015). https://doi.org/10.1016/j.comnet.2015.09.030
14. M. Yu, J. Rexford, M.J. Freedman, J. Wang, Scalable Flow-based networking with DIFANE, in *SIGCOMM' 10*, New Delhi, India, 2010
15. M. Karakus, A. Durresi, A survey: control plane scalability issues and approaches in software-defined networking (SDN). Comput. Netw. (2016). https://doi.org/10.1016/j.comnet.2016.11.017
16. L. Wang, Q. Li, R. Sinnott, Y. Jiang, J. Wu, An intelligent rule management scheme for software defined networking. Comput. Netw. (2018)
17. OpenFlow Switch Specification. https://www.opennetworking.org/software-defined-standards/specification. Accessed 13 June 2019

18. I. Arsovski, T. Chandler, A. Sheikholeslami, A ternary content- addressable memory (TCAM) based on 4T static storage and including a current-race sensing scheme. IEEE J. Solid-state Circuits **38**(1), 155–158 (2003)
19. A.R. Curtis, J.C. Mogul, J. Tourrilhes, P. Yalagandula, P. Sharma, S. Banerjee, Devoflow: scaling flow management for high-performance networks. ACM SIGCOMM Comput. Commun. Rev. **41**(4) (2011)
20. Y. Kanizo, D. Hay, I. Keslassy, Palette: distributing tables in software-defined networks. Proc. IEEE INFOCOM **900**, 545–549 (2013)
21. B.A. Nunes, M. Mendonca, X.N. Nguyen, K. Obraczka, T. Turletti, A survey of software-defined networking: past, present, and future of programmable networks. IEEE Commun. Surv. Tutor. **16**(3) (2014)
22. O. Rottenstreich, R. Cohen, D. Raz, I. Keslassy, Exact worst-case TCAM rule expansion, IEEE Trans. Comput. (2012)
23. Z.A. Qazi, C.C. Tu, L. Chiang, R. Miao, V. Sekar, M. Yu, Simplifying middlebox policy enforcement using *sdn*, in *ACM SIGCOMM'2013* (2013), pp. 27–38. https://doi.org/10.1145/2486001.2486022
24. M. Soliman, B. Nandy, I. Lambadaris, P. Ashwood-Smith, Source routed forwarding with software defined control, considerations and implications, in *Proceedings of the 2012 ACM Conference on CoNEXT Student Workshop, ACM2012* (2012), pp. 43–44. https://doi.org/10.1145/2413247.2413274
25. MPLS in the SDN Era. https://www.oreilly.com/library/view/mpls-in-the/9781491905449/
26. MPLS. https://searchnetworking.techtarget.com/definition/Multiprotocol-Label-Switching-MPLS
27. P. Ashwood-Smith, M. Soliman, W. Tao, Sdn state reduction (IEFT draft)
28. Segment Routing Architecture. IETF Draft. https://datatracker.ietf.org/doc/rfc8402/?include_text=1
29. S. Agarwal, M. Kodialam, T.V. Lakshman, Traffic engineering in software defined networks. Proc. IEEE INFOCOM **40**(10), 118–124 (2002)
30. J. Xie, D. Guo, Z. Hu, T. Qu et al., Control plane of software defined networks: a survey. Comput. Commun. (2015)
31. A. Voellmy, J. Wang, Scalable software defined network controllers. ACM SIGCOMM Comput. Commun. Rev. **42**(4), 289–290 (2012)
32. D. Erickson, The beacon openflow controller, in *Proceedings of ACM on Hot Topics in Software Defined Networking (HotSDN)* (Hong Kong, China, 2013)
33. A. Tavakoli, M. Casado, T. Koponen, S. Shenker, Applying NOX to the datacenter, in *Proceedings of ACM on Hot Topics in Networks (HotNets)*, New York City, NY, USA, 2009
34. A. Tootoonchian, S. Gorbunov, Y. Ganjali, M. Casado, R. Sherwood, On controller performance in software-defined networks, in *Proceedings of USENIX Hot-ICE*, San Jose, CA, 2012
35. S.H. Yeganeh, Y. Ganjali, Kandoo: a framework for efficient and scalable offloading of control applications, in *Proceedings of ACM on Hot Topics in Software Defined Networking (HotSDN)*, Helsinki, 2012
36. T. Koponen, M. Casado, N. Gude, J. Stribling, L. Poutievski, M. Zhu, R. Ramanathan et al., Onix: a distributed control platform for large scale production networks, in *Proceedings of USENIX Operating Systems Design and Implementation (OSDI)*, Vancouver, BC, Canada, 2010
37. K. Phemius, M. Bouet, J. Leguay, Disco: distributed multi-domain sdn controllers, in *Proceedings of IEEE/IFIP Network Operations and Management Symposium (NOMS)*, Krakow, Poland, 2014
38. A. Tootoonchian, A.Y. Ganjali, Hyperflow: a distributed control plane for openflow, in *Proceedings USENIX Internet Network Management INM/WREN*, San Jose, CA, 2010
39. Opendaylight. http://www.opendaylight.org/. Accessed on 23 June 2019
40. Opencontrail. http://www.opencontrail.org/. Accessed on 16 June 2019

41. P. Berde, M. Gerola, J. Hart, Y. Higuchi, M. Kobayashi, T. Koide, B. Lantz, B. O'Connor, P. Radoslavov, W. Snow et al., Onos: towards an open, distributed SDN OS, in *Proceedings of ACM on Hot Topics in Software Defined Networking (HotSDN)*, Chicago, IL, USA, 2014
42. ONOS Software Module. https://onosproject.org/features/. Accessed on 17 June 2019
43. M. Santos, B. Nunes, K. Obraczka, T. Turletti, B. de Oliveira et al., Decentralizing sdn's control plane, in *IEEE 39th Conference on Local Computer Networks (LCN)* (2014), pp. 402–405. https://doi.org/10.1109/lcn.2014.6925802
44. D. Marconett, S. Yoo, Flowbroker: a software-defined network controller architecture for multi-domain brokering and reputation. J. Netw. Syst. Manag. **23**(2), 328–359 (2015). https://doi.org/10.1007/s10922-014-9325-5
45. K.S. Atwal, A. Guleria, M. Bassiouni, A scalable peer-to-peer control plane architecture for Software Defined Networks, in *IEEE 15th International Symposium on Network Computing and Applications (NCA)*. https://doi.org/10.1109/nca.2016.7778609
46. B. Gorkemli, A.M. Parlakisik, S. Civanlar et al., Dynamic management of control plane performance in software-defined networks, in *IEEE NetSoft Conference and Workshop (NetSoft)*
47. B. Heller, R. Sherwood, N. McKeown, The controller placement problem, in *Procedding 1st Workshop Hot Topics Software Defined Network* (2012), pp. 7–12
48. J. Kempf et al., Implementing a 3G packet core in a cloud computer with Openflow data and control planes. U.S. Patent 8,762,501, issued June 24, 2014
49. M.A. Sharkh, M. Jammal, A. Shami, A. Ouda, Resource allocation in a network-based cloud computing environment: design challenges. IEEE Commun. Mag. **51**(11), 46–52 (2013)
50. M. Banikazemi, D. Olshefski, A. Shaikh, J. Tracey, G. Wang, Meridian: an SDN platform for cloud network services. IEEE Commun. Mag. **51**(2), 120–127 (2013)
51. Y.N. Hu, W.D. Wang, X.Y. Gong, X.R. Que, S.D. Cheng, On the placement of controllers in software-defined networks. J. China Univ. Posts Telecommun. **19**(2), 92–171 (2012). http://dx.doi.org/10.1016/S1005-8885(11)60438-X
52. M. Obadia, M. Bouet, J.L. Rougier, L. Iannone, A greedy approach for minimizing SDN control overhead, in *1st IEEE Conference on Network Softwarization (NetSoft)* (2015), pp. 1–5. https://doi.org/10.1109/netsoft.2015.7116135
53. H. Bo, W. Youke, W. Chuan'an, W. Ying, The controller placement problem for software-defined networks, in *2nd IEEE International Conference on Computer and Communications (ICCC)*
54. Y. Hu, W. Wang, X. Gong, X. Que, S. Cheng, On reliability-optimized controller placement for software-defined networks. Commun. China **11**(2), 38–54 (2014)
55. Y. Hu, W. Wendong, X. Gong, X. Que et al., Reliability-aware controller placement for software-defined networks, in *2013 IFIP/IEEE International Symposium on Integrated Network Management (IM 2013)* (2013), pp. 672–675
56. Y. Jimenez, C.C. Pastor, A.J. Garcia, On the controller placement for designing a distributed SDN control layer, in *Proceedings of Networking Conference*, IFIP Trondheim, Norway, 2014
57. H.K. Rath, V. Revoori, S.M. Nadaf, A. Simha, Optimal controller placement in Software Defined Networks (SDN) using a non-zero-sum game, in *Proceeding IEEE International Symposium on World Wireless, Mobile Multimedia Network*, Sydney, NSW, Australia, June 2014
58. B. Soper, J. Musacchio, A non-zero-sum, sequential detection game, in *Proceedings 53rd Annual Allerton Conference Communication, Control, and Computing (Allerton)*, Monticello, IL, USA, May 2015, pp. 361–371
59. S. Russell, P. Norvig, *Artificial Intelligence (A Modern Approach)*, 3rd edn. (Prentice Hall, New Jersey, 1995)
60. S.B. Kotsiantis, I. Zaharakis, P. Pintelas, Supervised machine learning: A review of classification techniques. Emerg. Artif. Intell. Appl. Comput. Eng. **160**, 3–24
61. J. Friedman, T. Hastie, R. Tibshirani, *The Elements of Statistical Learning*, vol. 1. Springer Series in Statistics (New York, 2001)

62. T. Cover, P. Hart, Nearest neighbor pattern classification. IEEE Trans. Inf. Theory **13**(1), 21–27 (1967)
63. J. Xie, F.R. Yu, T. Huang, R. Xie, J. Liu, C. Wang, Y. Liu, A Survey of machine learning techniques applied to software defined networking (SDN): research issues and challenges. IEEE Commun. Surv. Tutor. 1–1
64. L. Breiman, Random forests. Mach. Learn. **45**(1), 5–32 (2001)
65. S.N. Haykin, Network, a comprehensive foundation. Neural Netw. **2**(2004), 41 (2004)
66. S. Timotheou, The random neural network: a survey. Comput. J. **53**(3), 251–267 (2010)
67. S. Basterrech, G. Rubino, A tutorial about random neural networks in supervised learning (2016). arXiv:1609.04846
68. H. Bakirciouglu, T. Koccak, Survey of random neural network applications. Eur. J. Oper. Res. **126**(2), 319–330 (2000)
69. Y. LeCun, Y. Bengio, G. Hinton, Deep learning. Nature **521**(7553), 436 (2015)
70. J. Baker, Artificial neural networks and deep learning (2015)
71. J. Schmidhuber, Deep learning in neural networks: an overview. Neural Netw. **61**, 85–117 (2015)
72. A. Krizhevsky, I. Sutskever, G.E. Hinton, Imagenet classification with deep convolutional neural networks, in *Advances in Neural Information Processing Systems* (2012), pp. 1097–1105
73. C. Li, Y. Wu, X. Yuan, Z. Sun, W. Wang, X. Li, L. Gong, Detection and defense of DDoS attack-based on deep learning in OpenFlow-based SDN. Int. J. Commun. Syst. (2018)
74. H. Sak, A. Senior, F. Beaufays, Long short-term memory recurrent neural network architectures for large scale acoustic modeling, in *Fifteenth Annual Conference of the International Speech Communication Association* (2014)
75. S. Hochreiter, J. Schmidhuber, Long short-term memory. Neural Comput. **9**(8), 1735–1780 (1997)
76. T. Mikolov, M. Karafiát, L. Burget, J. Černockỳ, S. Khudanpur, Recurrent neural network based language model, in *Eleventh Annual Conference of the International Speech Communication Association* (2010)
77. X. Li, X. Wu, Constructing long short-term memory based deep recurrent neural networks for large vocabulary speech recognition, in *Proceedings of IEEE ICASSP '15*, Brisbane, QLD, Australia, April 2015, pp. 4520–4524
78. A. Patle, D.S. Chouhan, SVM kernel functions for classification, in *Proceeding IEEE ICATE '13*, Mumbai, India, January 2013, pp. 1–9
79. I. Steinwart, A. Christmann, *Support Vector Machines*. (Springer Science & Business Media, 2008)
80. B. Yekkehkhany, A. Safari, S. Homayouni, M. Hasanlou, A comparison study of different kernel functions for SVM-based classification of multi-temporal polarimetry SAR data. Int. Arch. Photogramm. Remote. Sens. Spat. Inf. Sci. **40**(2), 281 (2014)
81. G.E. Box, G.C. Tiao, *Bayesian Inference in Statistical Analysis*, vol. 40 (John Wiley & Sons, 2011)
82. J. Bakker, Intelligent traffic classification for detecting DDoS attacks using SDN/OpenFlow (Victoria University of Wellington, 2017), pp. 1–142
83. P. Holgado, V.A. Villagra, L. Vazquez, Real-time multistep attack prediction based on hidden markov models. IEEE Trans. Dependable Secur. Comput. **99**, 1 (2017)
84. L.R Rabiner, A tutorial on hidden markov models and selected applications in speech recognition. Proc. IEEE **77**(2), 257–286 (1989)
85. E. Alpaydin (2014) Introduction to machine learning (MIT Press, 2014)
86. J. Friedman, T. Hastie, R. Tibshirani (2001) *The Elements of Statistical Learning*, vol. 1. Springer Series in Statistics (New York, 2001)
87. T. Kanungo, D.M. Mount, N.S. Netanyahu, C.D. Piatko, R. Silverman, A.Y. Wu, An efficient k-means clustering algorithm: analysis and implementation. IEEE Trans. Pattern Anal. Mach. Intell. **24**(7), 881–892 (2002)

88. M.M. Van Hulle, Self-organizing maps, in *Handbook of Natural Computing* (Springer, Berlin, 2012), pp. 585–622
89. T. Kohonen, The self-organizing map. Neurocomputing **21**(1–3), 1–6 (1998)
90. X. Zhu, Semi-supervised learning literature survey, in *Computer Sciences TR 1530* (University of Wisconsin, Madison, 2008)
91. H. Wu, S. Prasad, Semi-supervised deep learning using pseudo labels for hyperspectral image classification. IEEE Trans. Image Process. **27**(3), 1259–1270 (2018)
92. L.P. Kaelbling, M.L. Littman, A.W. Moore, Reinforcement learning: a survey. J. Artif. Intell. Res. **4**, 237–285 (1996)
93. R.S. Sutton, A.G. Barto, *Reinforcement Learning: An Introduction*, vol. 1, no. 1 (MIT Press, Cambridge, 1998)
94. K. Arulkumaran, M.P. Deisenroth, M. Brundage, A.A. Bharath, Deep reinforcement learning: a brief survey. IEEE Signal Process Mag. **34**(6), 26–38 (2017)
95. S. Russell, P. Norvig, *Artificial Intelligence: A Modern Approach*, 3rd edn. (Prentice Hall, New Jersey)
96. Y. He, C. Liang, R. Yu, Z. Han, Trust-based social networks with computing, caching and communications: a deep reinforcement learning approach. IEEE Trans. Netw. Sci. Eng. 1 (2018)
97. O. Narmanlioglu, E. Zeydan, Learning in SDN-based multitenant cellular networks: a game-theoretic perspective, in *Proceeding IEEE INM'17*, Lisbon, Portugal, May 2017, pp. 929–934
98. S. Ranadheera, S. Maghsudi, E. Hossain, Mobile edge computation offloading using game theory and reinforcement learning (2017). arXiv:1711.09012
99. S. Doro, L. Galluccio, S. Palazzo, G. Schembra, A gametheoretic approach for distributed resource allocation and orchestration of softwarized networks. IEEE J. Sel. Areas Commun. **35**(3), 721–735 (2017)
100. J. Xie, R. Xie, T. Huang, J. Liu, F.R. Yu, Y. Liu, Caching resource sharing in radio access networks: a game theoretic approach. Front. Inf. Technol. Electron. Eng. **17**(12)
101. J. Xie, F.R. Yu, T. Huang, R. Xie, J. Liu, C. Wang, Y. Liu, A survey of machine learning techniques applied to software defined networking (SDN): research issues and challenges. IEEE Commun. Surv. Tutor. 1
102. M. Glick, H. Rastegarfar, Scheduling and control in hybrid data-centers, in *Proceedings IEEE PHOSST'17*, San Juan, Puerto Rico, July 2017, pp. 115–116
103. P. Xiao, W. Qu, H. Qi, Y. Xu, Z. Li, An efficient elephant flow detection with cost-sensitive in SDN, in *Proceedings IEEE INISCom'15*, Tokyo, Japan, March 2015, pp. 24–28
104. P. Amaral, J. Dinis, P. Pinto, L. Bernardo, J. Tavares, H.S. Mamede, Machine learning in software defined networks: data collection and traffic classification, in *Proceedings IEEE ICNP'16*, Singapore, November 2016, pp. 1
105. A. Nakao, P. Du, Toward in-network deep machine learning for identifying mobile applications and enabling application specific network slicing. IEICE Trans. Commun. 2017CQI0002 (2014)
106. P. Wang, S.C. Lin, M. Luo, A framework for QoS-aware traffic classification using semi-supervised.machine learning in SDNs, in *Proceedings IEEE SCC'16*, San Francisco, CA, USA, June 2016
107. L. Yanjun, L. Xiaobo, Y. Osamu, Traffic engineering framework with machine learning based meta-layer in software-defined networks, in *Proceedings IEEE ICNIDC'14*, Beijing, China, September 2014
108. A. Azzouni, R. Boutaba, G. Pujolle NeuRoute: predictive dynamic routing for software-defined networks (2017). arXiv:1709.06002
109. S. Sendra, A. Rego, J. Lloret, J.M. Jimenez, O. Romero, Including artificial intelligence in a routing protocol using software defined networks, in *Proceeding IEEE ICC Workshops'17*, Paris, France, May 2017
110. S.C. Lin, I.F. Akyildiz, P. Wang, M. Luo, QoS-aware adaptive routing in multi-layer hierarchical software defined networks: a reinforcement learning approach, in *Proceeding IEEE SCC'16*, San Francisco, CA, USA, June 2016, pp. 25–33

111. G. Stampa, M. Arias, D. Sanchez-Charles, V. Muntes-Mulero, A. Cabellos, A deep-reinforcement learning approach for software-defined networking routing optimization (2017). arXiv:1709.07080
112. Y. He, F.R. Yu, A. Boukerche, Deep reinforcement learning based resource management in software-defined and virtualized vehicular adhoc networks, in *Proceedings of ACM DIVANet'17*, Miami Beach, FL, November 2017
113. M.A. Salahuddin, A. Al-Fuqaha, M. Guizani Software-defined networking for RSU clouds in support of the internet of vehicles, IEEE Internet of Things J **2**(2), 133–144 (2015)
114. R. Sherwood, G. Gibb, K.K. Yap, G. Appenzeller, M. Casado, N. McKeown, G.M. Parulkar, Can the production network be the testbed? OSDI **10**, 1–6 (2010)
115. A. Al-Shabibi, M.D. Leenheer, M. Gerola, A. Koshibe, G. Parulkar, E. Salvadori, B. Snow, OpenVirteX: make your virtual SDN's programmable, in *Proceedings of ACM HotSDN'14*, Chicago, Illinois, USA (2014), pp. 25–30
116. ETSI, Network function virtualization: an introduction, benefits, enablers, challenges, & call for action
117. Y. Li, M. Chen, Software-defined network function virtualization: a survey. IEEE Access **3** (2015)
118. J.A. Guerrero-ibanez, S. Zeadally, J. Contreras-Castillo, Integration challenges of intelligent transportation systems with connected vehicle, cloud computing, and internet of things technologies. IEEE Wirel. Commun. **22**(6), 122–128 (2015)
119. L. Hu, M. Qiu, J. Song, M.S. Hossain, A. Ghoneim, Software defined healthcare networks. IEEE Wirel. Commun. **22**(6), 67–75 (2015)

Learning Rich Features from Software-as-a-Service Cloud Computing for Detecting Trust Violations

Mahreen Saleem, M. R. Warsi and Saiful Islam

Abstract We are witnessing a transition era of cloud security, as cloud computing paradigm is shifting its focus from provider to the consumer. Cloud service trust manipulation detection is different from traditional on-site service trust detection because cloud performs data operations at diverse geographically remote data centers; thus diminishing consumer control over the kind of service to be utilized. When the Cloud user submits a particular job to cloud, user has to rely upon the good behavior of the cloud to perform the task without violating the user trust in services utilized. However, due to lack of transparency in cloud, consumers find it hard to evaluate trust. Inspired by the recent progress of Spatial Rich Models (SRM) in image forensics domain, we propose to employ SRM and Machine learning approach to verify the trusted behavior of cloud by analyzing the rich features of the output produced by cloud service. We investigated noise distributions in data for violation detection. The approach is based on the hypothesis that every data processing task leaves certain distinct traces on the data. We identify those digital footprints to analyze whether cloud service provider has utilized the legitimate software-as-a-service for processing consumer requests. The inconsistency between authentic and obtained output acts as a proof-of-work for trust violation detection. The experimental results for the standard image dataset demonstrate that noise distributions in spatial domain can be successfully utilized to detect Cloud service trust violations.

Keywords Cloud trust · Trust violation · Rich features · Machine learning · SRM

Supported by TEQIP-III.

M. Saleem (✉) · M. R. Warsi · S. Islam
Aligarh Muslim University, Aligarh 202002, UP, India
e-mail: mehkhan27@gmail.com

M. R. Warsi
e-mail: warsimr@yahoo.com

S. Islam
e-mail: saifulislam@zhcet.ac.in

M. Dutta et al. (eds.), *Proceedings of International Conference on IoT Inclusive Life (ICIIL 2019), NITTTR Chandigarh, India*, Lecture Notes in Networks and Systems 116,
https://doi.org/10.1007/978-981-15-3020-3_38

1 Introduction

Adopting cloud services instead of deploying in-house applications has seen a remarkable growth in recent years. Organizations and individual consumers frequently outsource high complexity and resource intensive computations to a cloud service provider, thereby minimizing upfront infrastructure and computation expenses. However, since the third-party service provider is untrusted, the clients are hesitant to trust the task output. In order to build confidence in cloud service adoption, there is a need to verify the servers' output so that the consumers rest assured that the cloud has accomplished their assigned jobs faithfully. Utilizing the power of Machine learning, one of the effective ways to solve the challenge of building trust in cloud is to employ feature extraction for violation detection [1]. Massa et al. [2] created 'Trustlet.org' to collect trust datasets and evaluate various trust metrics in social network trust. In the study [3], we discuss the notion of trust management and its prominent issues in cloud computing ecosystems. In this work, we consider image processing service as a use-case for verifying trust in cloud. Since image processing tasks are usually computationally, and storage wise quite expensive and require sophisticated resource intensive platform to cater to large-scale processing needs, it becomes inevitable to outsource the computation intensive tasks to a trusted cloud service provider. A set of image features learned from the processed output using an authentic or certified data processing application would vary from the features learnt from an application that is counterfeit or unlicensed version of the legitimate one. Fundamentally, assembling the individual models is a feature selection problem as we strive for best detection accuracy for a given feature dimensionality. Therefore, the area of feature learning needs to be explored for solving issues in cloud trust violation detection as Machine learning has a potential to provide promising results in the area of cloud trust evaluation.

1.1 Related Work

Sherchan et al. [4] presented a Hidden Markov trust prediction model for predicting future reputation of web services. Pan et al. [5] proposed a model for service selection in cloud systems based on user interaction frequencies to identify similar trusted neighbors. Alhamad et al. in their work [6], described various cloud trust metrics and proposed a trust evaluation model for infrastructure-as-a-service cloud based on those trust metrics using fuzzy-set theory. Recently, spatial rich models (SRM) that are based on extraction of local noise features have been extensively used for steganalysis, and have proved significant for providing additional evidence for image-forensic applications [7, 8]. These techniques utilize the local noise distribution from neighboring pixels, to capture the tampering by diagnosing its inconsistency from the authentic regions. In the work proposed by Cozzolino et al. [9], the authors employed SRM feature model to distinguish between tampered regions and authentic ones. To localize the manipulations, they perform quantization and truncation to integrate SRM features with a CNN model [10]. Rao et al. [11] utilize SRM filter to improve detection accuracy. Han et al. [12] proposed a two-stream R-CNN model to uncover tampering artifacts in image manipulations using an RGB-stream and a SRM based noise-stream. Zhou et al. [8] employed noise residuals to detect tampered faces using a two-stream CNN. No prior work has utilized noise features to investigate cloud trust detection from the servers' output data. In this paper, we use two-stream noise residual data generated from image processed data and extract features to train the ensemble classifier for trust violation detection.

2 Proposed Model

As a use-case, we consider image processing service being invoked by a cloud consumer. The user submits a job of performing 'Edge detection' over a set of images to a cloudlet. The service provider however may either faithfully return the results processed with a legitimate application (say Matlab) or utilize a low-priced unlisensed version of the application (say Octave) to process the images. Even after careful observation, consumers find it hard to recognize whether the output was produced using the legitimate application or otherwise. Our initial philosophy of feature extraction for trust detection is inspired by the capturing of slightly different embedding artifacts in the data processed from different applications, as it occurs in the strategy of building steganographic detectors in digital images [12]. The proposed approach of trust feature extraction can be perceived as a stride towards providing a roadmap for automatizing trust evaluation to promote speedy trust estimation for various computation services over Cloud. Various approaches are adopted to obtain noise residual features of data. In our experiments, we employed SRM approach for feature detection analogous to its application in steganalysis. In SRM approach, only the noise distribution associated with the image (noise residual) is modeled instead of the image content. This proposed work is novel as no prior work has investigated cloud trust detection from noise distributions in the output data. The SRM technique being focused in this experiment is based on the concept of spatial-domain steganography. Rich models are built in spatial domain as the desirable detection is realized by construting models directly in the spheres wherein the embedding modulations are confined and thus more pronounced. The rich models consist of a large number of discrete submodels and each of the submodels takes interrelationships among neighboring noise residual samples into consideration. The submodel residuals are constructed using high-pass filters of the form:

$$R_{ij} = \hat{X}_{ij}(N_{(ij)}) - cX_{ij} \tag{1}$$

For a pixel X_{ij}, N_{ij} are its local neighboring pixels and $c \in N$ denotes the residual order. $\hat{X}_{ij}(.)$ is a local pixel predictor of cX_{ij} determined from N_{ij}. The union of all

106 spatial rich submodels inclusive of their disparately quantized variants has a total dimensionality of 34671. These noise features are extracted by linear and non-linear filters. SRM performs truncation and quantification operation on the filter output to proximate co-occurrence as the final features also termed as local noise descriptors. These noise descriptors are then fed to the neural network for model learning to detect whether the output was produced using the genuine application for which the user had been charged with.

For the learning part, we adopted ensemble classifier [13] constituting of arrays of base learners for model training, because fast machine learning is required. Because of the low intricacy of the classifier and its capacity to handle features of high dimensions and big training datasets, the training and testing time is remarkably reduced.

3 Experiments and Results

High dimensional SRM features were obtained for the two data streams corresponding to Matlab and Octave processed data similar to the corresponding stego and cover sources in steganography. The spatial rich features obtained from the two data streams are constructed in the spatial domain for detection of localized changes where they are most pronounced. The extracted features are observed for deriving the inconsistencies between the two data streams. For the experimentation, the standard BOSSbase dataset containing 10,000 images was used as the source data to generate required datasets for analysis. Image processing algorithm, 'Canny Edge Detection' at threshold value 0.2 was applied on the source dataset using Matlab and Octave, to generate two substreams of data for further classification, as shown in Fig. 1. The feature set obtained for the dataset is represented in the following form:

$$\mathrm{F} : 10000 \times 34671\,\mathrm{double}$$
$$\mathrm{Names} : 10000 \times 1\,\mathrm{cell}$$

where F corresponds to the SRM feature set obtained from 10000 BOSSbase dataset images each having dimensionality of 34671 features and Names are the corresponding image labels. High dimensional SRM features (F), corresponding to Matlab and Octave processed data, were analyzed to find out how different residual models engage in detection of differences in obtained feature sets.

The rich model is constructed by assembling the individual submodels selected on the basis on their detected error estimate utilizing the out-of-bag (OOB) error evaluated over the training set. The model training is carried out by random samples selected from the corresponding matlab and octave feature sets. Ensemble classifier [13] constituting of random forest of L binary base learners ($B^{(L)}$) is employed to assemble the model. Each base learner gets trained on a random subspace dimensionality from the extracted noise residual features. The ensemble fuses all individual decisions of L base learners to reach at its decision using majority voting.

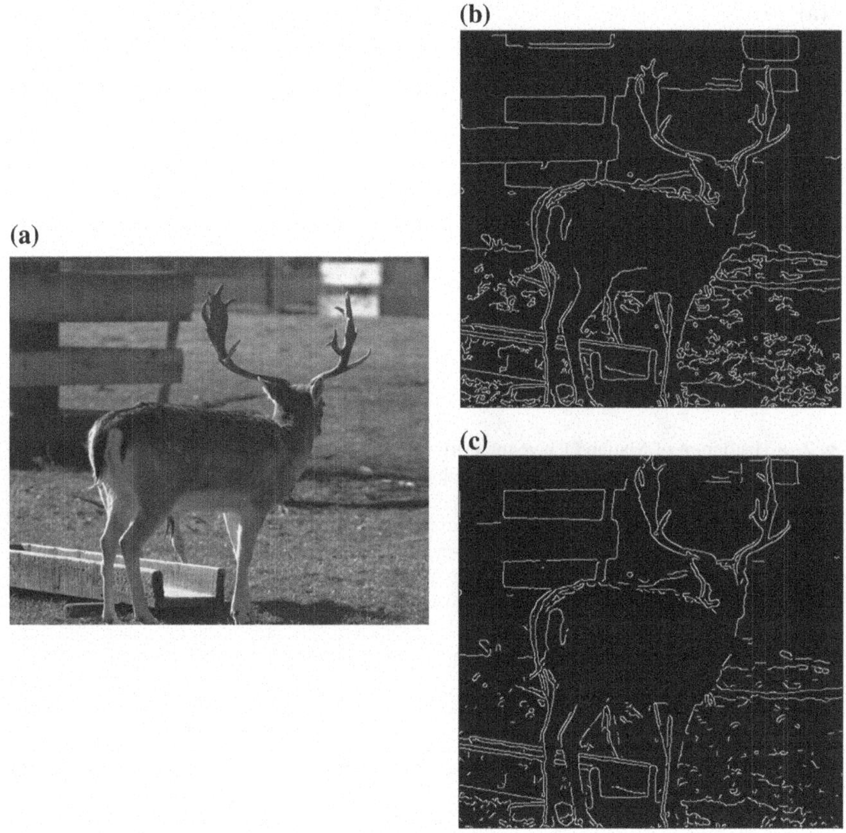

Fig. 1 **a** Original BOSSbase image. **b** Matlab processed image using Canny Edge Detection algorithm at threshold = 0.2. **c** Octave processed image using Canny Edge Detection algorithm at threshold = 0.2

Denoting the training set features obtained from Matlab and Octave as $x^{(m)}$ and $\bar{x}^{(m)}$ respectively, where m is an instance of training feature set of total size N^{trn}. We started by calculating the OOB error estimates, as given in Eq. 2, for training and testing sets at different settings.

$$E_{\text{OOB}}^{(L)} = \frac{1}{2N^{\text{trn}}} \sum_{m=1}^{N^{\text{trn}}} \left(B^{(L)}(x^{(m)}) + 1 - B^{(L)}(\bar{x}^{(m)}) \right) \tag{2}$$

Figure 2 depicts the histogram of majority voting values of base learners obtained using Ensemble Classifier. No overlapping histograms were detected as the classifier predicted classes accurately. Figure 2b shows how the out-of-bag error estimate is affected by the change in number of base learners. Number of base learners is varied to check the decrease in OOB, once OOB saturates, number of base learners is fixed.

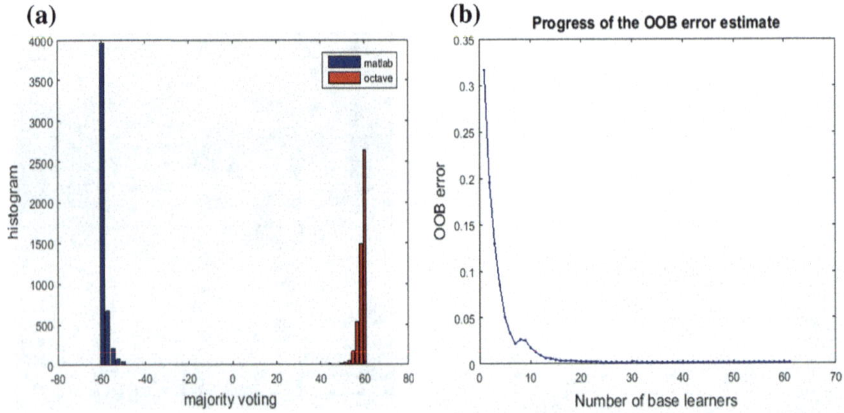

Fig. 2 **a** Histogram of Matlab/Octave majority voting results using Ensemble Classifier for individual base learners. **b** OOB error progress with the variation in the count of base learners

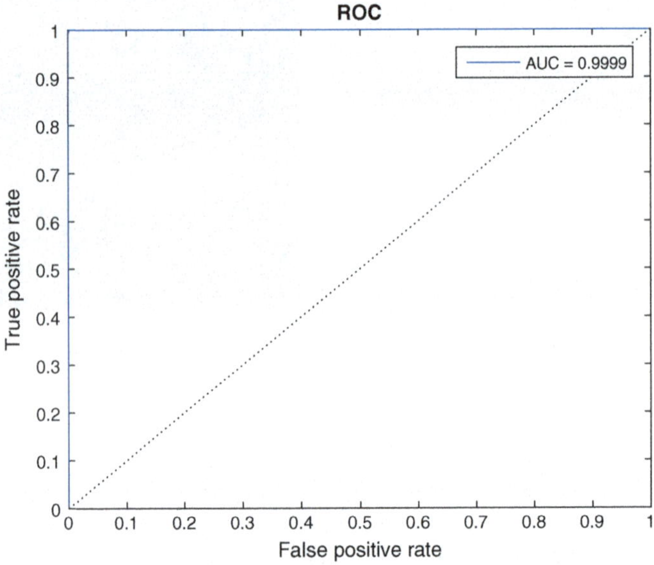

Fig. 3 ROC curve to plot the diagnostic ability of the classifier at various threshold settings

Figure 3 is the plot of ROC curve that is true positive rate or sensistivity or recall as a function of false positive rate or fall-out. The area-under-curve (AUC) measures the classification performance of the model. The results prove that spatial noise residual analysis was successful in detecting the legitimacy of data processing application that was employed by the cloud server for processing the requested tasks. Average testing error over 10 splits was calculated as 0.1458 ($\pm$0.0056). Overall classification accuracy recorded is 99% as ploted by ROC curve (refer to Fig. 3).

4 Conclusion

In order to build trustworthy cloud ecosystems, consumer-centric trust evaluation policies need to be enforced so that the consumer faith in cloud service utilization is no longer based on assumption of belief in service providers' good behavior, but is backed up with trusted evidence. The experimental results for the standard image dataset demonstrate that noise distributions in spatial domain can be successfully utilized to detect Cloud service trust violations. In the extension of this work, it will be interesting to find out how different residual models engage in detection of differences in obtained feature sets and reveal the underlying computational processes applied on the test data. Our further experimentation on general feature extraction would involve modeling of domain specific features to find how the generic domain-based features engage in uncovering the inconsistencies in the output data streams obtained from cloud service provider.

References

1. T.S. Dybedokken, Trust management in fog computing. Futur. Gener. Comput. Syst. **5**(June), 15619–15629 (2017)
2. P. Massa, K. Souren, Trustlet, open research on trust metrics. CEUR Work. Proc. **333**, 31–44 (2008)
3. M.S. Khan, M.R. Warsi, S. Islam, Trust management issues in cloud computing ecosystems, in *Elsevier SSRN Series* (Elsevier, 2019), pp. 2233–2238
4. W. Sherchan, S. Nepal, A. Bouguettaya, *Proceedings of 10th IEEE International Conference on Trust Security Privacy Computing Communication* (2011), pp. 258-265
5. Y. Pan, S. Ding, W. Fan, J. Li, S. Yang, Trust-enhanced cloud service selection model based on QoS analysis. PLoS One **10**(11), 1–14 (2015)
6. M. Alhamad, T. Dillon, E. Chang, A trust-evaluation metric for cloud applications. Int. J. Mach. Learn. Comput. **1**(4), 416–421 (2013)
7. J. Fridrich, J. Kodovsky, Rich models for steganalysis of digital images. IEEE Trans. Inf. Forensics Secur. **7**(3), 868–882 (2012)
8. P. Zhou, X. Han, V.I. Morariu, L.S. Davis, Two-stream neural networks for tampered face detection, in *IEEE Computer Society Conference on Computer Vision and Pattern Recognition Workshops*, July 2017, pp. 1831–1839
9. D. Cozzolino, G. Poggi, L. Verdoliva, Splicebuster: a new blind image splicing detector, in *2015 IEEE International Workshop on Information Forensics and Security WIFS 2015 - Proceedings*, IEEE, Nov 2015, pp. 1–6
10. D. Cozzolino, G. Poggi, L. Verdoliva, Recasting residual-based local descriptors as convolutional neural networks: an application to image forgery detection, in *IH and MMSec 2017 - Proceedings of the 2017 ACM Workshop on Information Hiding and Multimedia Security*, Mar 2017, pp. 159–164
11. Y. Rao, J. Ni, A deep learning approach to detection of splicing and copy-move forgeries in images, in *8th IEEE International Workshop on Information Forensics and Security WIFS 2016*, IEEE, Dec 2017, pp. 1–6

12. X. Han, L.S. Davis, Learning rich features for image manipulation detection RPN layer RGB stream input RGB RoI features bilinear noise stream input noise conv layers noise RoI features, in *IEEE Computer Society Conference on Computer Vision and Pattern Recognition Workshops* (2018), pp. 1313–1328
13. J. Kodovský, J. Fridrich, V. Holub, Ensemble classifiers for steganalysis of digital media. IEEE Trans. Inf. Forensics Secur. **7**, 432–444 (2012)

Improved Symbiotic Organism Search Based Approach for Scheduling Jobs in Cloud

Deepika Srivastava and Mala Kalra

Abstract Optimal task scheduling plays an important role in improving the performance of cloud computing. As it has been classified as a NP-complete problem, researchers are not able to get an exact solution for this issue. Symbiotic Organism Search (SOS) algorithm is the latest meta-heuristic technique that is widely used for finding the solution of optimization problems. Improved SOS (ISOS) algorithm is imitated from symbiotic relationships that exist among different organisms of an ecosystem. This work presents a scheduling algorithm based on ISOS Algorithm for the best possible mapping of various tasks on available cloud resources. The proposed algorithm is aimed to minimize two-objective functions that are makespan and cost. To validate the performance of presented work, it is compared with PSO algorithm. Simulation results show that ISOS algorithm gives 19.71–49.50% improvement in terms of makespan and 27.65–42.73% improvement in terms of cost over PSO algorithm when the number of tasks is varied from 100 to 500.

Keywords Cloud computing · Task scheduling · Improved symbiotic organism search · Makespan · Cost

1 Introduction

Cloud Computing has gained wider recognition among business enterprises in the last decade. Efficient task scheduling can benefit cloud service providers as well as customers. Many algorithms have been applied for solving task scheduling problem like Ant Colony Optimization, Genetic Algorithm, Particle Swarm Optimization and Symbiotic Organism Search (SOS), etc.

SOS is a recent algorithm inspired from nature. The term Symbiosis means "living together". This term is first used by D. Barry in 1878 to define cohabitation behavior

D. Srivastava (✉) · M. Kalra
Department of Computer Science & Engineering, NITTTR, Chandigarh, India
e-mail: sh.deepa1987@gmail.com

M. Kalra
e-mail: malakalra2004@gmail.com

M. Dutta et al. (eds.), *Proceedings of International Conference on IoT Inclusive Life (ICIIL 2019), NITTTR Chandigarh, India*, Lecture Notes in Networks and Systems 116,
https://doi.org/10.1007/978-981-15-3020-3_39

of different organisms. Nowadays this term is used to explain a relationship between any two dissimilar organisms of an ecosystem.

The most general symbiotic relationships that exist in environment are "mutualism, commensalism, and parasitism". Mutualism signifies a relation of two dissimilar organisms in which both organisms benefit each other in order to get better survival in ecosystem such as a relationship between flowers and bees. In this relationship, the bee gets benefitted as they are getting food from flowers and flowers are benefited as they are pollinated. Commensalism represents a relationship between two dissimilar organisms in which the first organism gets some benefit from the second organism while the second organism is unaffected like a relation of barnacles and whales. Barnacles attach themselves to whales and get benefit of transportation and more feeding ability while whale does not have any positive or negative effects. Parasitism represents a relationship between two dissimilar organisms in which the first organism gets some benefit from the second organism while the second organism is actively harmed such as the relation of bacteria and human beings. In this relationship, bacteria get benefited while human is negatively affected.

Cheng and Prayogo [1] have simulated symbiotic relationship of organisms in ecosystem and implemented this method for finding solution of mathematical and engineering design problems.

Symbiotic organism search algorithm starts with an initial ecosystem (initial population). Ecosystem consists of a group of randomly selected organisms, which signifies possible solutions of the problem. Each organism has a fitness value, which represents degree of adaptation to the required objective function. All three symbiotic relationships act as operators to modify the existing solution. Each iteration of algorithm produces new modified solutions. Near optimal solution is produced when termination criteria are satisfied.
The main steps of algorithm are

Step 1: Ecosystem initialization
Step 2: Repeat

(a) Phase 1 (Mutualism)
(b) Phase 2 (Commensalism)
(c) Phase 3 (Parasitism)

Step 3: If termination condition is not true go to step 2 else stop.

The detail of each phase in context of the proposed work is explained in Sect. 3.

Nama et al. [2] presented an improved SOS algorithm in which a new symbiotic relationship called predation is introduced. Predation represents a relationship between two organisms in which the first organism kills and consumes the second organism. A random weighted reflective parameter (RWRV) is also used so that the algorithm's ability to search the best solution can be enhanced.

The remaining paper is structured as follows. Section 2 reviews existing algorithms focused on task scheduling in cloud and SOS algorithm. Section 3 presents problem definition and Sect. 4 details proposed algorithm along with flow chart. Section 5 presents the results whereas Sect. 6 concludes the paper.

2 Related Work

Chawla and Bhonsle [3] proposed a technique that uses a cost-based scheduling algorithm which can be optimized dynamically. The algorithm combines two strategies one is cost-based task scheduling and other is dynamically optimized resource allocation method. The first strategy is beneficial to the user and second one is beneficial to service provider. The tasks are grouped, prior to resource allocation so that computation/communication ratio and resource utilization can be improved.

Panda and Jana [4] presented an algorithm which is a multi-objective algorithm designed for multiple clouds. The algorithm attempted to make a balance between execution time and computation cost of tasks. Three objective parameters, total cost, total execution time, and average utilization of cloud are used. Two other algorithms are used for comparison of results, by taking all parameters into consideration. The experiments concluded that this work achieves better results when compared with existing algorithms.

Bey et al. [5] presented a heuristic technique based on improved Min-Min algorithm, which uses a balancing procedure for all available resources with the method of task exchange. The objective is to decrease the execution time and enhance the resource utilization in cloud ecosystem. A heterogeneous environment is considered. The experimental results prove the enhanced performance of suggested method as compared to Sufferage and Min-min algorithms.

Cheng and Prayog [1] applied an algorithm called SOS (Symbiotic Organisms Search) algorithm for finding solutions of different problems related to optimization, statistics, and engineering design. SOS imitates the symbiotic relation of organisms which they maintain to stay alive and grow in the ecosystem. Few well-known optimization methods are used for comparison of results. Results confirm that the algorithm gives the outstanding performance in solving a variety of difficult numerical problems.

Abdullahi et al. [6] implemented Symbiotic Organism Search (SOS) algorithm in cloud environment for finding optimal schedule for task on available cloud resources. A discrete version of SOS (DSOS) is proposed. Results obtained demonstrate that algorithm has better performance over PSO and is suitable for large size scheduling problems.

Namaa et al. [2] proposed a new version of Symbiotic organism search called ISOS algorithm to obtain solutions of complex unconstrained optimization problems. For this purpose, they added a new phase called predation phase. They have also used a random weighted reflective parameter. The experimental outcomes of improved SOS are compared with various algorithms. Results prove the better efficiency of presented algorithm over other optimization strategies.

Tejani et al. [7] proposed three modified versions of Symbiotic organism search. Adaptive benefit factors are introduced and effective combination of these adaptive benefit factors is considered. When compared with SOS and some other algorithms, results prove its better reliability and efficiency.

Eki et al. [8] implemented SOS algorithm for solving Capacitated Vehicle Routing Problem (CVRP). Solution representation decoding method is used along with SOS. The results show that SOS can be used as a better alternative for the efficient solution of CVRP.

3 Problem Definition

This work presents a discrete algorithm based on Improved Symbiotic Organism Search (ISOS) for scheduling of independent tasks. The main aim is to reduce the completion time and computation cost for execution of tasks on available virtual machines. Here we are using an objective function as the weighted sum of both objective functions.

$$f(x) = w \times f_1(x) + (1 - w) \times f_2(x)$$

where w is weighing factor, i.e., $0 \leq w \leq 1$.

4 Proposed Work

The details of each step are as follows:

Step 1: Parameter initialization and Initial ecosystem generation

Initialize following parameters
n which represents total number of organisms in ecosystem.
max_it which represents maximum number of iterations.
Generate initial ecosystem (X) randomly $X = \{x_1, x_2, \ldots x_n\}$.

Step 2: Calculate ETC Matrix and Cost Matrix

ETC matrix is a $t \times m$ matrix which stores time taken by each task on each VM. It is calculated as
ETC(a, b) = length of Task(a)/capacity of VM(b)
Cost matrix is a $1 \times n$ matrix which comprises of cost of each virtual machine.
Where *t* is total number of tasks and *m* is total number of virtual machines.
The values of ETC matrix and cost matrix are normalized by dividing with respective maximum values [4].

Step 3: Repeat steps 4–9 until number of iterations is reached to maximum value and $i < n$.

Step 4: Fitness function evaluation

Evaluate fitness function $f(x_a)$ for each organism x_a (Solution).

$$f(x_a) = w \times f_1(x_a) + (1 - w) \times f_2(x_a)$$

where w is weighing factor, i.e., $0 \leq w \leq 1$.

$f_1(x)$ is objective function for makespan defined as

$$f_1(x) = \max\{T_{kl}\ \forall \text{task k mapped on VM l}$$

$f_2(x)$ is objective function for computation cost that can be defined as

$$f_2(x) = \sum_{i=1}^{m} C_l \times UT_l\ \forall\, \text{VMs l}$$

Here T_{kl} is the execution time of all virtual machines, C_l is cost of virtual machine l. UT_l is the utilization time of virtual machine l. m is number of available virtual machines.

Utilization time (UT) = End Time (ET) – Start Time (ST)

Step 5: Identify best organism x_{best}

Find organism x_b for which $f(x_b)$ is lowest. Assign this x_b to x_{best}.

Step 6: Mutualism Phase

(a) Select a random organism, x_b, where $x_a \neq x_b$
(b) Calculate mutual relationship vector (Mut_Vec) and benefit factor (b_f)

$$Mut\_Vec = (x_a + x_b)/2 \tag{1}$$

b_f_1 and b_f_2 = either 1 or 2 (selected randomly)

(c) Modify organism x_a and x_b using Eqs. 4 and 5

$$S_1 = x_a + rwrv \times \left(x_{best} - Mut\_Vec^{*} b\_f_1\right) \tag{2}$$

$$S_2 = x_b + rwrv \times \left(x_{best} - Mut\_Vec^{*} b\_f_2\right) \tag{3}$$

$$x_a^{'} = \lceil S_1 \rceil \bmod m + 1 \tag{4}$$

$$x_b^{'} = \lceil S_2 \rceil \bmod m + 1 \tag{5}$$

where RWRV is a weighted random number called random weighted reflection vector which is calculated as using Eq. 6

$$rwrv = 1 - 0.5 \times (1 + rand(0, 1)) \tag{6}$$

(d) Calculate Fitness Value of $x_a^{'}$ and $x_b^{'}$

(e) if $f(x_a') < f(x_a)$
$x_a \leftarrow x_a'$
$f(x_b') < f(x_b)$
$x_b \leftarrow x_b'$

Step 7: Commensalism Phase

(a) Select a random organism x_b, where $x_b \neq x_a$
(b) Modify organism x_a using Eq. 8

$$S_3 = x_a + rwrv \times (x_{best} - x_b) \tag{7}$$

$$x_a' = \lceil S_3 \rceil \bmod m + 1 \tag{8}$$

(c) Calculate Fitness Value of x_a'
(d) if $f(x_a') < f(x_a)$
$x_a \leftarrow x_a'$

Step 8: Parasitism Phase

(a) Select a random organism, x_b, where $x_b \neq x_a$
(b) Create a parasite vector (x_p) from organism x_a using Eq. 10

$$S_4 = rwrv \times x_a \tag{9}$$

$$xp = \lceil S_4 \rceil \bmod m + 1 \tag{10}$$

(c) Compute Fitness Value of the parasite vector
(d) if $f(x_p) < f(x_b)$
$x_b \leftarrow x_p$

Step 9: Predation Phase

(a) Select a random organism, x_b, where $x_b \neq x_a$
(b) Create a predation vector (x_{pr}) from organism X_a using Eq. 12

$$S_5 = x_a + rwrv \times (x_a^{max} - x_a^{min}) \tag{11}$$

$$xpr = \lceil S_5 \rceil \bmod m + 1 \tag{12}$$

where x_a^{max} and x_a^{min} are the maximum and minimum value of dimension of organism x_a
(c) Compute Fitness Value of the predation vector
(d) if $f(x_{pr}) < f(x_b)$
$x_b \leftarrow x_{pr}$.

5 Result and Discussion

Table 1 shows the parameters used for the proposed algorithm and ISOS. Table 2 shows parameters settings for CloudSim toolkit.

We have assumed single data center with 20 virtual machines. The capacity of virtual machines is varied from 250 to 1250 MIPS and cost is varied from 0.05 to 0.25$. The machines with higher capacity are costlier than the machines with lower capacity. The average bandwidth of the datacenter is supposed to be 1000 Mbps. Space-shared policy is used for virtual machines.

The efficiency of the proposed algorithm is evaluated by comparing it with PSO algorithm. Execution time and cost are considered as performance metrics. The number of tasks is varied from 100 to 500. Figure 1 shows the average makespan obtained for considered algorithms. Results displayed are the average of values obtained by executing both algorithms 10 times.

From the Figure, it is clear that Improved Symbiotic Organism Search (ISOS) gives lesser values of average makespan as compared to Particle Swarm Optimization Algorithm (PSO) when the range of number of tasks is 100–500.

For 100 tasks, proposed algorithm gives approximately 49.5% smaller makespan than PSO. For 200 tasks, our algorithm achieves 35.03% better makespan. It achieves 39.79%, 24.5%, and 19.71% lesser makespan than PSO algorithm for 300, 400, 500 tasks, respectively.

Figure 2 shows that the improved symbiotic organism search algorithm gives lesser cost than PSO algorithm when the range of number of tasks is 100–500.

Table 1 Parameter settings of algorithm

Algorithm	Parameters	Values
PSO	Population size	50
	Maximum iteration	50
	Self recognition coefficient (c1)	2
	Social effect (c2)	2
ISOS	Ecosystem size	50
	Maximum iteration	50
	Weighing factor	0.5

Table 2 Parameter settings of CloudSim

Datacenter	Number of datacenters	1
	Type of manager	Space-shared
VM	Number of VMs	20
	VM memory(RAM)	512
	Bandwidth	1000
	Policy	Space-shared
Task	Number of tasks	100–500

Fig. 1 Average makespan with respect to no. of tasks

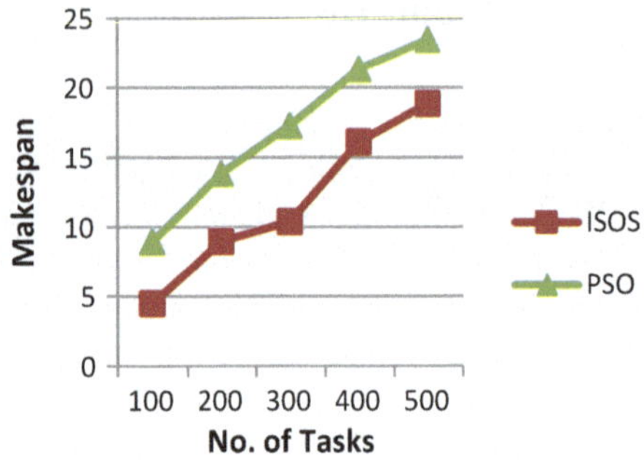

Fig. 2 Average costs obtained by varying no. of tasks

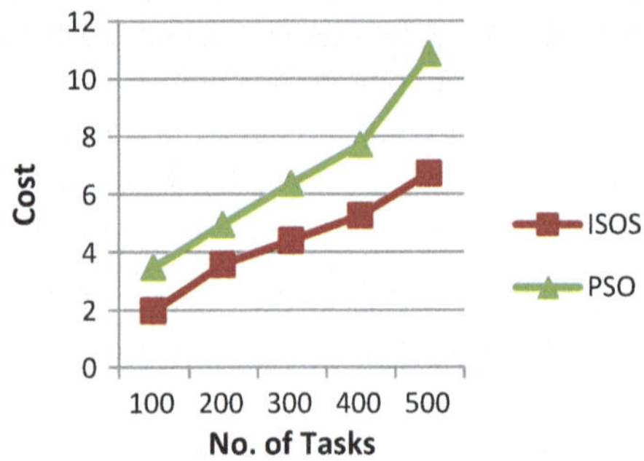

For 100 tasks, proposed algorithm gives approximately 42.73% lesser cost than PSO. For 200 tasks, our algorithm achieves 27.65% lesser cost. It achieves 30.78%, 31.61%, and 38.16% lesser cost than PSO algorithm for 300, 400, and 500 tasks, respectively.

6 Conclusion and Future Scope

This work presents a scheduling approach based on Improved SOS algorithm which is motivated by symbiotic relationship of organisms. These relationships can be defined in terms of mutualism, commensalism, parasitism, and predation. This algorithm has good explorative and exploitative capability. It works in different phases where each phase gives it ability to search better solution in search space. Moreover, the algorithm requires fewer parameters.

The given approach is intended to reduce execution time and cost for executing tasks on available cloud resources. For the validation of algorithm's performance, it is compared with PSO algorithm. Simulation results show that the proposed algorithm gives 19.71–49.501% improvement in makespan and 27.65–42.73% improvement in cost over PSO. In future, the suggested technique can be used for scheduling workflows in real cloud. Further using this approach, we can optimize additional objectives while scheduling.

References

1. M. Cheng, D. Prayogo, Symbiotic organisms search: a new metaheuristic optimization algorithm. An Int. J. Comput. Struct., Article in Press (2014). Elsevier
2. S. Nama, A. Saha, S. Ghosh, Improved symbiotic organisms search algorithm for solving unconstrained function optimization. Decis. Sci. Lett. **5**, 361–380 (2016)
3. Y. Chawla, M. Bhonsle, Dynamically optimized cost based task scheduling in cloud computing. Int. J. Eng. Trends Technol. Comput. Sci. **2**(3) (2013)
4. S. Panda, K.P. Jana, A multi-objective task scheduling algorithm for heterogeneous multi-cloud environment, in *International Conference on Electronic Design, Computer Networks & Automated Verification (EDCAV)*, Jan 2015, pp 82–87
5. K. Bey, F. Benhammadi, R. Benaissa, Balancing heuristic for independent task scheduling in cloud computing, in *12th International Symposium on Programming and Systems (ISPS)*, Apr 2015, pp 1–6
6. M. Abdullahi, M. Ngadi, S. Abdulhamid, Symbiotic organism search optimization based task scheduling in cloud computing environment. Futur. Gener. Comput. Syst. (2015)
7. G. Tejani, V. Savsanim, V. Patel, Adaptive symbiotic organisms search (SOS) algorithm for structural design optimization. J. Comput. Des. Eng. **2**(1) (2016)
8. R. Eki, V. Yu, S. Budi, P. Redi, Symbiotic Organism Search (SOS) for solving the capacitated vehicle routing problem. Int. J. Mech. Aerosp. Ind. Mechatron. Manuf. Eng. **9**(5) (2015)

The manufacturer's authorised representative in the EU is Springer Nature Customer Service Centre GmbH, Europaplatz 3, 69115 Heidelberg, Germany. If you have any concerns regarding our products, please contact ProductSafety@springernature.com

Printed and bound by CPI Group (UK) Ltd, Croydon, CR0 4YY
30/09/2026
02225556-0014